Studies in Logic
Volume 78

Logic and Philosophy of Logic
Recent Trends in Latin America and Spain

Volume 67
Many-Valued Logics: A Mathematical and Computational Introduction
Luis M. Augusto

Volume 68
Argument Technologies: Theory, Analysis, and Applications
Floris Bex, Floriana Grasso, Nancy Green, Fabio Paglieri and
Chris Reed, eds

Volume 69
Logic and Conditional Probability. A Synthesis
Philip Calabrese

Volume 70
Proceedings of the International Conference. Philosophy, Mathematics, Linguistics: Aspects of Interaction, 2012 (PhML-2012)
Oleg Prosorov, ed.

Volume 71
Fathoming Formal Logic: Volume I. Theory and Decision Procedures for Propositional Logic
Odysseus Makridis

Volume 72
Fathoming Formal Logic: Volume II. Semantics and Proof Theory for Predicate Logic
Odysseus Makridis

Volume 73
Measuring Inconsistency in Information
John Grant and Maria Vanina Mrtinez, eds.

Volume 74
Dictionary of Argumentation. An Introduction to Argumentation Studies
Christian Plantin. With a Foreword by J. Anthony Blair

Volume 75
Theory of Effective Propositional Paraconsistent Logics
Arnon Avron, Ofer Arieli and Anna Zamansky

Volume 76
Argumentation and Inference. Proceedings of the 2nd European Conference on Argumentation. Volume I
Steve Oswald and Didier Maillat, eds.

Volume 77
Argumentation and Inference. Proceedings of the 2nd European Conference on Argumentation. Volume II
Steve Oswald and Didier Maillat, eds.

Volume 68
Logic and Philosophy of Logic. Recent Trends in Latin America and Spain
Max A. Freund, Max Fernández de Castro and Marco Ruffino, eds.

Studies in Logic Series Editor
Dov Gabbay   dov.gabbay@kcl.ac.uk

Logic and Philosophy of Logic
Recent Trends in Latin America and Spain

Edited by
Max A. Freund
Max Fernández de Castro
and
Marco Ruffino

© Individual author and College Publications, 2018
All rights reserved.

ISBN 978-1-84890-293-0

College Publications
Scientific Director: Dov Gabbay
Managing Director: Jane Spurr

http://www.collegepublications.co.uk

Printed by Lightning Source, Milton Keynes, UK

All rights reserved. No part of this publication may be reproduced, stored in a retrieval system or transmitted in any form, or by any means, electronic, mechanical, photocopying, recording or otherwise without prior permission, in writing, from the publisher.

	Introduction	vii
	List of Contributors	xvii
	Logic and Ontology	1
1	**The Problem of Relations** *Roderick Batchelor*	3
2	**On formal aspects of the epistemic approach to paraconsistency** *Walter Carnielli, Marcelo Coniglio, Abilio Rodrigues*	49
3	**Limitaciones expresivas y fundamentación: la teoría ingenua de conjuntos basada en LP** *Luis Estrada González*	77
4	**Sobre la función de la teoría de modelos y el significado de la teoría de conjuntos** *Mario Gómez-Torrente*	89
5	**Which is the least complex explanation? Abduction and complexity** *Fernando Soler-Toscano*	103
6	**On the naturalness of new axioms in set theory** *Giorgio Venturi* . .	121
	Logic and Epistemology	147
7	**Cómo pensar sobre otra cosa** *Axel Arturo Barceló Aspeitia*	149
8	**Algunos problemas de epistemología social: Probabilidades grupales dinámicas** *Eleonora Cresto*	169
9	**Remarks on the use of formal methods** *Décio Krause y Jonas R. Becker Arenhart*	187
10	**Explanation, Understanding, and Belief Revision** *Andrés Páez* . . .	209
	Logic and Language	229
11	**Yablo's Paradox and omega-Paradoxes** *Eduardo Alejandro Barrio*	231
12	**Kripke's Interpretation of the Theories of Frege and Russell in Comparison with Mill's Views** *Luis Fernández Moreno*	253
13	**Logics of programs as a fuelling force for semantics** *Francisco Hernández-Quiroz*	273
14	**¿Puede un personaje literario estar durmiendo en una playa? Ficción, referencia y verdad** *Eleonora Orlando*	281

Philosophical Logic and Philosophy of Logic in Latin America and Spain

Introduction

In the last decades, logic and philosophy of logic have increasingly become areas of research and great interest for mathematicians and philosophers in Latin America and Spain. Important work has been done and continues to be done in those areas. As examples, we can mention the development of paraconsistent logic in Brazil, abductive logic in México and Spain, and epistemic logic in Colombia and Argentina. The goal of this volume is to call attention to the above developments by means of a collection of original and unpublished papers by a number of specialists from Latin America and Spain carrying out serious and important work in either philosophy of logic or philosophical logic. No volume of this sort has been published so far and in this way our aim is to fill a gap. Given the nature of the papers included in this volume, the book might be suitable as a companion to a graduate course in logic or philosophy of logic as well as to related disciplines. Specialist might find the volume relevant to their work as well. Some of the papers are written in Spanish and the others in English. The idea of having papers in two different languages was to attract the interest and use of the volume in different linguistic communities. The volume is divided in three sets of papers dealing with related topics. The first group contains contributions dealing with general aspects of logical relations (inference, consequence, etc.) and ontology (relations, sets, etc.). The second group contains contributions dealing with applications of logic to epistemology (epistemic logic, epistemic paradoxes and formalization of pre-theoretical notions). And the third contains those contributions dealing with logic and language (reference, propositions, etc.)

Logic and Ontology

Roderick Batchelor's "The Problem of Relations" is a long and detailed discussion concerning the nature of the connection between relations and their *relata* in what he calls *relational situations*. He describes his contribution as an attempt to carry further the critical part of Kit Fine's "Neutral Relations" (2000), although Batchelor is not in agreement with Fine's positive proposal. He reviews a number of alternative explanations for the connection between relations and *relata*, and argues that each of them is unsatisfactory (for different reasons). There is one favored perspective that he calls Symmetricalism, which sees the nature of the connection between a relation and its *relata* as being independent from any order or direction etc. in which the latter are taken (e.g., the situation *a is parallel to b* is the same as *b is parallel to a*). But the main point of the paper is neither to develop nor to offer positive reasons for the favored position, but rather to deepen the negative reasons against the rival positions. Batchelor discusses (in many cases improving on Fine's criticisms) the following views of relational situations: *Orderism*, i.e., the view that a relational situation consists in a relation applied to its terms in a certain order (he distinguishes, going beyond Fine, a weak and a strong version of this view); *Directionalism*, i.e., the view that a relational situation consists in a relation applied to its terms in a certain direction or sense, or as "proceeding" from one term to the other; *Positionalism*, i.e., the view that relations contain "positions" (argument-places) and the relational situation consists in *relata* occupying these positions; *Variablism*, i.e., the view that relations contain *Objective Variables*, which are the correlate in the world of variables 'x', 'y', etc. in language, and in the relational situation *relata* replace these Objective Variables; *Symmetricalism*, i.e., the view that there is only one relational situation given by the application of a relation to suitable *relata*. Batchelor considers yet another set of different positions (without any special name for it) which he describes only negatively as those which account for the possibility of multiple completions of the same relation by the same *relata* without employing notions like order, direction, position or variable. The paper ends with a list of *desiderata* for a theory of relational predication (and the indication that none of the positions and theories reviewed in it satisfy all of the *desiderata*).

Walter Carnielly, Marcelo Coniglio and Abílio Rodrigues' "On formal aspects of the epistemic approach to paraconsistency?" is a study on foundational aspects of paraconsistency. It gives continuity to several previously published articles by the same authors on this topic. They search both for a reasonable philosophical interpretation of the admissibility (in some special circumstances) of contradictions in scientific theories without getting explosion (and thereby triviality) and for a sound (complete) semantics for a formalized paraconsistent system. According to the authors, contradiction in science occurs more frequently than one might expect and it can be seen in those many situations in which there is evidence both for a proposition P and for its negation. Hence giving a central role to the notion of evidence (which is weaker than truth) might be a good way of treating contradiction in scientific theories in a natural way without accepting explosion. The authors' leading idea is to treat contradictions in science as basically an epistemic phenomenon, differently from the perspective known as *dialetheism* (e.g., Priest and Berto 2013) according to which contradictions are in the nature of things, i.e., part of the world itself. Carnielli, Coniglio and Rodrigues present and compare two formal systems, the Basic Logic of Evidence (BLE), which is a system of natural deduction designed to preserve evidence (or, as the authors describe it, as "suited to the reading of contradictions as conflicting evidence"), and the Logic of Evidence and Truth (LETj), which is a system that adds to BLE some mechanisms of recovering classical logic for those formulas that have been established as true or false. As the authors describe it, LETj is a system that can both express preservation of evidence and preservation of truth. It results from a combination of the so called Logics of Formal Inconsistency (i.e., logics for which a contradiction might be accepted without explosion) with Logics of Formal Undeterminateness (i.e., logics for which, for some formula A, $A \wedge \neg A$ is not a theorem) formally represented by the introduction of a special one-place operator ○ that represents both consistency and definiteness. (○A means that the truth of A has been conclusively established.) Towards the end of the article they develop an algebraic approach to both and BLE and LETj, which culminates with a proof that the latter is both sound and complete with respect to the so-called Fidel structures.

Luis Estrada González' "Limitaciones expresivas y fundamentación: la teoría ingenua de conjuntos basada en LP" discusses the prospects of naïve set theory based on paraconsistent logical systems. More specifically, based on systems resembling LP, extensively used by Priest. More broadly, and insofar as big chunks of mathematics are based on set theory, he discusses the prospects of an inconsistent mathematics (in which the principle of explosion does not hold) based on this sort of paraconsistent set theory. Central to Estrada González discussion are Thomas' (2014) results that seem to be pretty discouraging for the prospects of such a theory: naïve set theory based on any system of a class of LP logics on the one hand cannot prove the existence of singleton sets, or of ordered pairs, or of ordered infinite series and, on the other, can prove something like $\forall x \forall y ((x \in y) \wedge (x \notin y))$, which makes it rather trivial. In particular, it cannot derive classical arithmetic. Estrada González wants to throw some skepticism both over the cogency of Thomas' technical results and to the negative conclusions about the prospects of paraconsistent mathematics. He argues, contrary to the view of some philosophers (e.g., Priest himself), that paraconsistent mathematics does not need to recover theorems of set theory or a "decent part" of classical mathematics in order to be taken as a serious alternative.

Mario Gómez-Torrente's "Sobre la función de la teoría de modelos y el significado de la teoría de conjuntos" is a discussion concerning the foundations of the usual definition of logical truth and logical consequence in terms of models. More precisely, he discusses the worry expressed by some philosophers of logic (e.g., Boolos, Williamson and McGee) that the tarskian definition of logical truth centered on the notion of truth in all models (and the corresponding definition of logical consequence) is enough to rule out the possibility that some sentence in a formal language might be true in all set-theoretical models and, nevertheless, false in the intended interpretation of set theory (*mutatis mutandis, the same for logical consequence*). Gómez-Torrente call this the *foundations' force criticism*. The strategy is to dissolve this criticism appealing to the role and transparency of what he calls (following, e.g., Shapiro) *reflection principles* for set theory. He argues that these principles do not presuppose nor lead to the conclusion that there is an intended interpretation of set theory. But this presupposition is essential for the foundation's force criticism. It follows that the latter is based on an unwarranted presupposition, and there is no point in claiming that model-theoretic definition of logical truth or consequence might not capture a more natural conception lying at the boton of an intented interpretation of set theory (because part of the morals of set theory is that there is no such intended interpretation).

Fernando Soler-Toscano's "¿Cuál es la explicación más simple? Abducción y complejidad" discusses criteria for choosing the simplest among several alternative explanations in an abductive problem, i.e., the problem of explaining a surprising event φ given a body of knowledge Θ. In one version of the problem (which he calls "novedoso" (surprising)), we have that neither the event nor its negation is expected on the basis of Θ i.e., $\Theta \not\vdash \varphi$ and $\Theta \not\vdash \neg\varphi$ In another version (which he calls "anómalo" (anomalous)), the event is actually an anomaly from the perspective of Θ, i.e., $\Theta \not\vdash \varphi$ but $\Theta \vdash \neg\varphi$. In both versions, the problem is to find a hypothesis Δ such its addition to Θ explains φ i.e., find Δ such that $\Theta, \Delta \vdash \varphi$. Normally there are several possible explanatory hypothesis that can do this job, and a natural *desideratum* is that the best explanatory hypothesis is the simplest one. This means that we need a criterion for the "simplest explanation". Soler-Toscano first discusses the prospects of using the formal notion of algorithm complexity formulated as Kolmogorov's complexity of a sequence s, i.e., the length of the shortest program that can reproduce s in a universal Touring machine. The second alternative discussed is the solution of abductive problem within the framework of plausibility models in belief-revision theory. He examines the merits and shortcomings of each alternative.

Giorgio Venturi's "On the naturalness of new axioms in set theory" is concerned with epistemic and methodological aspects of the justification of new axioms in mathematics, more specifically in set theory. He wants to find an alternative to the standard strategy of justification of new axioms that has strong ties to mathematical realism. Central to Venturi's discussion is the notion of *naturalness* (to which Gödel famously appeals in his remarks on the introduction of new axioms in view of the independence of the Continuum Hypothesis with respect to ZFC, but which also had an increasing use in the XXth century, as Venturi points out.) He starts by pointing out some difficulties for the standard strategy of justification which is, according to him, based on two "dogmas". The first dogma is that there is a sharp division between intrinsic and extrinsic reasons justifying new axioms, understanding by "intrinsic" the capacity of the axiom for capturing or analyzing the essence of the formalized notion, and by "extrinsic" the capacity of the axiom for yielding new interesting results. Venturi explores a parallel between intrinsic reasons and analytic statements about sets on the one hand, and extrinsic reasons and synthetic statements on the other. But, as he argues, this parallel seems to make the intrinsic/extrinsic distinction pray to something resembling Quine's criticism of the analytic/synthetic distinction. The second "dogma" is that new axioms can only be justified if they deal with a fully general notion of set and, hence, "local" axioms, i.e., axioms that have consequences only for the lowest levels of the cumulative hierarchy, are not justifiable. Venturi formulates his own proposal of a new notion of naturalness based on two hypotheses concerning the real mathematical practice. The first is that the increasing use of terms like "natural" and "naturally" in the mathematical literature derives from the attribution of relevance to axioms relatively to theoretical contexts that include other pieces of mathematics seen as equally relevant. In other words, the attribution of the terms 'natural' and 'naturally' are contextual (hence something "natural" in one scientific context might not be natural in other one). The second is that some or most fundamental mathematical concepts have an open texture (in Weismann's sense), being open to new axioms as statements that are meant to complete the shaping of the same concepts. This approach corresponds better, according to Venturi, to the real mathematical practice, in particular to the development of set theory in the XXth century.

Logic and Epistemology

Axel Barceló's "Cómo pensar sobre otra cosa" is a proposal for solving some classical epistemic paradoxes such as the lottery, the preface and sorites. Normally such paradoxes have this form: from true premises there is a valid inference to a false or contradictory conclusion. Solutions to these paradoxes normally proceed by either (i) showing that at least some premise is not really true, or (ii) showing that the inference is not really valid or (iii) showing that the conclusion is not really false or contradictory. Barceló's strategy is of the second kind, i.e., he wants to develop a way of looking at these paradoxes as invalid inferences from true premises (under special assumptions) to false conclusions. The strategy is based on the recognition of what he calls *normality conditions*: each of the premises is accepted as true having as background its normality conditions; but these normality conditions cannot be taken jointly in a consistent way. Barceló first introduces this strategy as a solution to a simpler paradox that he calls the *Paradox of the Three Coins*, and then extends it to the other classic paradoxes. This *Paradox* is presented in the following story: the sophist Proclo, who carries three coins of a dracma, is supposed to pay back one dracma that he borrowed from Glaucón before. But then he shows the three coins to Glaucón and asks of each one if that is the one that must be paid, and of course the answer is 'no' for each of them because there is no particular coin that he owes to Glaucón. At the end Proclo draws the inference that, since none of the particular coins is the one that should be paid back, there is no coin at all that he should pay back, and poor Glaucón does not receive his payment. Barceló's solution for this paradox states that the normality condition for the acceptance of each premise of the form 'Coin x is not the one that must be given to Glaucón' as true is that one of the other two coins must be paid instead. But all three premises cannot be accepted as true simultaneously without a conflict in their normality conditions. Barceló shows how to extend this solution in a natural way to the other (harder, according to him) paradoxes. In the case of the preface paradox, the author of a book might accept each of the particular statements as true under the normality condition that some other statements might be false; in the case of the lottery paradox, one considers that each particular ticket is unlikely to win under the normality condition that some other ticket will win. In the sorites case, according to Barceló, each premise of the form 'If n grains does not make a heap, then $n + 1$ grains does not make a heap either' is accepted under the normality condition that there must be a limit between non-heap and heap, but the limit lies elsewhere (i.e., for some other $m \neq n$).

Eleonora Orlando's "¿Puede un personaje literario estar durmiendo en una playa? Ficción, referencia y verdad" proposes a semantic account of statements containing fictional names. Her starting point is the recognition that such statements are intuitively true in a great number of occasions, despite of the fact that they are not about real entities. In the first part of the paper she reviews some of the most important theoretical alternatives such as neo-meinongianism (e.g., Parsons, Priest), possibilism (e.g., Lewis), abstraccionism (e.g., Kripke, Salmon, Thomason and Predelli) and descriptivism (e.g., Frege, Russel, Walton). She concludes that we must face a dilemma regarding fictional names: either we adopt a *realist* position and accept fictional entities in our ontology (thereby accepting singular propositions including them as well) or we adopt an *anti-realist* position and provide some sort of paraphrase of fictional statements in which singularity disappears (thereby renouncing to the claim that fictional statements are of the form subject-predicate). Her own choice is the first one: fictional names are, for her, not only proper names of abstract entities, but also rigid, i.e., they designate the same abstract entity in every possible world (in which that entity exists). She does not regard the resulting ontological inflation as a big problem because, as she says, philosophers have always been committed to one or another sort of abstract entities. She immediately faces a problem, namely, explaining the content of statements such as 'Ulysses sleeps in the beach of Ithaca', since an abstract entity that corresponds to 'Ulysses' cannot have an ordinary property such as sleeping in a beach. This is the question taken up in the second part of the paper. Orlando follows the broad lines of an account suggested, e.g., by Predelli (1997, 2005 and 2008) and by Recanati (2010) according to which fictional uses of proper names in a statement require a change in the context of evaluation, i.e., change in the circumstances under which a statement is evaluated as true or false without the use of intensional operators. But she makes clear some important differences between her proposal and Predelli's. First, according to her, indexicals and definite descriptions take their value from the context of utterance (while, for Predelli, they take it from the context required by the fictional use), and only later the content (determined in the context of utterance) is evaluated in the fictional context, and the latter is determined not by any semantic requirement but by our imagination. Second, since according to Orlando fictional characters are abstract entities, and their names are rigid designators, and it is part of their essence that they cannot have ordinary properties, it follows that, for her, the context of evaluation is always a metaphysically impossible (although epistemically possible) world.

Décio Krause and Jonas Arenhart's "Remarks on the use of formal methods" discuss some fundamental methodological and epistemological aspects of formalization and axiomatization (in logic, mathematics and empirical theories) of informal or pre-theoretic notions. They center their discussion on three leading questions: first, what are the resources required for developing a formal system; second, what are the criteria of adequacy between the formalized notions and their informal counterparts; third, what is the adequate evaluation of discrepancies between the formalized notions and their informal counterparts. In dealing with these questions, Krause and Arenhart make extensive use of examples taken from the history of mathematics (e.g., Peano's arithmetic and set theory), logic (e.g., Tarski's formal notion of logical consequence), and from attempted formalization of empirical theories (e.g., Mary William's axiomatization of Darwin's theory and MacKinsey, Sugar and Suppes's axiomatization of classical particle mechanics). Regarding the first question (the starting-point of formalization), they consider whether there is a kind of circularity in the very notion of formalization, since there must be already at least some knowledge of the theoretical body that goes under formalization or axiomatization for the process to get off the ground. They advocate the view (following a suggestion from da Costa) that, at this stage, only a minimum core of rationality is required (encapsulated in what they call the "pragmatic principles of reason", which include, e.g., the resources from intuitionistic mathematics and the underlying intuitionistic logic) and this minimum is present only at the metalinguistic level, while there is almost unlimited freedom in the shaping of the linguistic level. (E.g., some paraconsistent logics admit violations of the principle of non-contradiction only at the linguistic level but not at the metalinguistic level.) Regarding the second question, they review several proposals for a criterion of adequacy of formalization. Among them are the so-called 'it *squeezing arguments*, i.e., arguments to the effect that an informal concept is extensionally equivalent with a formally defined one. Krause and Arenhart see some limitation in the applicability of these arguments such as, e.g., in those cases in which the informal concept is too vague (or "rough", as they say). Finally, concerning the third question, Krause and Arenhart discuss how the formal concept can explain the informal one given the fact that the former does not clearly matches the latter, and also that the former often brings new knowledge not foreseeable in latter. As one would expect, Carnap's notion of *explication* is in the center of their discussion. They consider several historical examples, among them the notion of model in Tarskian semantics and its unfolding into non-standard models, countable models of analysis, etc.

Andrés Páez's "Expanation, Understanding and Belief Revision" is an attempt to show that (and how) the notion of understanding (in the sense of a particular epistemic state of individuals involved in the scientific enterprise) can be relevant for a theory of explanation. More broadly, Paez wants to find an approach to the notion of explanation that is somewhere in between purely pragmatic approaches (such as van Fraassen's) and more usual approaches in which pragmatic considerations have no serious role. According to Paez, the agent's acquisition of new information depends on a balance between the explanatory power of this information and the risk of error. The paper focuses primarily on the explanation not of general laws but of particular facts. Three theses are central to his strategy (and a substantial part of the paper is a defense of each of them): first that the determination of all possible explanations of a fact is a non-pragmatic matter (Páez follows a probabilistic model of explanation); second that it is possible to determine the epistemic value of most potential explanations of a fact in a non-arbitrary way, although such value is the result of the evaluation of individual researchers; third that the criteria of acceptance of an explanation in the corpus of beliefs of researchers are based on their joint assessment of the credibility and epistemic value of potential explanations. Paez also discusses the reasons for there being a widespread opinion that probability values are essential for explanation (contrary to his own opinion). He examines Hempel's, Salmon's and Humphey's theory and argues that they avoid the epistemic relativity of explanation (something that Paez takes to be necessary) but only at a very high price. Paez takes up a suggestion by Friedman (1974) and tries to develop a way of looking at understanding in terms of common doxastic states and goals of participants in scientific activity. The epistemic framework that he uses is Isaac Levy's versión of Peirce's belief-doubt model in which the understanding of a fact expressed by a proposition P depends on how well P fits into the agent's cognitive system. There are some conditions that a set of statements must satisfy in order to provide an explanation of a given fact, and these conditions are related to the previously existing beliefs of the agent or of a community of agents. In the last section Paez analyses (and rejects) several proposals for judging the epistemic value of an explanation and, finally proposes his own solution.

Logic and Language

Eduardo Barrio's "La Paradoja de Yablo y las Omega-Paradojas" discusses Yablo's Paradox, which is generated by an infinite sequence of sentences each one stating that the following sentences in the list are all false. According to Yablo, this paradox does not involve self-reference and, hence, does not depend on some sort of circularity. In the first part of the paper Barrio analyses some attempts to formalize the paradox using first order arithmetic. He shows that Priest's formalization generates a theory that is consistent but omega-inconsistent and argues that this is a serious defect. On the other hand, in a second order language two principles necessary to generate the paradox lead to a theory that does not generate a formal contradiction but that has no models either. Barrio's conclusion is that these formalizations are inadequate. In the second part he discusses whether the paradox really does not involves some form of circularity (as Yablo claims) and considers some criticism of this claim. One of them was raised by Priest (who argues that the existence of Yablo's list in a formalized version does require some implicit circularity). Another one by Beall (who understands that such a principle involves epistemic circularity). The last part of the paper examines formalizations that do not use arithmetic but rather, for instance, an infinitary logic or several satisfaction-predicates. Barrio's conclusion is that it is doubtful whether these attempts of formalizing of Yablo's paradox are successful.

Luiz Fernández Moreno's "Kripke's Interpretation of the Theories of Frege and Russell in Comparison with Mill's Views" discusses Kripke's interpretation of Frege's and Russell's theories of proper names (and of what they have in common) in *Naming and Necessity* (1980). He also wants to establish how these theories resemble or diverge from Mill's theory (as stated by Kripke). Fernández Moreno first reviews the details of Frege's doctrine of the sense of a proper name and touches on the interpretative question of whether the latter regards the sense of all names as equivalent to the sense of some definite description. (Contrary to Dummett, he endorses the affirmative view.) Next he reviews the details of Russell's theory of definite descriptions having as background the distinction between knowledge by acquaintance and knowledge by description. He contrasts both Frege's and Russell's views with Mill's, and checks where they coincide and where they differ. Finally, he explains the assumptions under which Kripke's assimilation of Russell's theory of proper names to Frege's doctrine of sense and reference can be seen as plausible.

Francisco Hernández' "Logic of Programs as a fuelling force for Semantics" studies the prospects of establishing equivalences between axiomatic semantics and other semantics for programming languages. He starts by pointing out that many different kinds of semantics for such languages have been proposed, as well as many dichotomies that are meant to give some orientation in this "jungle" (e.g., denotational versus operational semantics, which roughly corresponds to the dichotomy between platonism and formalism). Axiomatic semantics was originally conceived as a series of formal tools for evaluating the reliability of softwares. One way in which this semantics can proceed is by means of a specification of formal inference rules of the conditions under which a program can move from one state to another. One can raise two objections against such proposal: a) that is not a semantics properly speaking but rather a verification tool, b) that, given the variety of programs that may be designed for a single language, one cannot suppose that each of them provide the semantics for that language. Hernández replies that the specification of conditions for a program that takes a given state to another is similar, for instance, to Davidson's program of providing the meaning of an assertion by means of the determination of its truth conditions. Regarding the second criticism, Hernández's reply appeals to a domain theory that allows the derivation of a logic of programs from a denotational semantics, thereby avoiding the ad hoc character of the systems. However, as the author remarks, this is not the ultimate solution for the problem since domain theory can only yield a relatively simple program logics.

Eleonora Cresto's "Algunos Problemas de Epistemología Social: Probabilidades Grupales Dinámicas" searches for an adequate definition of group probability. This is a particular case of a more general problem, i.e., of combining propositional attitudes of different members of a group so that the group itself can be seen as the agent of a collective attitude. Cresto first faces the more basic problem of the combination mechanism, i.e., what are the group probabilities that one must aggregate. If we decide to add posterior probabilities, we have to make clear how they depend on individual attributions. Cresto argues that they are individual conditional probability functions having the collective knowledge as condition. This is usually understood in at least two different ways: common knowledge and distributed knowledge. The first is, roughly speaking, the kind of knowledge shared by all members of the group and such that they know that they share. The second is that kind of knowledge that a subject would have by collecting everybody else's individual knowledge. Cresto's proposal is a dynamic concept in which ideal members of a group follow the ideal of incorporating the distributed knowledge until they transform it into shared knowledge. The logic of public announcements provides the tools to follow this path. Among the conceptual tools that she uses is the logic of public announcements in which the accessibility relation between worlds change successively as the agents make public their knowledge. The models that she uses include, besides the traditional models, probability functions for each agent in each world and paths that lead from distributed knowledge to common knowledge at a given moment (although the individuals may acquire information by other means). At the end she discusses the consequences of combining individual probabilities using a geometric aggregation function.

The particular selection of topics included in this volume is, of course, not exhaustive. It is only meant as representative of research done in a family of issues that currently occupy philosophers (and, in some cases, mathematicians) in Latin America and Spain. We hope it can trigger further research and development in these and related issues.

We would like to thank Mauricio Andrade for his help in the edition of this text.

List of Contributors

Jonas R. Becker Arenhart is Associate Professor of Philosophy at the Federal University of Santa Catarina, in Florianópolis, Brazil. He received his PhD in Philosophy from the Federal University of Santa Catarina in 2011, under the supervision of prof. Décio Krause. His main areas of interest are the philosophy of quantum mechanics and philosophy of logic, with special interest in the metaphysical aspects of such investigations. Currently, he is Associate Editor of Principia: an international journal of epistemology.

Axel Arturo Barceló Aspeitia holds a research position at the National University of Mexico's Institute for Philosophical Research where he studies human representations (words, formulas, pictures, diagrams, etc.) and their use, especially in inference and argumentation. He got his PhD at Indiana University, Bloomington. He was awarded the National University Recognition of Distinction Award for Young Researchers in the Humanities, and throughout his academic career, he has published more than 40 journal articles and book chapters in Mexico and abroad. Recent publications include "Las Imágenes como Herramientas Epistémicas" (*Scientiae Studia* 14(1) 2016) "What makes quantified truths true?" (*Quantifiers, quantifiers, quantifiers*, edited by Alessandro Torza, Synthese Library 2015), "Words and Images in Argumentation" (*Argumentation* 26(3) 2012) and "Semantic and Moral Luck", in *Metaphilosophy* (43(3) 2012).

Eduardo Alejandro Barrio (Ph.D. - University of Buenos Aires) is Full Professor (Tenure Truck) at University of Buenos Aires and Principal Researcher at National Scientific and Technical Council (Conicet). He works in the philosophy of logic and his main research interests include the notion of *truth*, especially, the semantic paradoxes and the expressive limits of formal languages. He has also done some work on substructural logics. Some of his papers appeared published in the *Journal of Philosophical Logic, Review of Symbolic Logic, Studia Logica, Analysis, Synthese, Journal of Applied Non-Classical Logics* and *Logical Journal of IGPL*. He has also published three books: *La Verdad Desestructurada* (Buenos Aires, EUDEBA, 1998), *Paradojas, Paradojas y más Paradojas* (Londres, College PU, 2014) and *La Lógica de la Verdad* (Buenos Aires, EUDEBA, 2014).

Roderick Batchelor is currently Lecturer in Philosophy at the University of São Paulo. He received his Ph.D. from the University of London in 2004 (thesis: *Investigations in Modal Logic*). He specializes in formal logic, philosophy of logic, and ontology. His publications include "Grounds and Consequences" (*Grazer Philosophische Studien*, 2010), "Topic-Neutrality" (*Mind*, 2011), and "Complexes and Their Constituents" (*Theoria* [Lund], 2013).

Walter Carnielli is professor of logic and philosophy at the Department of Philosophy, and member and former Director of the Centre for Logic, Epistemology and the History of Science of the University of Campinas (Brazil). He received his PhD in Mathematics from the University of Campinas and held research positions at the University of São Paulo, University of California Berkeley, University of Münster, University of Bonn, Superior Technical Institute of Lisbon, and Université du Luxembourg. He is the author of circa 100 scientific papers and several books on theory and application of contemporary logic and set theory, focused on foundations of reasoning.

Marcelo E. Coniglio is Full Professor in Logic at the Philosophy Department of the Institute of Philosophy and the Humanities (IFCH) of the University of Campinas (UNICAMP), and he is member and Director of the Centre for Logic, Epistemology and the History of Science (CLE) of UNICAMP (period 2016/2018). He was President of the Brazilian Logic Society (SBL) from 2014 to 2017. He obtained his PhD in Mathematics from the University of São Paulo (Brazil) in 1997. He has published more than sixty refereed articles and published and edited six books on different subjects of Non-Classical Logic, with emphasis on Paraconsistent Logic and Combination of Logics.

Eleonora Cresto (Ph.D. Columbia University, 2006) is a Permanent Researcher at the CONICET (National Council for Scientific and Technical Research, Argentina), and she is also Full Professor at UNTREF and Universidad Torcuato Di Tella (Argentina). Her main areas of research are formal and mainstream epistemology, epistemic logic, decision and game theory, and social choice theory. Some of her papers include "Lost in Translation: Unknowable Propositions in Probabilistic Frameworks" (*Synthese*, 2017); "Confirmational Holism and the Amalgamation of Evidence" (with M. del Corral, D. Tajer, A. Cassini y J. Nascimbene, *Recent Developments in the Philosophy of Science*, Springer, 2016); "A Defense of Temperate Epistemic Transparency" (*The Journal of Philosophical Logic*, 2012); and "On Reasons and Epistemic Rationality" (*The Journal of Philosophy*, 2010).

Luis Estrada-González, a proud member of the Logicians' Liberation League, is a permanent research fellow at the Institute for Philosophical Research at the National Autonomous University of Mexico (UNAM), having previously worked at the Universities of Tartu (Estonia) and Groningen (The Netherlands). He holds a Ph.D. in Contemporary Philosophy from the Autonomous University of the State of Morelos, which he earned under the supervision of Ivonne Pallares-Vega. His Erdös number is not small enough to brag about. He has taught courses in logic and related areas almost uninterruptedly since 2009. His main research interests all lie within the areas of logic and its philosophy and the philosophy of mathematics. Some highlights of his published work include "Models of possibilism and trivialism" (*Logic and Logical Philosophy*, 2012), "Remarks on some general features of abduction" (*Journal of Logic and Computation*, 2013) and "Through Full Blooded Platonism, and what Paraconsistentists could find there" (*Logique et Analyse*, 2016).

Luis Fernández Moreno got his PhD at the Free University of Berlin and he is a Professor of Logic and Philosophy of Science at the Complutense University of Madrid. He was the Head of the Department of Logic and Philosophy of Science at that university from 2005 to 2013, and from 2016 to 2017. He has had stays as a researcher at the Department of Philosophy of the *Trinity College* of the University of Dublin, at the Department of Philosophy of Harvard University and at the *Center for the Study of Language and Information* (CSLI) of Stanford University. He has published the books *Wahrheit und Korrespondez bei Tarski* (Königshausen & Neumann, 1992), *La referencia de los nombres propios* (Trotta, 2006) and *The Reference of Natural Kind Terms* (Peter Lang, 2016). He has edited the books *Ensayos sobre lenguaje, naturaleza y ciencia* (2007), *Para leer a Wittgenstein: lenguaje y pensamiento* (2008) and *Language, Nature and Sciencie: New Perspectives* (2009), having been co-editor of the book *Cuestiones filosóficas. Ensayos en honor de Eduardo Rabossi* (2008). He is likewise author of over eighty papers in specialized journals and collective works, mainly on subjects of Philosophy of Language and logical-philosophical issues.

Mario Gómez-Torrente (Lic. Fil., Barcelona, 1990; Ph.D. in Philosophy, Princeton, 1996) is a researcher at the Instituto de Investigaciones Filosóficas of the Universidad Nacional Autónoma de México (UNAM). He works mainly on topics in the philosophy of language and the philosophy of logic. His papers include "Tarski on Logical Consequence" (Notre Dame Journal of Formal Logic, 1996), "The Problem of Logical Constants" (Bulletin of Symbolic Logic, 2002), "Rigidity and Essentiality" (Mind, 2006), "How Quotations Refer" (Journal of Philosophy, 2013) and "Perceptual Variation, Color Language, and Reference Fixing. An Objectivist Account" (Nous, 2016).

Francisco Hernández Quiroz has a B.A. in Mathematics and Philosophy from UNAM and a Ph.D. in Computer Science from Imperial College (London). He is currently professor of the Department of Mathematics and a collaborator of the Graduate Program in Philosophy of Science, both at UNAM. He supervises students both in Computer Science and Philosophy of Science. His fields of interest are modal logic and its applications (both in Computer Science and Philosophy) and computation theory and applications, and he has several publications in these fields.

Décio Krause is professor of logic and philosophy of science at the Department of Philosophy of the Federal University of Santa Catarina. He is retired as professor of foundations of mathematics of the Department of Mathematics of the Federal University of Paraná. His got his PhD by the Department of Philosophy of the University of São Paulo with a thesis in the logical foundations of quantum mechanics. His main interests are in the logic and the metaphysics of quantum theories, the philosophy of logic, and the applications of non-classical logics to the foundations of scientific theories.

Eleonora Orlando got a MA Degree from the University of Maryland and a Ph.D. from the University of Buenos Aires. She is currently a professor at the Philosophy Department of the University of Buenos Aires and a researcher at the National Research Council of Argentina (CONICET). She specialises in different topics in the philosophy of language, more specifically, general term rigidity, the semantics of fictional terms, the contextualist and relativist debates and the relation between semantics and ontology. Recently, she has also focused on the semantic analysis of aesthetic judgements and its relation to aesthetic experience. Her publications include the books *Concepciones de la referencia* (Eudeba, 1999), *Significados en contexto y verdad relativa. Ensayos sobre semántica y pragmática* (Título, 2015) and a variety of essays on general terms, fictional names, contextualism and relativism. She has been the president of the Argentinian Society for Philosophical Analysis (SADAF, 2015/2017) and the Latin-American Association of Analytic Philosophy (ALFAn, 2010/2012).

Andrés Páez (Ph.D., The City University of New York, Graduate Center) is Associate Professor of Philosophy at the Universidad de los Andes in Bogotá, Colombia, director of the Research Group in Logic, Epistemology, and Philosophy of Science. His work focuses on the philosophy of science and on formal, social, and legal epistemology. Among his publications are the book Explanations in K (Oberhausen: Athena Verlag, 2007) and the articles: "Artificial Explanations: The Epistemological Interpretation of Explanation in AI," Synthese 170: 131-146, 2009; "Probability-Lowering Causes and the Connotations of Causation," Ideas y Valores 151: 43-55, 2013; "The prediction of future behavior: The empty promises of expert clinical and actuarial testimony." Teoria Jurídica Contemporánea 1: 75-101, 2016; and "Reputation and Group Dispositions." Review of Philosophy and Psychology

Abilio Rodrigues Filho is Associate Professor of Philosophy at the Federal University of Minas Gerais, Belo Horizonte, Brazil. He received his PhD from the Pontifical Catholic University of Rio de Janeiro, Brazil in 2007. He is the author of several scientific papers and his main areas of interest are logic, philosophy of logic, paraconsistency and intuitionism.

Fernando Soler-Toscano is Associate Professor at the University of Seville, Spain. His background is both in Philosophy and Computer Science. His research focuses on formal models of abductive reasoning in dynamic epistemic logic. He also works on algorithmic complexity measures and mathematical models of the brain. He is a foreign collaborator of the Lisbon Center for Philosophy of Sciences.

Giorgio Venturi is Professor of Logic at the State University of Campinas (UNICAMP), since 2017. He earned his PhD in philosophy at the Scuola Normale Superiore (SNS) and his PhD in mathematics at the Université Paris Diderot (Paris 7), in 2014. His main areas of interest are set theory and the foundations of mathematics, but he also contributed to the fields of modal logic and the history of logic. Since 2012 he is member of the Association for Symbolic Logic (ASL) and he regularly publishes research papers in philosophy, mathematics and logic. He serves as referee for many international journals and for the Mathematical Reviews of the American Mathematical Society (AMS).

Logic and Ontology

The Problem of Relations

Roderick Batchelor
University of São Paulo

Contents

1	**Introduction**	**4**
2	**Identical symmetry**	**7**
3	**Orderism and Directionalism**	**8**
4	**Positionalism and Variablism**	**12**
	4.1 Fine's objections	12
	4.2 Svenonius	15
	4.3 Three further objections to Positionalism and/or Variablism	16
	4.3.1 Failure of Uniqueness of Substitution under Positionalism	16
	4.3.2 Reflexion	16
	4.3.3 The Love of Variables	18
	4.4 Open Situationism and the Incompleteness Thesis	18
	4.5 Nuclear and Anti-Nuclear Open Situationism	19
5	**Open Situationism, complex attributes, and higher-order attributes**	**22**
	5.1 Some terminology and some preliminary remarks	22
	5.2 Connexions between Open Situationism and the questions of existence of complex and higher-order attributes	24
	5.3 Further arguments for Anti-O.S. + Simplicity Thesis + Elementarism	26
	5.3.1 An argument for Elementarism	27
	5.3.2 An argument against Open Situationism + Higher-Orderism	28
	5.3.3 An argument for the Simplicity Thesis, via an argument for Atomism	29
	5.4 Attributes, conditions, and situational functions	32
6	**Miscellaneous theories**	**35**
	6.1 Fine's theory	35
	6.2 Russell 1913	37

| 7 | Symmetricalism | 40 |
| 8 | Concluding remark | 45 |

1   Introduction

The present paper is about the following question: *What is the nature of the combination between a relation and its terms in a relational situation?* Such combination may be called 'relational predication'; so our question is that of *the nature of relational predication*.

We will make certain *assumptions* here. We will assume that there *are* relations, as *sui generis* entities (rather than as sets of sequences, or functions from worlds to sets of sequences, or anything else). We will also assume that there are situations (also known as 'states of affairs'), again as sui generis entities. Such situations we take to include both *factual* situations – situations which obtain, are the case – and *counterfactual* situations – situations which do not obtain, are not the case.

We use also the term *attribute*, to encompass both properties (also taken here as sui generis entities) and relations. And whatever is not an attribute may be called a *substance*.

Our question then is how a relation combines with its terms (relata) in a relational situation. This is what I call 'the problem of relations'. I call it a *problem* because it is not clear to me what the correct answer to the question is.

Now when it is not clear to one what the correct answer to a question is, this may be for one of two reasons. It may be because one can think of no answer to the question which is at all plausible. Or it may be because one can think of many different answers to the question, each of which is not without some plausibility but none of which is clearly superior to all the others. The present case is of this second kind. We will distinguish no less than six types or groups of theories, most of which sub-divide into some, and sometimes *many*, alternative versions.

Several of these types of theory (though not all, and also not many relevant sub-divisions) are discussed in Kit Fine's paper 'Neutral relations', to which it will be obvious that I am much indebted. Indeed the present work may be regarded as an attempt to carry further the themes of that paper. Fine's own positive theory, however, I do not find congenial (for reasons which I explain in §6 below).

The *first* of our six types of theory is what I call *Orderism*: this says that, in a relational situation, the relation applies to the relata *in a certain order* – first, second, third, etc. Such order is supposed to be a fundamental feature of relational situations, not analyzable in other terms.

The *second* type of theory is what we will call *Directionalism*: this says that the relational situation contains a certain (basic) *directionality* or *sense*: the relation goes 'from' one term 'to' another (and 'through' others if the relation is more than binary).

The *locus classicus* of this view is Russell's *Principles of Mathematics*. 'It is a characteristic of a relation of two terms that it proceeds, so to speak, *from* one *to* the other. This is what may be called the *sense* of the relation ... The sense of a relation is a fundamental notion, which is not capable of definition' (§94).

There is a certain affinity between Orderism and Directionalism, but surely they are not identical: order is one thing, and direction another. (Curiously the two views even evoke different notations for [binary] relational situations: Orderism evokes Rab, whereas Directionalism evokes aRb.) In Fine's paper, however, (and in various other places in the literature) the two are conflated: what Fine calls 'the standard conception of relations' he formulates now in terms of order, now in terms of 'sense' or 'direction'. I should say also that I rather doubt that either Orderism or Directionalism or even their disjunction deserves to be called 'standard conception'.

The *third* type of theory is what we will call, following Fine, *Positionalism*: according to this, it is a basic feature of relations that they contain 'argument-places' or 'positions', which in the relational situation are then occupied by the relata. Of course if this holds for *relations*, then it no doubt holds for *properties* as well.

The *fourth* type of theory is what I will call *Variablism*: this appeals to the idea of Objective Variables, alias Variable Objects (supposed sui generis worldly correlates of the symbolic variables 'x', 'y' etc.), and says that it is a fundamental feature of relations (or at least of some relations) that they contain such Objective Variables, which in the relational situation are then replaced by the relata. Again, if this holds for relations then it no doubt holds for properties as well.

As with Orderism and Directionalism, here too with Positionalism and Variablism we have two closely kindred theories, though again not identical. (For one thing, when a Variable is replaced it disappears, but when a position is occupied it is still there though now no longer empty but occupied.) However, it will be convenient to have a special term to mean the disjunctive thesis 'Positionalism or Variablism'; and we will use *Open Situationism* as that term – since under both views attributes are naturally regarded as worldly correlates of open sentences (just as situations are worldly correlates of [closed] sentences), whether written in the style '__ loves ...' or in the style 'x loves y'.

I will argue in this paper that (although this may sound surprising) there is a close connexion between (i) Open Situationism and (ii) the thesis that there are *complex* attributes and (iii) the thesis that there are *higher-level* attributes (attributes of attributes). Namely: if one believes in one (whichever one) of these three theses, then one ought to believe also in the other two! Thus the problem of relations turns out to have wider repercussions than might have been expected.

In my *fifth* group of theories (which in this case is really a mere 'group', not a very cohesive 'type') I have included miscellaneous theories which attempt to account for the supposed possibility of multiple completions of the same relation by the same relata (like [John loves Mary] and [Mary loves John]) *without* resorting to the above-mentioned ideas of order or direction or positions or Variables. This essentially negative characterization leaves room of course for many possible views; but here we will consider only two specific views of this kind – viz. the respective theories of Fine 2000 and Russell 1913.

The *sixth* and last type of theory is what I call *Symmetricalism*. It is the radical view that actually there is always only *one* situation which results from the application of a given relation to suitable relata, in which situation there is no asymmetry whatever among the relata with respect to the situation (just as e.g. there is no asymmetry whatever among the elements of a set with respect to the set). In other words all relations are what Fine calls *strictly symmetric*, though I myself prefer to say *identically symmetric*: they are all like parallelism, adjacency, etc., in that they have only one 'completion' or 'value' for any given terms (the situation [a is parallel to b] = the situation [b is parallel to a]). The application of a relation to the terms in a relational situation is not then essentially different from the application of a *property* to its subject in a property–object situation: in both cases we have one or more things (the subjects), and in the situation the attribute applies to *those things* – e.g. a and b, *that is* b and a. – Prima facie, of course, one would have thought that there are non--identically-symmetric as well as identically symmetric relations; but according to the Symmetricalist this is an illusion. He then may – or may not – try to give some sort of 'reduction' of non--identically-symmetric relations in terms only of properties and/or identically symmetric relations.

As I have already indicated above, I do not *know* what the solution to this problem of relations is. I will not however hide the fact that to one of the possible views I am more inclined, or less disinclined, than to the others: that one is Symmetricalism. But this is a preference, not a firm conviction. The purpose of the present paper is not to 'defend' the 'claim' that Symmetricalism is the true view. Rather my principal aims have been (i) to classify the main alternative views and sub-views – to draw as it were a map of dialectical space –, and (ii) to make various remarks about the various possible theories.

We begin with a preamble (§2) on the notion of identical symmetry. Then in §3 we discuss the two related views Orderism and Directionalism; in §§4–5 we discuss Positionalism and Variablism (§4 containing more 'direct' considerations and §5 developing the aforementioned connexion between Open Situationism and the questions of existence of complex and higher-order attributes); in §6 we discuss the other, miscellaneous theories; and in §7 Symmetricalism. The final §8 is a brief concluding remark.

2 Identical symmetry

We may say that a relation R is *materially symmetric* if it never has two values, for the same objects as arguments, which are not *materially* equivalent. We may say that R is *strictly symmetric* if it never has two values, for the same objects as arguments, which are not *strictly* equivalent (i.e. necessarily materially-equivalent). (By 'necessity' in this paper I always mean so-called *metaphysical* necessity.) And we may say that R is *identically symmetric* if it never has two values, for the same objects as arguments, which are not *identical*.

Thus if R is identically symmetric then for any given objects in its 'range of significance' (or better 'range of application') there will be a single situation which results from the application of R to such objects. And if R is materially, or strictly, symmetric, there may be several such situations but they will all be materially, respectively strictly, equivalent.

It should be noted that we avoid here formulations like

$$\forall x \forall y (Rxy \leftrightarrow [\text{or } \backepsilon, \text{ or } =] Ryx).$$

(By the way the formulas with $\leftrightarrow$ and $\backepsilon$ are easily seen to be equivalent to the more usual ones with $\rightarrow$ and $\dashv$ respectively.) For the notation Rxy (or xRy) has really no neutral meaning antecedent to an account of the nature of relational predication; whereas the formulation adopted above is more neutral in this sense. Once Rxy is given a suitable meaning, however, the present formulas will normally be equivalent to our formulations above.

(For the sake of simplicity I give throughout this paper a 'possibilist' interpretation to quantifiers, whether verbal or symbolic. But it is usually easy to reformulate things in terms of 'actualist' quantifiers. Here e.g., for the notions with $\backepsilon$ or $=$, it would be enough to replace $\forall x \forall y$ by $\Box \forall x \Box \forall y$.)

There is also an abnormal possibility, however, viz. always only one of Rab and Rba might make sense – e.g. only [Socrates lived in Athens] not [Athens lived in Socrates]. Such are the relations called 'heterogeneous' in Russell 1913. They count thus as identically symmetric by our official definition, since there is only one value for any suitable arguments.

Since it is *identically symmetric* that is the crucial notion for the present discussions, it is tempting to appropriate the simple term 'symmetric' for it. But I think this would be too grave a trespass, since usage of the term in the material sense is very firmly established. We will however avoid circumlocution by using the abbreviations 'i.s.' for 'identically symmetric' and 'non-i.s.' for 'non--identically-symmetric'.

We should note that there are of course also 'partial' notions of symmetry, i.e. relative to *certain* 'argument-places'. This has no *neutral* general definition, but will have its respective (and usually straightforward) formulation within each conception of relations.

3 Orderism and Directionalism

We come now to Orderism and Directionalism. Since the considerations concerning the two are often analogous, we will concentrate on Orderism and then later point out the similarities and differences with respect to Directionalism.

Two versions of Orderism may be distinguished:

Strong Orderism: *Every* relational situation consists of a relation attributed to its terms in a certain order (first, second, etc.).

Weak Orderism: *Some*, but *not all*, relational situations consist of a relation attributed to its terms in a certain order.

The ones that don't will be precisely the ones where the relation is identically symmetric.

(Note that despite the names Strong Orderism does not imply Weak Orderism: on the contrary, the two are incompatible theses.)

Strong Orderism is of course a more 'uniform' theory than Weak Orderism; which is a virtue. On the other hand, there *seem* to be some identically symmetric relations, and so Weak Orderism might be thought to be superior to Strong in that it allows for such relations. (Notice also that a duplication of atomic situations Rab, Rba, for R strictly symmetric, would violate the Principle of Independence of Atomic Situations.)

However, things are not so simple: there are also cases of *partial* symmetry, as in 'x is between y and z'. Here the Weak Orderist cannot say that there is no order at all, since we must distinguish e.g. [a is between b and c] from [b is between a and c]; and if he says that it is a normal case of order then he will have to distinguish [a is between b and c] from [a is between c and b], which is no less counterintuitive than distinguishing e.g. [a is parallel to b] from [b is parallel to a], the avoidance of which was precisely the motivation for going for Weak Orderism in the first place instead of the more uniform Strong Orderism. Perhaps a Weak Orderist might say that, in [a is between b and c], there is one first term, a, and *two second terms*, b and c, i.e. c and b. But this is an idea of doubtful intelligibility. (And to make things worse there are also the 'more complicated symmetries' which we will see later in §4; for such cases even this desperate move of more than one n-th term will not do.) – I surmise from these considerations that Strong Orderism seems to be a *better* theory than Weak Orderism.

(Fine formulates the *Strong* version of Orderism as what he calls 'the standard conception of relations'. Although he adduces *against Positionalism* the consideration that it seems incapable of accommodating i.s. relations, in fact the same consideration applies just as much against the 'standard conception'. – Armstrong 1997 [§§6.4 and 8.2] recommends Weak Orderism / Directionalism. Like Fine, he speaks now of order, now of direction, as if they were the same thing. He goes even further, however, and throws some Positionalism into the mixture. He seems to think [and gives also a quotation from Grossmann 1983 §67 to similar effect] that, in relational situations where the relation is non-i.s., there is order, or direction, *among the argument-places*!

– This is theoretical over-shooting if anything is.)

Objections to (Weak or Strong) Orderism: –

(1) There is a sort of 'intuitive' objection: What *is* 'first', 'second', etc.? These do not seem to be fundamental notions. We have eliminated order in the form of *sequences* (as a supposed fundamental) from the foundations of mathematics (through such devices as the Wiener–Kuratowski definition of ordered pairs in terms of sets); we would not like to re-introduce it in the metaphysics of relations.

(2) Under Orderism, it seems to be at least *conceivable* that there should be two ontically simple relations, R and its 'converse' $\check{R}$. Then no doubt the situation Rab would be different from (though strictly equivalent to) the situation $\check{R}ba$. But this is intuitively implausible, as Fine explains at length. (Cp. also Russell 1913, p. 87: 'It seems ... so obvious as to be undeniable that, when we think of what actually takes place rather than of its verbal expression, we cannot find a vestige of difference between x preceding y and y succeeding x. The two are merely different names for one and the same time sequence.' – The theme has an older history than one might perhaps expect: see Hansen 2016 and references therein. Williamson 1985 quotes even a [very apt] aphorism from Heraclitus, viz.: 'The way up is the way down'.) And moreover it gives a facile violation of the Principle of Independence of Atomic Situations; there would be a 'brute' necessary connexion between R and $\check{R}$.

There are several *coherent* responses which the Orderist can give to this objection concerning converses, but none of them seems to be really *satisfactory*.

First, the Orderist can take a *Monist* view, that for each pair of supposed converse relations like before/after, above/below etc., there is always really only *one* relation. This is certainly an 'economical' view; but also sort-of arbitrary. After all, *which* one? (Is it *before*, or is it *after*? Etc.) – We may distinguish two sub-views. *Skeptical Monism* says that basically we never know which one it is. (A melancholy condition.) And *Cognitivist Monism* is the more optimistic view that at least in some cases we *do* know which one it is, or at least have good justification for belief. One might appeal to what seems to us intuitively more 'natural'. E.g. *loves* seems more 'natural' than *is loved by*, *hates* than *is hated by*, *kicks* than *is kicked by*, *materially implies* than *is materially implied by*, *before* than *after*, *smaller than* than *greater than*, *belongs* (to set) than *has as element*, *is subset of* than *has as subset*, etc. etc. Though it is true that there are more 'difficult' cases (*above* or *below*? *left* or *right*?). – But of course even in the 'easier' cases the Skeptic may accuse the Cognitivist of excessive confidence in the reliability of superficial psychological impressions.

Secondly, the Orderist can take a *Pluralistic* view, that there are always (for any non-i.s. relation R) both R and $\check{R}$. (It is no doubt absurd to think that *sometimes* there is just one of the two, and sometimes both!) We can then distinguish between the more straightforward *Monopredicativistic Pluralism* according to which always $Rab \neq \check{R}ba$ (a given relational situation can only be the value of *one* of R, $\check{R}$), and the less straightforward *Polypredicativistic Pluralism* according to which always

$Rab = \check{R}ba$ (the same situation a value of *both* R and $\check{R}$). (Again no doubt it is absurd to think that *sometimes* $Rab = \check{R}ba$ and sometimes not!) Monopredicativistic Pluralism is subject to the objections already indicated above; though of course it is not an *incoherent* view. Polypredicativistic Pluralism on the other hand goes against the apparently obvious facts that (1) a relational situation must be the completion of a *single* relation (at any rate of a single *simple* relation); and (2) under Orderism, in a situation Rab (at least for R non-i.s.) the object a is *the* first term of the situation and b *the* second – the given order, and no other, should be written into the nature of the situation. (Incidentally, this [dubious] 'poly-orderist' idea might be used also by the Strong Orderist to try to accommodate i.s. relations: he might say that [e.g.] Adjacency(a, b) = Adjacency(b, a) – there is always order but not always a *single* order!) And although Polypredicativistic Pluralism avoids a counterintuitive proliferation of *situations*, still it allows a counterintuitive proliferation, more specifically (in the binary case) *duplication*, of *relations*.

– It is worth mentioning here that the very characterization of the concept of converse under Orderism is a delicate matter. (The literature is full of careless formulations. E.g.: 'If a non-symmetric binary relation R applies to a and b (in that order), a converse R^* may be defined as the relation that applies to b and a (in that different order).' [MacBride 2007, p. 26.] Even in Fine's paper [p. 3, fn.1] we find: 'A converse [of a given relation] is one that differs from the given relation merely in the order of its arguments.' But of course there is no order of arguments *inside a relation*, only inside relational situations.) Perhaps the most plausible attempt is something along the lines of: the converse of R $=_{\text{df}}$ the relation S such that, for any objects x and y, the situation Sxy has the same *import* as the situation Ryx. This idea of 'import' however is of course far from completely pellucid. (We do not want to say *identical*, as we do not want to prejudge the case in favour of Polypredicativistic Pluralism.) Moreover the presumption of uniqueness (*the* relation S such that ...) might be questioned. An eccentric but curious view is that ˇ is an ontic-complexity--producing item, so that $(\check{R})\check{ } \neq R$ (like $\neg\neg p \neq p$ under a structural notion of situation). Would then not e.g. the *triple* converse of R give the same 'import' – so that the presumption of uniqueness would fail? – Or for that matter the double negation of $\check{R}$?

– Most of the considerations above apply equally to *Directionalism*: – We can distinguish Weak from Strong Directionalism, the Strong version being superior because it is more uniform and because the motivation for the Weak version cannot in any case be satisfied in cases of 'partial' (let alone 'more complicated') symmetries. 'Direction' does not intuitively seem to be a fundamental notion. And the problem of converses arises in the same way, and the similar reactions are possible.

(Russell 1903 advocates *Strong* Directionalism for his primary notion of relation [for 'propositional functions' on the other hand he has a Variablist view]. V. §§218 and 94; in the latter of which however a quaint exception is made for 'are two'. [Russell in fact refuses to call that a 'relation', but that is just his bad terminology.] Cf. also

3 Orderism and Directionalism 11

Gaskin & Hill 2012, esp. fn. 5 and corresponding text.)

One might complain also, against both Orderism and Directionalism, that they give no *explanation* of fixed degree ('arity'). What *makes* an e.g. binary relation binary? One possible answer is that it is simply the specific *contentual nature* of a relation that makes it have the degree it has. A more radical answer, which might be given either by the Orderist or the Directionalist (or others – e.g. the Symmetricalist), is that actually *all* attributes are multigrade and indeed can apply to *any* number of arguments (≥ 1). Is this so absurd?

– As with Orderism, also with Directionalism it is not clear how exactly one might characterize the notion of converse. Again, (under Directionalism) the direction is surely not inside the relation but only inside the relational situation: in aRb, R goes from a to b; *inside* R, surely there is nothing going from or to anything. So again it is nonsense to say that the converse of R is the relation which differs from R only in its direction. Not that there *might* not be something, inside the relation, which is connected somehow to (perhaps even accounts for) the capacity of the relation to apply to objects with a certain direction. And should this inner something be susceptible to 'inversion', then this would allow a characterization of the notion of converse. (Here, and below, the same applies to Orderism.) But the Directionalist *as such* is not committed to the further existence of this inner something, in addition to the directionality in relational situations. And nor can this inner something *replace* the directionality in the relational situation: whatever may be supposed to be *inside* the R, we still must be told what is the *mode of combination* of R with the relata in the relational situation. So Directionalism *with* the postulation of such 'inner something' is positing *two* things: the 'inner something', and the directionality in the situation. Thus the inner something by itself is *not sufficient*, and *nor* is it *necessary* to the Directionalist view as such. And of course one may then say that, by Ockham's razor, if it is not necessary then we should not postulate it.

(In Fine's paper however there is often suggestion of direction or 'bias' as an inner something, and that as if that was sufficient. See e.g. the penultimate paragraph of the paper [p. 32], where he speaks of the 'built-in bias' and thus 'internal complexity' of relations under the 'standard conception'; also page 16: 'biased relations appear to possess a genuine complexity, which only becomes disentangled once we distinguish their "content" and their "bias"'; also the picture of *the relation* as an arrow [p. 10], suggesting of course some kind of directionality as an aspect of the *inner* nature of the relation.)

– We note here the following perplexity, specific to Directionalism. The Directionalist says that, in aRb, the relation R 'goes' 'from' a 'to' b. But what does that really mean? If a person (or a ship, etc.) goes from one place to another, this is of course a temporal process, with beginning, middle and end in time, involving movement of the person across a certain stretch of space. A relational situation, on the other hand, like *any* situation, is no doubt a fixed, static thing, with the constituents in a fixed, as it were frozen, configuration. But then how are we to understand this idea that in the relational situation the relation 'goes' 'from' a 'to' b? Has it already *gone*? (Then it is *at b*??) Is it still going? Where is it? Is it moving? (If it is an infinitary relation, it keeps going forever?) – I find this puzzling. But I have called it a 'perplexity' and perhaps should not call it an 'objection': perhaps my temperament is over-Parmenidean, and a philosopher of more Heraclitean proclivities might find these things an 'attractive feature' of Directionalism.

4 Positionalism and Variablism

4.1 Fine's objections

Fine's main objection to Positionalism is simply that it is implausible to think that there are such entities as argument-places among the 'ultimate furniture of the world'. This is a very simple but also very compelling objection.

The similar objection applies to Variablism – i.e. it is implausible to think that there are Objective Variables among the 'ultimate furniture of the world'. – Fine himself does not consider what we are calling here Variablism; but I think he would agree with this objection to Variablism: despite his (in a sense) 'defence of arbitrary objects', he does clearly state at least in one place that he does not think that there really are arbitrary objects in the ontologically fundamental sense, among the 'ultimate furniture of the world': '...If now I am asked whether there are arbitrary objects, I will answer according to the intended use of "there are". If it is the ontologically significant use, then I am happy to agree with my opponent and say "no". I have a sufficiently robust sense of reality not to want to people my world with arbitrary numbers or arbitrary men.' (Fine 1985, p. 7.)

Another consideration which Fine adduces against Positionalism is that it cannot accommodate identically symmetric relations (nor multigrade relations). E.g. the adjacency relation according to the Positionalist would contain two argument-places, but then the situation where a occupies one of these argument-places and b the other cannot be identical with the situation where it is the other way round (supposing of course $a \neq b$). (Cf. Grossmann 1992 p. 57, which suggests a version of Positionalism and explicitly admits this consequence. Another curious peculiarity of this passage is that it introduces *positions* as a supposed precisification of the idea of *direction*!)

Note that *this* objection does *not* transfer to Variablism. For the Variablist can perfectly coherently say that the situation which results from the relation [x is adjacent to y] by replacement of the Objective Variable x by the object a and the Objective Variable y by the object b is *identical* with the situation which results from [x is adjacent to y] by replacement of x by b and y by a. (Just as e.g. the set which results from the set $\{a, b\}$ by replacement of a by c and b by d is *identical* with the set which results from $\{a, b\}$ by replacement of a by d and b by c.) (Note however that, if

4.1 Fine's objections

the Variablist reifies the adjacency 'nucleus' or 'primary relation' [on which see 4.5 below], then in order for him to coherently uphold the present identity claim he must combine it with a *Weak* form of Orderism or Directionalism for primary relations.)

Returning to Positionalism: – One naturally thinks of trying to escape the above objection either by (i) allowing positions which can be occupied by more than one argument (including variable numbers of arguments), or else by (ii) going for 'Weak' Positionalism rather than 'Strong', i.e. saying that relations *sometimes* contain positions, *not always*. – (i) seems better, because the 'hybridness' of Weak Positionalism sounds somehow even more grating than that of Weak Orderism / Directionalism.

Fine does briefly consider this answer (i), and gives the following interesting reply (which actually can be modified to apply to (ii) as well). (This is in his footnote 10. My own formulation does not follow Fine's very literally, but gives what I take to be his point. – Cf. also Svenonius 1987, p. 56.) Take the relation R expressed by the open sentence 'x, y, z are arranged in a circle (clockwise, and in that order)'. Corresponding to the variables x, y, z in the open sentence, there will be *at most* three argument-places in the relation, which we may call respectively α, β, γ. And given names n_1, n_2, n_3, we may write '$Rn_1n_2n_3$' to represent the completion of R where the objects denoted by n_1, n_2, n_3 occupy the positions α, β, γ respectively. But $\alpha \neq \beta$, since $Rabc$ need not even be materially equivalent to $Rbac$ (indeed the two situations are incompatible); and by similar reasoning $\alpha \neq \gamma$ and $\beta \neq \gamma$. So there are in fact *exactly* three positions in R; and so e.g. the situation $Rabc$ must be distinct from the situation $Rbca$. But intuitively these situations should be identical, as identical as [a is adjacent to b] and [b is adjacent to a].

Or to give a more homely example, take the quaternary relation S expressed by the open sentence 'x and y are playing tennis against z and u'; and call $\alpha, \beta, \gamma, \delta$ the corresponding argument-places of S. Again it is easily seen that we must have $\alpha \neq \gamma$, $\alpha \neq \delta$, $\beta \neq \gamma$, $\beta \neq \delta$ (though now *not* that $\alpha \neq \beta$ or $\gamma \neq \delta$). And so $Sabcd$ must, counterintuitively, be distinct from $Scdab$. (Cf. Fine 2007 fn. 2.)

What distinguishes these 'more complicated symmetries' from the simpler cases like 'x is parallel to y', or 'x is between y and z', is that now the permissible permutations of variables are not only within neatly segregated disjoint clusters.

(The following more precise characterization may be of interest to some readers. We may say that a permutation σ of $\{1, \ldots, n\}$ [$n \geq 2$] is a *strict symmetry* for the open sentence $\varphi(x_1 \ldots x_n)$ if $\forall x_1 \ldots \forall x_n \ (\varphi(x_1 \ldots x_n) \leftrightarrow \varphi(x_{\sigma(1)} \ldots x_{\sigma(n)}))$. A class G of permutations of $\{1, \ldots, n\}$ may be said to be *elementary* if there is a quasi-partition [i.e. something like a partition except that the empty set is allowed as an element] $\{A_1, \ldots, A_k\}$ of $\{1, \ldots, n\}$ such that, for every permutation σ of $\{1, \ldots, n\}$, we have: $\sigma \in G$ iff both (1) $\sigma(m) = m$ for all $m \in A_1$, and (2) $\sigma[A_i]$ (i.e. the set of values of σ for arguments in A_i) $= A_i$ for $i = 2, \ldots, k$. [Intuitively, $A_2 \ldots A_k$ correspond to the disjoint clusters of symmetry, and A_1 to the other variables.] The vague statement that φ has 'more complicated symmetries' can then be replaced by the precise statement: the class of all strict symmetries for φ is not elementary.)

Three Positionalist reactions to this argument are possible. *First*, the Positionalist might say that the argument shows that the idea of multiple occupancy does not work for these 'more complicated symmetries' and so is best dropped altogether: we should stick to the simple view 'one position, one object'. *Secondly*, the Positionalist might say fair enough, I'll distinguish e.g. $Rabc$ and $Rbca$ since I *have* to, but *not* [a is adjacent to b] and [b is adjacent to a], since I *don't* have to – *here* I'll use the multiple occupancy idea. And *thirdly*, the Positionalist might propose 'wedding positionalism with a sparse theory of relations' (considered in this connexion in MacBride 2007, p. 41) and Fine's argument may be taken to suggest that there really *are* no such supposed relations as this R.

The first Positionalist says two counterintuitive things but is uniform. The second says only one counterintuitive thing but is *not* uniform. (Between these two the first view, i.e. 'Basic Positionalism' – 'one position, one object' –, seems preferable. Uniformity [and hence simplicity] of theory seems a weightier consideration than being able to give what is only a limp half-solution to the symmetry issue.) The third is probably worst of all: as will become clear below (§5), Positionalism and a 'sparse' theory of attributes are a most ill-suited couple.

– There is a further curious question for a proponent of multiple occupancy in the case of multigrade relations, viz.: How would one distinguish a relational situation, e.g. [a and b raised the piano], from a corresponding 'partial completion' by the same objects, e.g. the property $\lambda x(x$ and a and b raised the piano)? No doubt in both cases a and b have been inserted into the position of the raised-the-piano relation; what else is different?

Indeed, a similar question can be posed even to the Positionalist who proposes multiple occupancy in the case of ordinary, non-multigrade relations: What is the difference between e.g. [a is parallel to a] and $\lambda x(x$ is parallel to $a)$? (Incidentally, on this view parallelism is not *quite* 'unigrade' since a can go alone into the position and produce a situation.)

– One last point I would like to make about Fine's discussion of Positionalism concerns his remark (p. 12): 'I doubt that there is any reasonable basis, under positionalism, for identifying an argument-place of one relation with an argument-place of another.' But surely e.g. if we take the ternary relation [x gave y to z] and fill its 'x-place' with say Socrates, we get a binary relation [Socrates gave y to z] with two argument-places in common with the earlier relation? Surely by going into the x-place Socrates does not destroy or modify the y- and z-places? – Perhaps however Fine was thinking only of *basic* relations at this point; for *that* what he says is much more plausible.

Finally, we should say something about *multigrade* relations under *Variablism*. Clearly Basic Variablism – 'one Variable, one object' – cannot accommodate multigrade relations. Corresponding to 'Multiple-Occupancy Positionalism', there is 'Plural Variablism', where some Objective Variables are allowed to be plural. But note that *here* the 'more complicated symmetries' do *not* cause trouble: e.g. with [X are playing tennis against Y] (where X and Y are Plural Objective Variables), the Plural Variablist can perfectly coherently say that what results from this relation by replacement of X by a and b and Y by c and d is *identical* with what results from it by replacement of X by c and d and Y by a and b.

4.2 Svenonius

The little-known but interesting paper Svenonius 1987 anticipates several of the themes of Fine's paper. Svenonius discusses a form of Directionalism / Orderism (§3); then a form of Positionalism (§4); and he gives the same two main objections against Positionalism (§5, part D, p. 45):

'... One such disturbing feature is that the theory seems to force us to treat the "empty places" ... as a new kind of "object". ... In the name of ontological economy, it seems that we should avoid the assumption of such strange entities, at least if we want to do something philosophically fundamental'. And then: 'There is a special reason against the assumption of these "places" as objects in the case of symmetries ...'.

Svenonius proposes to overcome the second, 'special' objection, by 'adding "indeterminacies"' and considering in particular the argument-places involved in a case of identical symmetry as 'indistinguishable'. (He is also aware of the special delicacy of the 'more complicated symmetries' as in Fine's fn. 10: he [Svenonius] says e.g. [pp. 51–2]: 'the indeterminacies that are the basis of the symmetries cannot always be reduced to pairwise indistinguishability of empty places, but consist in the fact that certain interchanges of empty places (which may have to involve more than two places) do not change anything'.) Regarding a binary case Svenonius says (p. 51): 'the number of "empty places" is two, but the places are *indistinguishable* (as two photons might be indistinguishable); because of that, there is really only one way of inserting two objects in the empty places (rather than two ways).' (Cf. also Leo 2008, §9.)

But really the objection, as can be seen from the clearer formulation of Fine's paper, depends only on there being *two* places, *not* on their being 'distinguishable' (whatever that means). – Also it is really the *first* objection that is more 'disturbing', and with this Svenonius's move to 'indistinguishable places' does not help at all – on the contrary, the 'strange entities' become even stranger.

Another point concerning this paper which may be worth mentioning is that it adopts an *intensional* criterion of identity for relations – which seems to me obviously unfortunate, as it is no doubt the finer 'hyper-intensional' or 'structural' level that is the proper environment for these issues.

4.3 Three further objections to Positionalism and/or Variablism

4.3.1 Failure of Uniqueness of Substitution under Positionalism

Positionalism violates what may be called the *Principle of Uniqueness of Substitution*: For any complex c, and for any occurrence o in c of a constituent a of c, and for any entity b of the same ontic category as a, there is a *unique* complex c' which results from c by the substitution of b for the occurrence o of a. This principle seems to me to have a certain inner plausibility, and it holds in all or almost all normal cases one can think of; but the Positionalist view violates it, since under this view there will not in general be unique results of substitution for relations in situations: when a relation is removed from a situation the relata as it were get scattered around, and there is not a unique way of reinserting them into the argument-places of a new relation. There is no general objective correspondence between the argument-places of different relations, to serve as ground for reassignment of terms.

(I have discussed this Principle of Uniqueness of Substitution, and proposed a treatment of the notions involved in its statement, in Batchelor 2013, to which interested readers may be referred.)

4.3.2 Reflexion

Let us use 'ghosts' to abbreviate 'positions (under Positionalism) or Objective Variables (under Variablism)'.

What is the Open Situationist to say about attributes or putative attributes like $\lambda x(x$ loves $x)$, where so to speak the ghosts of $\lambda xy(x$ loves $y)$ have been 'identified'?

There seem to be four main options:

Coordinative Open Situationism: In some attributes there is *coordination* of ghosts (yielding 'requirement' that they be filled or replaced by same object).

Coalescential Open Situationism: Some attributes are obtained from others by *merging* or *coalescence* of ghosts.

Open Situationism with Repetition of Ghosts: Some attributes contain *multiple occurrences* of the same ghost.

Bare Open Situationism: There is no such funny business (neither 'coordination' nor 'coalescence' nor 'ghost-repetition'), and really there are no such attributes as $\lambda x(x$ loves $x)$, only $\lambda xy(x$ loves $y)$.

But none of these options is without its difficulties. Coordinative Open Situationism and Coalescential Open Situationism are subject to a compounded form of the basic intuitive objection to Open Situationism: the coordination or coalescence of argument-places or Variables would here be supposed to be part of the basic account of the ultimate nature of relations (more generally attributes); but although the ideas of coordination or coalescence of argument-places or Variables, as well as the idea of argument-place or Variable itself, are no doubt *in some sense* legitimate, yet they seem

4.3 Three further objections to Positionalism and/or Variablism

to bear all the appearance of notions which ought to be *defined* somehow, rather than notions which enter into the most basic description of reality.

Also it would seem that both the Coordinative Open Situationist and the Coalescential Open Situationist would have to (counterintuitively) distinguish e.g. the situation $\lambda x(x$ loves $x)$(Socrates) from the situation $\lambda xy(x$ loves $y)$(Socrates, Socrates). For the Coalescential Open Situationist $\lambda x(x$ loves $x)$ has a single ghost, which in the situation $\lambda x(x$ loves $x)$(Socrates) is occupied or replaced by Socrates; so no doubt Socrates occurs only once here, but twice in $\lambda xy(x$ loves $y)$(Socrates, Socrates), and so the two situations cannot be identical. For the Coordinative Open Situationist on the other hand $\lambda x(x$ loves $x)$ does have two (coordinated) ghosts, so that presumably Socrates *will* occur twice in $\lambda x(x$ loves $x)$(Socrates). But in this situation the predicate is an attribute involving this 'coordination', which one would have thought does not magically disappear when Socrates enters the positions or replaces the Variables; and so this situation would again be distinct from $\lambda xy(x$ loves $y)$(Socrates, Socrates) where no such 'coordination' figures.

As for Open Situationism with *Repetition of Ghosts*, one puzzling question concerns the identity of the repeated ghost: is e.g. the ghost figuring twice in self-love one of the two ghosts of love? If so, *which*? (And if ghosts can be shuffled around, why is there not a result of *interchanging* the two ghosts of e.g. the amatory relation? – On this see further below.) If not, if it is third ghost, does it have some kind of relation to the original ghosts, or is it an entirely new ghost? – The *Positionalist* version of this view suffers from additional difficulties. *First*, again we would have to distinguish $\lambda x(x$ loves $x)$(Socrates) from $\lambda xy(x$ loves $y)$(Socrates, Socrates). And *secondly*, although one might think that there is nothing especially problematic about an *Objective Variable* occurring multiple times in a complex (beyond what may be problematic about *any* object occurring multiple times in a complex), on the other hand an argument-*place*, a *position*, seems more like a kind of *region* in the complex, so that its occurring several times in a complex seems a problematic idea – it seems like saying e.g. that the Earth could contain several North Poles. – I will add, however, that even with *Objective Variables*, the permission of multiple occurrences seems dubious: it seems to give to Objective Variables a dangerous degree of 'autonomy'. Socrates, Plato etc. are 'autonomous' objects and so can indeed be repeated in a situation, can be substituted for or by other objects, etc. If then Objective Variables can be repeated, why can they not be say interchanged? Where [x loves y] is the amatory relation, there would then be another relation, [y loves x], resulting from the interchange of the two Objective Variables. (Surely this would indeed be an*other* relation, since no doubt e.g. substitution of Socrates for x and Plato for y in the *former* relation would give the situation [Socrates loves Plato] but the same substitution in the *latter* relation would give *not* [Socrates loves Plato] but [Plato loves Socrates].) No doubt these two distinct relations would be 'converses' of each other – so that the 'Problem of Converses' is reinstated. Indeed here it would seem to appear in an especially puzzling form, as it is not easy to see which of the two relations should be considered *loves* and which *is loved by*. (Surely there isn't a tablet in Platonic – or Platonic–Finean – heaven on which is inscribed an Alphabetic Ordering of Objective Variables.) – Nor is it clear why we would have to stop at the inter-substitution of the two given Objective Variables and there could not be results of substituting *other* Objective Variables for these.

The *Bare* Open Situationist too is not without his awkwardnesses. Typical Open Situationism models attributes upon open sentences, i.e. sentences with free variables. But now suddenly each variable can only occur *once*. Thus e.g. quantification cannot be given a straightforward Fregean account: $\exists x(Px)$ is okay as the attribution of a second-level attribute to a first-level one, but not so $\exists x(Px \wedge Qx)$, or $\exists x Rxx$, which at first sight might seem equally legitimate.

4.3.3 The Love of Variables

Under *Variablism*, why is e.g. [x loves y] not a *situation*? If the Objective Variable x is a certain entity, and the Objective Variable y is a certain entity, why should [x loves y] be thought to be relevantly different from a situation like [John loves Mary]? Of course [x loves y] would presumably be a *counterfactual* situation: for although the Variable y might be loved (e.g. by a metaphysician), still the Variable x, or for that matter any other Variable, presumably does not love anything, nor has indeed any kind of feeling at all. But then e.g. [x does not love y] would be a fact – if one wants a fact rather than a counterfactual situation. Anyway the point here is: why are such things not *situations* as opposed to attributes?

This is a question which the Variablist would have to address. I will refrain (wisely, I think) from a detailed discussion of this question here, but merely mention two possible lines of response for the Variablist: –

(1) He might say that Variables, despite being entities, are entities of a special sort which cannot figure as subjects of situations at all. So if [$\dots a \dots b \dots$] is a situation, then [$\dots x \dots y \dots$] *must* be an attribute.

(2) He might distinguish Objective Variables *qua* Objects from Objective Variables *qua* Variables, and say that e.g. in the *fact* [Kit Fine loves x] the Objective Variable x occurs qua Object, whereas in the *property* [Kit Fine loves x] (i.e. the property of being loved by Fine) it occurs qua Variable.

4.4 Open Situationism and the Incompleteness Thesis

Let us use the name *Incompleteness Thesis* for the thesis that attributes (both properties and relations) are in some sense 'incomplete', 'unsaturated'.

If Open Situationism is true, then no doubt this Incompleteness Thesis is true too: containing positions or argument-places (also known by the less stately name of gaps), or containing Variables, is no doubt one (reasonable) sense of being 'incomplete', 'unsaturated'.

The converse passage, from the Incompleteness Thesis to Open Situationism, is much less compulsory: it seems perfectly coherent to accept the Incompleteness Thesis and nevertheless reject Open Situationism. All the same Open Situationism does provide a natural way – indeed *the* most natural way so far as I can see – of specifying the sense in which attributes are supposed to be 'incomplete', 'unsaturated' (should they *be* supposed to be that).

Frege was of course an explicit proponent of the Incompleteness Thesis. But although one naturally thinks of him as a kind of proto-Positionalist, indeed as the proto-Positionalist-maximus, still he cannot really be said to be an *Explicit* Positionalist: he never says anything as definite as that concepts and relations *contain Argument-Places* as *ENTITIES*; and indeed one imagines that he would *deny* this if asked.

Perhaps also Williamson, who has been reckoned by Fine and others as an Explicit Positionalist (on account of the paper Williamson 1985), would deny this. (Although he does in one place [p. 261, fn. 20 *ad fin.*] go so far as to allow argument-places to occur themselves as relata to other relations; which would appear to suggest that argument-places are not only being reified but even being taken as 'self-sufficient' entities which can go about their business elsewhere, rather than mere 'dependent parts'.)

I suppose *Explicit* Positionalists are as uncommon as authors of more or less casual Positionalist-sounding remarks are common: cf. the long list of references in Gilmore 2013 fn. 3 (to which an interesting addition is Weyl 1949). (Gilmore himself though says in that paper that he is 'attracted' to [Explicit] Positionalism, which he calls by the name – of uneven dignity – 'Slot Theory'.)

4.5 Nuclear and Anti-Nuclear Open Situationism

The Open Situationist says that, in the relation corresponding to the open sentence say 'x loves y', there is a ghostly constituent (Variable or position) corresponding to the linguistic variable 'x', as well as a ghostly constituent corresponding to the linguistic variable 'y'. But now what about the linguistic verb 'loves' (which after all seems to be 'the main thing' in the open sentence 'x loves y')? Does the relation also contain a (non-ghostly) constituent corresponding to *that* – thus a sort of *nucleus* of the relation? Or is rather 'the love' the whole 'texture' of the relation (into which the ghosts are 'immersed') rather than a third constituent on a par with, only more 'fleshy' than, the ghosts?

Both alternatives seem to be at least superficially coherent – and that whether within Positionalism or Variablism. Thus there are here four views, which we may call Nuclear Positionalism (i.e. Positionalism with reification of nuclei), Anti-Nuclear Positionalism (without such reification), Anti-Nuclear Variablism, and Nuclear Variablism. Let us briefly consider each of these in turn.

(1) *Nuclear Positionalism*. On this view, the atomic situation aRb will have, despite appearances, *six* constituents: the a, the b, the relation – and that relation in turn has three constituents (the nucleus and the two positions), which are therefore also constituents of the whole situation.

We have been considering 'the Problem of Relations' – 'How does the relation combine with the relata in the relational situation?'. But now here there will be another, similar question, 'the Problem of Nuclei' – 'How does *the nucleus* combine with *the positions* in *the relation*?'! So it might at first sight seem that the Nuclear Positionalist would have to postulate some order or direction (surely not meta-ghosts), only *inside the relation*! (So that Armstrong's 'order among positions' would not be theoretical over-shooting after all.) However, fortunately for the Nuclear Positionalist, he does not really have to do that. He can perfectly well say that the relation itself is a *symmetric complex* (even if it is a 'non-symmetric relation' like the amatory relation) in the sense that there is no asymmetry among the immediate constituents of the complex (here the nucleus and the positions) w.r.t. the complex. (Just as e.g. there is no asymmetry among the elements of a set w.r.t. the set. Though of course the relation will not *be* a set but a supposedly sui generis sort of symmetric complex.) The desired difference between aRb and bRa is sufficiently accounted for by difference in which positions are occupied by which objects, with no need for asymmetry in the constitution of the relation itself from its constituents.

– We may define: entity e is an *inseparable constituent* of entity e' $=_{df}$ e is a constituent of e', and every entity which has e as a constituent also has e' as a constituent, or else is identical with e'. Open Situationists may with some plausibility say that the ghosts of a basic attribute are always in this sense inseparable constituents of the attribute. (Though they may also with some plausibility deny it – e.g. they might think that everything has a unit set, even a ghost.) And the Nuclear Positionalist may say the same of the nucleus. Or at any rate if the nucleus can separate itself from the relation and go do something else (such as e.g. belong to a unit set, or even help constitute a 'reflected' attribute like self-love), at least this something else cannot be *predication* of the nucleus to some subjects, which no doubt here should be taken to always require mediation through associated positions. So the nucleus should not here be regarded as another kind of attribute (i.e. predicable), but rather as the core of attributes (of the only kind there is).

(2) *Anti-Nuclear Positionalism*. Now aRb will have *five* constituents: the a, the b, the R, and the two positions of R.

Again the relation may – and so should – be taken as a mere symmetric complex, the constituents now being only the positions. However, if this complex (the relation) consists of its constituents combined in a certain way (and isn't that what a complex always is?), then the *empirical content* of the complex (in the case of an empirical relation) surely would not be located in the constituents, which are mere ghosts, but

4.5 Nuclear and Anti-Nuclear Open Situationism

would have to be located then in their mode of combination. This gives a violation of what may be called the Principle of Logicality of All Form, viz. the principle (which clearly has some plausibility) that empiricality is always located in constituents of situations, not in their mode of combination which is always logical (e.g. [perhaps viz.] predication).

Also, a basic *monadic* attribute (property) would have a single constituent (viz. the position), in violation of plausible 'Principles of Supplementation' for constituency (on which see Batchelor 2013, §2).

(3) *Anti-Nuclear Variablism.* Now aRb has *two* constituents, namely a and b. For it results from xRy, whose constituents are just the Objective Variables x and y (there being on this view no nucleus), by replacement of x by a and y by b. Note that here the relation actually never occurs in the 'relational' situation!

Again the ghostly Variables may combine to produce an empirical relation, thus violating the Principle of Logicality of All Form. (And again monadic attributes will yield violation of Supplementation Principles.) But now also the complex which the relation is *cannot be taken to be symmetric*, otherwise we could not here distinguish aRb from bRa. (If xRy were a symmetric complex, the result of substituting a for x and b for y in it would have to be identical with the result of substituting b for x and a for y in it.) Hence presumably either order or direction (surely not 'meta-ghosts') will have to be postulated here, inside the relation.

(4) *Nuclear Variablism.* By this we do not mean that there must *always* be a nucleus, but only that in the simplest cases there is the nucleus. Thus there might be a nucleus for xRy but not for $\neg xRy$.

Now aRb has finally *three* constituents – and the a, the R and the b at that. The nucleus or 'primary relation' R is thus a *separable* constituent of the Variabled relation xRy.

Now how does the nucleus R combine with the Objective Variables x and y in xRy? Again the complex cannot be taken to be a symmetric complex. Thus presumably the most reasonable view here is that the combination is effected by order or direction, and then of course the exact same mode of combination occurs in aRb, as aRb results from xRy by the appropriate replacements. (Here of course there is no violation of the Principle of Logicality of All Form [the empirical content is in the R], nor of Supplementation Principles [as Px has now two constituents].)

The R here is a sort of autonomous entity, and can be directly predicated of a and b, without previous intervention of Variables. (aRb can indeed be correctly described as the result of substituting a for x and b for y in xRy, but this is a sort of indirect, inessential description, not one that is directly revelatory of the inner nature of the situation aRb.) Thus the 'nucleus' here is if anything *more* deserving of the appellation 'attribute' than the Variabled attribute – which is however still deserving enough, since the replacement of Variables by specific objects is surely naturally called 'predication' of a kind. So here we have a hybrid theory, with two kinds of (sui generis) attributes – the primary attributes (presumably always simple) and the derivative (but still *sui generis entities* rather than certain equivalence classes or the like) Variabled attributes.

(This was essentially the view of Russell's *Principles of Mathematics*; and is also very close to Dummett's Frege.)

5 Open Situationism, complex attributes, and higher-order attributes

(Although it is a continuation of the discussion of Open Situationism, the present section leads us rather far afield into other issues. So it may be advisable [and it is possible] to skip it on a first reading, and return to it after reading the rest of the paper.)

5.1 Some terminology and some preliminary remarks

By a *complex* entity I mean an entity which has at least one other entity as a constituent; and by *simple* I mean not complex. An *attribute of attributes* is an attribute which, in some situation, is predicated of some entity or entities at least one of which is itself an attribute. I use also the terms 'higher-order attribute' or 'higher-level attribute' for attribute of attributes. (Note that on this usage an attribute of *situations* is *not* a higher-order attribute.)

We will use *Anti--Open-Situationism*, or the abbreviation *Anti-O.S.*, for the rejection of Open Situationism. And we introduce also the following terms for the positive and negative answers to the questions of existence of complex and higher-order attributes:

Simplicity Thesis: All attributes are simple.
Complexity Thesis: There are complex attributes.
Elementarism: There are no attributes of attributes. (I.e. all attributes are attributes of substances only.)
Higher-Orderism: There are attributes of attributes.
('Elementarism' has been used in something like this sense by G. Bergmann [v. e.g. Bergmann 1957] and others.)

If a situation is the value of some attribute (for some arguments), we may call it an *attributive* situation. (Arguably of course *every* situation is attributive in this sense; but whether this is or is not the case is not particularly relevant now.) There is then the question whether a single situation can be the value of more than one attribute; again we introduce names for the positive and negative answers:

Polypredicativism: Some situation is the value of more than one attribute.
Monopredicativism: No situation is the value of more than one attribute. (I.e. every attributive situation is the value of a *unique* attribute.)

5.1 Some terminology and some preliminary remarks

If all attributes are simple (Simplicity Thesis), then no doubt Monopredicativism is true. (Ignoring Polypredicativistic Pluralism about [supposedly simple] converses, which as we have seen there is good reason to reject.) It is if one thinks that there are complex attributes that the Polypredicativism vs. Monopredicativism question becomes more relevant.

Thus a polypredicativistic complex-attribute theorist might wish to say that e.g. the situation Rab is the value of R for a and b as arguments (in the given order, or occupying specified positions, or whatever may be his account of relational predication), but also e.g. the value of $\lambda x(Rxb)$ for a as argument. Or again, that $\neg\neg p$ is the value of $\neg$ (a situational property – 'counterfactuality') for $\neg p$ as argument, but also the value of $\neg\neg$ for p as argument.

On the other hand a monopredicativistic complex-attribute theorist might wish to distinguish the situation Rab, where the principal attribute (i.e. the attribute of which the situation is a value) is R, from the situation $\lambda x(Rxb)(a)$, where now the complex attribute $\lambda x(Rxb)$ is the principal attribute. Or again between $\neg(\neg p)$ and $(\neg\neg)(p)$ (or in lambda-notation $\lambda q(\neg\neg q)(p)$).

But now, even in a polypredicativistic theory, no doubt there cannot be two distinct *simple* attributes such that the same situation is the result of applying one of them to some terms and also the result of applying the other to some terms. Thus, if we suppose that R is a simple relation, the *only* decomposition of Rab into a *simple* attribute applied to some terms is that into R applied to a and b; all other decompositions must be into a *complex* attribute and its terms. Just as a natural number can have multiple decompositions into arbitrary factors but must have a unique decomposition into *prime* factors, so similarly an attributive situation can have multiple decompositions into arbitrary predicates and their subjects but must have a unique decomposition into a *simple* predicate and its subjects. This may be called the *canonical decomposition*. Thus the canonical decomposition of Rab is into R applied to a and b, and the canonical decomposition of $\neg\neg p$ is into $\neg$ applied to $\neg p$ (supposing that negation is a simple property).

(Cf. Dummett 1973/1981 Ch. 2 and Dummett 1981 Chs. 15 & 16. Dummett's articulation of these issues seems to me excellent in itself, but probably [and less importantly] not very faithful to Frege. E.g. the following remark from Frege 1919 [third par.] seems pretty unequivocal: 'I do not begin with concepts and put them together to form a thought or judgement; I come by the parts of a thought by analyzing the thought. This marks off my *Begriffsschrift* from the similar inventions of Leibniz and his successors, despite what the name suggests; perhaps it was not a very happy choice on my part.' – I myself am utterly 'Leibnizian' on this, and find this thesis of 'priority of the thought' [or situation] almost mystical.)

A similar distinction can be made within a monopredicativistic theory: now not between two different types of decomposition of situations, but between two different types of situation. Namely: between attributive situations where the predicate is simple – like $\neg(\neg p)$ – and attributive situations where the predicate is complex – like $(\neg\neg)(p)$. The former may be called *canonical situations*. Every attributive situation has its *canonical correlate*: e.g. $\neg(\neg p)$ is the canonical correlate of $(\neg\neg)(p)$, and $R(a,b)$ is the canonical correlate of $\lambda x(Rxb)(a)$. (Note also that, since situations may have other situations as constituents, this notion of canonicity can of course be extended to all the layers of the structure of the situation.)

5.2 Connexions between Open Situationism and the questions of existence of complex and higher-order attributes

We will argue in what follows that there are certain connexions between the question whether Open Situationism is true and the question of existence of complex attributes and the question of existence of higher-order attributes. The Simplicity Thesis naturally leads to Elementarism, and vice versa; or equivalently, the Complexity Thesis naturally leads to Higher-Orderism, and vice versa. (When we say here that a certain thesis 'naturally leads' to another, what we mean is that it would be implausible, or at least disharmonious, to think that the former thesis is true and the latter false.) Moreover, Anti-O.S. naturally leads to the Simplicity Thesis, and vice versa; or equivalently, Open Situationism naturally leads to the Complexity Thesis, and vice versa. From these facts it then follows that, indirectly, Elementarism leads to Anti-O.S. and vice versa, or equivalently, Higher-Orderism leads to Open Situationism and vice versa. (Supposing that the 'natural leadings' are strong enough to sustain one application of transitivity.) Therefore, of the eight ($= 2^3$) combinatorial possibilities of answers to our three questions, the only harmonious ones are two: Anti-O.S. plus Simplicity Thesis plus Elementarism, and Open Situationism plus Complexity Thesis plus Higher-Orderism. This is a dialectically powerful fact, because with it any argument for one of the three theses in each group becomes by consequence an argument for the others; and indeed even an argument for the *disjunction* of the three theses becomes an argument for their *conjunction*. Notice moreover that, given one specific thesis of one of the groups, only *two* of our four 'natural leading' claims will be needed to pass to the other two theses in the group. E.g. given that we accept Anti-O.S. (which as we have seen there is ample reason to do), one step will take us to the Simplicity Thesis, and one more to Elementarism (so that the claims that Simplicity Thesis naturally leads to Anti-O.S. and that Elementarism naturally leads to Simplicity Thesis are not used).

That the Complexity Thesis naturally leads to Higher-Orderism can be seen as follows. The postulation of complex attributes occurring *predicatively* seems to be, in itself, wholly superfluous. For why suppose – as in a *monopredicativistic* theory of complex attributes – that, in addition to the canonical situation $\neg(\neg p)$, there is also this peculiar non-canonical situation $(\neg\neg)(p)$? Why not just recognize the canonical situation $\neg(\neg p)$ and leave it at that? What *need* is there for a situation in which this

5.2 Connexions between Open Situationism and the questions of existence of complex and higher-order attributes

supposed complex attribute double negation is applied to p, in addition to the situation where negation is applied to the situation where negation is applied to p?

Again, why suppose – as in a *polypredicativistic* theory of complex attributes – that, in addition to the attributes figuring as predicates in the canonical decompositions of situations, there are also attributes figuring as predicates in non-canonical decompositions? Why postulate a separate entity double negation, figuring as a lower-rate constituent in $\neg(\neg p)$? What *need* is there for such an assumption?

The assertion of existence of complex attributes seems then to be quite gratuitous in so far as we consider only their *predicative* occurrence. It is only when they are taken as subjects to other predicates, as in Fregean quantification, that their postulation acquires a really substantial motivation. (Again cf. Dummett.) These other predicates, like the Fregean quantifier, are of course attributes of attributes: so in this sense it may be said that it is only in the presence of attributes of attributes that complex attributes as it were come to life. Complex attributes as predicates can easily be dispensed with; but not complex attributes as subjects, as in Fregean quantification. (But it is true that, *once* complex attributes are admitted because of their use as subjects, it is natural enough to allow that they may also occur as predicates.) Thus complex attributes would be useless without attributes of attributes, and so it would be implausible to hold the Complexity Thesis without also holding Higher-Orderism.

The converse proposition, that Higher-Orderism naturally leads to the Complexity Thesis, can be justified as follows. If one thought that there were attributes of attributes, one would hardly think that their application was restricted to only simple attributes. E.g. if one thought that there was an attribute of attributes Universal Quantification, $\forall$, one would hardly think that its application was restricted to simple attributes, so that there was a situation $\forall x \varphi x$ whenever φ was simple, but not say $\forall x(xRb)$ and so on. Also, more generally, Higher-Orderism often tends to be motivated by a liberal attitude in postulation of worldly items corresponding to meaningful linguistic constructions; and this of course leads to complex attributes too. Thus it would be implausible, or at least disharmonious, to hold the Higher-Orderist view without also holding the Complexity Thesis.

We pass on now to the connexions between the Complexity Thesis and Open Situationism.

That Open Situationism leads to the Complexity Thesis is quite clear. Indeed there is a trivial sense in which Open Situationism implies not only that there are complex attributes but that *all* attributes are complex: for according to Open Situationism every attribute contains at least one ghost as constituent, as well as some 'surrounding material'. (The only possible exception to this would be a situational attribute $\lambda p(p)$, which would perhaps *be* a ghost, without any surrounding material.) But of course this kind of complexity is not so significant: what matters more is the complexity or simplicity of the 'surrounding material'.

Now, perhaps the easiest way of seeing the implausibility of Open Situationism without complex attributes is by considering *partial completions*. Take say the relation [x loves y], as conceived of by the Open Situationist, with its two ghosts; and suppose John is inserted into the x-place (under Positionalism) or substituted for the x Variable (under Variablism). Surely this produces the attribute [John loves y], alias λy(John loves y), alias being-loved-by-John. And that, of course, is a complex attribute. – I can see no decent reason for the Open Situationist to deny the existence of such partial completions.

The following more general (but also I suppose more resistible) point is also relevant. Typical Open Situationism, as we have said before, is a theory which models attributes upon their supposed linguistic counterparts, namely *open sentences*; and just as open sentences contain 'gaps' or 'slots' or linguistic free variables, so likewise their supposed objective correlates, the attributes, will be thought to contain corresponding 'gaps' or 'slots' or Objective Variables. (A theory where attributes are 'complete' and simple, on the other hand, may model attributes not on open sentences but rather on grammatically simple *verbs*, like 'smokes', 'loves', etc.) But open sentences may of course be of arbitrarily great complexity, and so Open Situationism, at least in its natural articulation, will countenance plenty of complex attributes.

That the Complexity Thesis leads to Open Situationism can be seen as follows. It is not difficult to regard a *simple* attribute as a self-contained, ghost-less entity. But a *complex* attribute one is much more apt to consider as a sort of quasi-situation, with ghosts at some points instead of specific objects. Take for instance a supposed complex relation of 'either mutual love or "no love lost"', i.e. the relation corresponding to the open sentence

$$(x \text{ loves } y \wedge y \text{ loves } x) \vee (\neg(x \text{ loves } y) \wedge \neg(y \text{ loves } x)).$$

It is easy to think of this as a quasi-situation with some ghosts helping to keep together the complex structure; but not easy to think of it in any other way. Probably we ought to admit that a theory might postulate some rather restricted class of complex attributes (e.g. perhaps negative ones) while remaining Anti-O.S.ist; but more extensive use of complex attributes would hardly be plausible without Open Situationism.

– So we have here what some philosophers like to call a 'package deal': we must either take Anti-O.S. plus Simplicity Thesis plus Elementarism (as I think we should) or else Open Situationism plus Complexity Thesis plus Higher-Orderism: we cannot pick and mix.

5.3 Further arguments for Anti-O.S. + Simplicity Thesis + Elementarism

We have already argued above against Open Situationism, i.e. in favour of Anti-O.S.; and we see now that this is indirectly argument for the conjunction of our three theses. We will now add three further arguments in favour of this package Anti-O.S. plus

5.3 Further arguments for Anti-O.S. + Simplicity Thesis + Elementarism 27

Simplicity Thesis plus Elementarism; of which the first is primarily an argument for Elementarism as against Higher-Orderism; the second primarily against the conjunction of Open Situationism with Higher-Orderism; and the third primarily against the Complexity Thesis.

5.3.1 An argument for Elementarism

Philosophers have often felt that there must be a radical difference between the nature of substances and the nature of attributes. Some have even gone so far as to say that attributes are not even properly speaking 'entities', or 'things', but merely *ways* for things to be, or *forms* of situations. Here we are supposing that there are such *entities* as attributes; but still it must surely be recognized that they are a kind of entity radically different from substances. Thus the Elementarist view may be thought to be superior to the Higher-Orderist view in so far as it makes substance and attribute more different in a regard that seems eminently suitable to satisfy that intuitive feeling of a radical difference between the two categories. Both the Elementarist and the Higher-Orderist will agree that substances must be subjects of situations and cannot be predicates; and that attributes *can* be predicates; and thus that there is this much of difference between substances and attributes – that attributes but not substances *can* be predicates. But the Elementarist will go further than this and will make the difference between substance and attribute even sharper, by saying that, just as substances are inexorably consigned to the rôle of subjects, so attributes are inexorably consigned to the rôle of predicates. Thus the radical difference between the natures of the two kinds of entities would be reflected in a sharp separation of their rôles in situations.

(Here I am ignoring the possibility of objects occurring in situations *neither* as subjects nor as predicates [e.g. as members of set]. Thus Elementarism becomes in effect the thesis that attributes can only occur predicatively. [Cf. Russell – 'A propositional function can only occur in a proposition through its values'.] In other contexts it might be proper to be more discriminative.)

It is a traditional idea, and perhaps a correct one, that attributes should be *defined* as the *predicables* (entities which are predicates in some situations), and substances as the *'subjectables' and non-predicables*. ('An attribute is what can be said *of* things, and a substance is that of which other things can be said but which cannot itself be said of anything'.) In fact we can improve on this a little by defining substances simply as the non-attributes, i.e. non-predicables. If all non-predicables are subjectables, as is probably the case, then the two definitions are equivalent and ours simpler; otherwise, i.e. if some entities are neither predicables nor subjectables, then our definition classes such entities as substances, which certainly seems best. With the present definition, then, the question whether there are subjectable attributes would become the question whether there are subjectable predicables. We might then say that the radical difference between attributes and substances, i.e. between predicables and non-predicables, is reflected in the fact that, just as substances are (now by definition) incapable of playing the rôle of predicates, so likewise attributes, i.e. predicables, are incapable of playing the rôle of subjects.

One ought not, however, to take the parallelism *so* far as to say that, just as attributes should be defined as the predicables, i.e., as we may say, the predicators of predications, so also likewise substances should be *defined* as the subjectables, i.e. the predicatees of predications. The question whether there are subjectable attributes could then also be phrased as the question whether there are entities which are both substances and attributes; and since if one attribute is subjectable then presumably all are, one might end up saying that 'all attributes are substances', which is perhaps proof enough that the terminology has been badly chosen. Moreover, although as we have said it is probable that all substances (properly so called) are subjectables, yet it is not completely inconceivable that this should fail to be the case for lack of suitable attributes: e.g. that there should be such entities as situations, which certainly should rank as substances, and yet no such entities as attributes of situations.

5.3.2 An argument against Open Situationism + Higher-Orderism

We come now to the second argument, viz. the argument against the conjunction of Open Situationism with Higher-Orderism. It is actually something more specific: an argument against the conjunction of Open Situationism with the view that *quantifiers* are simple higher-order attributes (*à la* Frege). To logically-minded metaphysicians quantifiers are the Prime Case of putative higher-order attributes, so that hitting *them* (quantifiers) is hitting Higher-Orderism; more empirically-minded metaphysicians may be less happy with this – but there it is. At any rate what follows seems to me to be an important point about the Fregean conception of quantifiers (though actually it applies just as much to any other case where there is also the kind of 'variable-binding' as with quantifiers).

For the sake of definiteness we will concentrate here on Positionalism. But the exactly similar considerations apply to Variablism.

Let us take first [Something is blue], which we can write also as $\exists x Bx$, or as $\exists(B)$. Here the position of the higher-level property $\exists$ has been filled by the lower-level property B, just like e.g. in [Socrates is mortal] the position of the property of mortality is filled by Socrates.

So far so good. But now take say the property [x loves something], or $\exists y(xLy)$. This property is no doubt involved e.g. in the situation $\forall x \exists y(xLy)$. And again no doubt the property $\exists y(xLy)$ consists in $\exists$ somehow applied to the relation xLy. But how? We cannot *just* say that the relation xLy is filling in the gap of $\exists$; we must be able to distinguish $\exists y(xLy)$ (the property of loving something) from $\exists x(xLy)$ (the property of being loved by something).

There is really only one sensible answer to this question (within the given frame-

5.3 Further arguments for Anti-O.S. + Simplicity Thesis + Elementarism

work). We must say that in $\exists y(xLy)$ the position of $\exists$ has been occupied by the relation xLy, with the existential quantifier *binding* the 'y-place'. The x-place of course is here neither 'bound' nor occupied by specific object, and so we have a one-place attribute (property). Then in the situation $\forall x \exists y(xLy)$ this property goes into the position of $\forall$, with the x-place being 'bound' by $\forall$. (Not that now there was any *choice* as to what $\forall$ might bind, but for the sake of uniformity of theory it seems recommendable to say that there is binding in this case too [as well as even in $\exists x Bx$].) Incidentally, I would suggest that in this framework one should allow also simultaneous binding of multiple positions, which gives essentially richer resources in the case of infinitary attributes. (E.g. we can have $\forall x \exists y_1, y_2, y_3, \ldots \varphi(x, y_1, y_2, y_3, \ldots)$, which could not be gotten with successive bindings of single positions.)

Now the *objection* is simply that thus the mechanism of predication must incorporate this 'binding' as something no doubt basic: it cannot now be *just* a matter of objects occupying positions (as with attributes of substances). The apparatus of positions by itself was already implausible as a supposed fundamental; with 'binding' as a necessary further primitive things become even worse.

(Similar objections can be made to quantification as a supposed higher-order attribute also in other theories of relational predication, in particular Orderism, Directionalism, and the theories to be considered in the next section.)

– We saw above (4.3.1) the failure of the Principle of Uniqueness of Substitution under Positionalism, in connexion with the *attributive* use of relations. We can now add that it is actually a *double* failure, since in such cases of binding there will not in general be unique result of substitution for the *subject* relation (for, to which position in the new relation should the earlier binding go?). Now, under *Variablism*, the *attributive* use of relations does *not* yield violation of the Principle of Uniqueness of Substitution, since under Variablism, as we have already said, the Variabled relation actually does not occur at all in the 'relational' situation. However the *subject* use, in cases of 'binding', *does* yield violation of this Principle (again, which Variable in the new relation should 'inherit' the earlier binding?). – And this is an objection to Variablism per se, since Variablism of course implies Open Situationism which, we have argued, leads to Higher-Orderism.

5.3.3 An argument for the Simplicity Thesis, via an argument for Atomism

We come finally to our third argument – an argument against the Complexity Thesis. We argue that: *If* there are complex attributes, *then* they are useless, and *so* by Ockham's razor they do not exist after all. The main body of the argument is actually an argument for a form of 'logical atomism'. In what immediately follows we state *this* argument (for atomism); and then after that we will indicate how this yields an argument against complex attributes.

By an *atomic situation* we mean a situation which consists of a simple attribute predicated of simple entities, i.e. either a simple property predicated of a simple entity or a simple relation predicated of some simple entities. (Given Elementarism these simple entities will always be substances, of course; but this is not needed for the present argument.) Then *Atomism* may be defined as the thesis that every situation is a truth-function – or better factuality-function – of atomic situations; i.e. every situation is made up out of some atomic situations by successive application of the basic truth-functors. (These might be e.g. negation and conjunction, or negation and *dis*junction, or joint denial, etc.) – This is at least close to the usual sense in which this word (or the variant 'logical atomism') is used in connexion with the early Wittgenstein, Russell etc. It should be noted however that the thesis as formulated here does not include the assertion of the modal independence of atomic situations.

Atomism, even in this relatively weak sense in which we are using the term here, is generally regarded nowadays as an extremely implausible view. I think therefore that the argument presented in what follows may be of some interest even apart from the questions of existence of complex attributes etc., since it derives Atomism, by fairly tight reasoning, from premises which I believe would be regarded as extremely plausible, or at least plausible.

I will state the argument at once, and then make some comments on it. (1) and (2) are the premises. In (2), 'the pattern of instantiation of attributes in the case of complexes' means: the truth-values – or better factuality-values (factuality or counterfactuality) – of the attributive situations where either the attribute is complex or (inclusive-or) at least one of the subjects is complex. And 'the pattern of instantiation of attributes in the case of simples' means: the factuality-values of the attributive situations where both the attribute and all the subjects are simple – i.e. atomic situations. Here then is the argument: –

(1) The totality of facts (i.e. which situations are factual and which counterfactual) supervenes on the pattern of instantiation of attributes (i.e. on which entities instantiate which attributes).

(2) The pattern of instantiation of attributes in the case of complexes supervenes on the pattern of instantiation of attributes in the case of simples (i.e. on the factuality-values of atomic situations).

∴ (3) The pattern of instantiation of attributes supervenes on the factuality-values of atomic situations. (From (2).)

∴ (4) For every situation p: the factuality-value of p supervenes on the factuality-values of atomic situations. (From (1) and (3).)

∴ (5) For every situation p: p is strictly equivalent (i.e. necessarily materially-equivalent) to a factuality-function of atomic situations. (From (4).)

5.3 Further arguments for Anti-O.S. + Simplicity Thesis + Elementarism

∴ (6) *Atomism* – i.e.: For every situation p: p *is* a factuality-function of atomic situations. (From (5).)

Here by saying that question Q *supervenes* on question R we mean that every answer to question R strictly implies some answer to question Q; or equivalently, that whenever two possible worlds agree on question R they must also agree on question Q. Thus e.g. premiss (1) means: If two worlds are exactly alike as regards which entities instantiate which attributes, then they are exactly alike as regards the factuality-values of all situations (from which presumably it follows that they are in fact one and the same possible world).

It is sometimes thought that every situation is attributive (i.e. consists of an attribute applied to some entity or entities). This would make premiss (1) automatically true, of course; but even otherwise, I think it is clear that (1) is at least highly plausible.

Premiss (2) also is clearly plausible. There may be complex attributes (if I am wrong in thinking otherwise) and other complex entities; but it is plausible to think that whether an attribution involving complexes is factual always supervenes on whether certain attributions involving only simples are factual (indeed attributions involving only simples which actually occur in the given complexes – though this is not needed for our argument).

Thus the premisses of the argument are clearly at least plausible (whether or not they are ultimately true). Let us now consider the steps leading to the final conclusion.

(3) is an immediate and inescapable consequence of (2). (If Q supervenes on R then Q-plus-R supervenes on R.)

Also (4) is an immediate and inescapable consequence of (1) and (3). (If Q supervenes on R and R supervenes on S, then Q supervenes on S.)

The passage from (4) to (5) is also inescapable, but it involves a point of supervenience theory which, although quite elementary, may be unknown to less logically-minded metaphysicians. (4) tells us that (for an arbitrary situation p) the factuality-value of p supervenes on the factuality-values of atomic situations: that is to say, if two worlds agree on the factuality-values of all atomic situations, then they must also agree on the factuality-value of p (i.e. p is factual in both or counterfactual in both). Thus certain (zero or more) attributions of factuality-values to the atomic situations (conjunctions of one out of each pair of an atomic situation and its negation) strictly imply p, whereas the other such attributions strictly imply the negation of p. Hence p is strictly equivalent to the disjunction of attributions in the former group; which of course is a factuality-function of atomic situations. (If that group is empty, p is equivalent to e.g. the conjunction of an arbitrary atomic situation with its negation.)

Finally, we come to the step from (5) to (6), which is the only one that is not quite inescapable. I take it simply as an application of Ockham's razor. There *are* the various factuality-functions of atomic situations – why postulate further situations which are strictly equivalent to such situations we already have?

A less razor-happy philosopher may question this last step. But it is worth noting that, if he accepts the premisses (1) and (2), he is compelled to accept (5), which is – not only in the graphical sense – one step away from Atomism. Indeed (5) may be called a weak form of Atomism.

– This then is the argument for Atomism. But then we may reason: If there are complex attributes then there is at least supervenience (premiss (2)) and so on, and we arrive at the atomistic view. But *then* complex attributes if they existed would occur only in the useless contexts of the kind we have seen before, where they are dispensable. So again by Ockham's razor there *are* no complex attributes.

5.4 Attributes, conditions, and situational functions

The rejection of complex and higher-level attributes (and indeed of non-i.s. relations as well) may seem to some philosophers rather counterintuitive. But it should seem at least *less* so once we make certain distinctions (which we ought to make anyway): namely between attributes (properties and relations, taken as sui generis entities) and what I call *situational functions* and what I call *conditions*.

A *situational function* is simply a function (in the usual sense of set theory) whose values are situations. There need not be any intrinsic connexion whatever between argument and value (although of course there *may* be such a connexion): thus Ramsey gives the example that such a function might assign, to Socrates as argument, the value [Queen Anne is dead], and to Plato as argument, the value [Einstein is a great man]. (Ramsey [1925, §4] considered what is essentially this notion of situational function; his own term for it was 'propositional function in extension'.)

By a *condition*, on the other hand, I mean the result of 'removing' certain occurrences of constituents from a situation. Thus e.g. the condition $\lambda x(x$ is a philosopher $\wedge\ x$ is Greek) is the result of removing (for instance) the indicated occurrences of Socrates from the situation [Socrates is a philosopher ∧ Socrates is Greek], or the indicated occurrences of Plato from the situation [Plato is a philosopher ∧ Plato is Greek]. More exactly, we define, in terms of substitution, a certain equivalence relation of 'congruence' holding between pairs consisting of a situation together with a class of occurrences of constituents; and then take the above-mentioned 'removal' as the corresponding equivalence class. (I omit details, which can be found in my paper Batchelor 2013, in particular §§4 and 7, along with accounts of the underlying notions of occurrence and substitution.)

(If, as is suggested in Batchelor 2013 [p. 342, fn. 1], there are no such entities as classes [= sets] but talk of classes is a mere *façon de parler*, then in particular conditions and situational functions, being defined as certain classes, are not entities; whereas attributes *are* entities. But even if we suppose that classes are entities, still conditions and situational functions would be, though now entities, not *sui generis* entities; and would be less fundamental entities than attributes; in particular, while

5.4 Attributes, conditions, and situational functions

attributes are *constituents* of the situations which are their 'values' ['completions'], conditions and situational functions [at least as defined above] *have* their 'values' *as* constituents. – Conditions in the present sense stand to the open situations of the Open Situationist exactly like the sensible Frege–Russell [or Frege–Russell–Scott; or von Neumann or Zermelo] numbers stand to the ghostly Cantorian [or Cantorian–Finean – see Fine 1998] numbers.)

Then, although there may be no complex attributes and no higher-level attributes (and no non-i.s. relations), there will be plenty of conditions and situational functions which are complex or higher-level (or non-i.s.) in the relevant senses. Thus e.g. although there may be no *attribute* of being a Greek philosopher, yet there is the *condition* which results e.g. from removing the indicated occurrences of Socrates from the situation [Socrates is a philosopher $\wedge$ Socrates is Greek], as well as the *situational function* which assigns to each object x (of suitable type) the situation [x is a philosopher $\wedge$ x is Greek]; and although there may be no attribute of being-instantiated-by-Socrates, yet there is the condition which results from removing the attribute of being wise from the situation [Socrates is wise], as well as a corresponding situational function; and so on.

Since the arguments of situational functions can be anything at all and in particular themselves situational functions, it follows that there are situational functions of arbitrarily high levels. The same does not hold, however, for conditions, because the arguments of conditions must be actual constituents of situations, and so (given that there are no attributes of attributes) either substances or attributes of substances; thus a condition on attributes of substances is as 'high' as we can get here.

For each attribute φ there is a corresponding condition $\varphi(-)$ and a corresponding situational function the set of all pairs $\langle x, \varphi x \rangle$. (Or at least *families* of conditions and situational functions, with 'type restrictions'.) And for each condition ... $(-)$ ... there is a situational function corresponding to *it*, viz. the set of all pairs $\langle x, \ldots x \ldots \rangle$. But a condition need not correspond to any attribute, and a situational function need not correspond to any condition (let alone attribute). In this sense it may be said that of the three notions that of situational function has the highest degree of generality.

Here we should point out a slight qualification for the case of polyadic conditions: namely that in this case there may not be a *single* corresponding situational function, because such a corresponding function may have to embody an ordering of the arguments, which is arbitrary and therefore not unique. Thus e.g. to the condition *material implication*, i.e. $\neg(-\wedge\neg(\ldots))$, there will correspond two situational functions: the set of all pairs $\langle \langle p, q \rangle, \neg(p \wedge \neg q) \rangle$, and the set of all pairs $\langle \langle p, q \rangle, \neg(q \wedge \neg p) \rangle$. (More exactly we should speak here of *families* of conditions and situational functions.)

Although I think that no situation is the value of more than one *attribute* (Monopredicativism), yet a situation may be, and indeed always is, the value of more than one *condition* (and a fortiori more than one *situational function*). In this sense it may be said that a situation does not have a unique decomposition into a condition and its arguments. Thus in cases like $\neg(\neg p)$ and Rab discussed before, we have indeed multiple legitimate decompositions of the situations into *conditions* and their arguments.

– One congenial and important use to which Ramsey (1925, §4) put this notion of situational function was to give a Leibnizian definition of *identity*. A definition quantifying over attributes or even conditions is of course problematic; but just quantify over situational functions and we are done! I.e. we define $a = b$ as $\forall \varphi(\varphi a \leftrightarrow \varphi b)$ where φ is a variable for situational functions. Thus, Ramsey says, (given also the Tractarian conception of quantification as truth-function of instances) if a is indeed the same entity as b, then this ($\forall \varphi(\varphi a \leftrightarrow \varphi b)$) is the conjunction of all situations of the form $p \leftrightarrow p$; and otherwise it is the conjunction of all situations of the form $p \leftrightarrow q$. (Actually in this another Tractarian element enters, viz. the intensional identity-criterion for situations. *Here* by contrast we are assuming a more fine-grained, structural notion of situation; so for us these conjunctions are 'too big'. But it is enough to consider situational functions with a restricted range of values [as well as a restricted domain]; indeed one 'canonical' tautology and one 'canonical' contradiction are enough.)

Note that on this view identity *itself* is not an attribute (relation), nor a condition, but only a (dyadic) situational function, viz. the situational function which assigns the conjunction of all situations of the form $p \leftrightarrow p$ (for p in a certain range) to pairs $\langle x, x \rangle$, and the conjunction of all situations of the form $p \leftrightarrow q$ (for p and q in the range) to pairs $\langle x, y \rangle$ where x is other than y. (Indeed it already follows from Elementarism that there cannot be an absolute relation of identity, i.e. one that is predicable of entities of any kind including attributes.) Identity would be, as we may say, a *proper* situational function, i.e. a situational function to which there corresponds no attribute and no condition; which shows that a proper situational function may be something quite significant and not just an odd curiosity like Ramsey's example of Socrates and [Queen Anne is dead] and Plato and [Einstein is a great man].

(It is perhaps worth pointing out here that if identity *were* an attribute, then to every situational function [of entities] there would after all correspond at least an equivalent condition: to a situational function f where $f(a) = p$, $f(b) = q$, …, there would correspond the condition $(x = a \wedge p) \vee (x = b \wedge q) \vee \ldots$ [And similarly for polyadic cases.] The same would hold if in some other way there were, for each entity e, a 'haecceitical' condition H^e, i.e. a condition which, necessarily, held of e only. For we could then replace, in the formula above, $x = a$, $x = b$, etc., by $H^a(x)$, $H^b(x)$, etc.)

Necessity (and possibility etc.) gives another excellent example of proper situational function, assuming Atomism (in the sense of 5.3.3 above) and the Principle of Independence of Atomic Situations. It is simply the function which assigns, to each situation p (in a given range), either (say) p itself if p is a tautology (tautological factuality-function of atomic situations), or (say) $p \wedge \neg p$ otherwise (i.e. if p is *not* a tautology). (Indeed even the Principle of Independence of Atomic Situations is not really essential here: we can re-interpret 'tautology' as situation factual under all *permissible* factuality-value assignments to atomic situations!)

6 Miscellaneous theories

We come now to theories which attempt to account for non-i.s. relations without resorting to the more familiar ideas of order or direction or positions or Variables. Of course many different theories may conform to this essentially negative description. Here we have selected for consideration only two theories, viz.: first, the theory of Fine 2000; and secondly, the theory of Russell 1913.

6.1 Fine's theory

In his paper 'Neutral relations', after dismissing Orderism / Directionalism and Positionalism, Fine proposes his own theory of relational predication, which he calls 'anti-positionalism'. The name is rather infelicitous: the theory is not the only theory which is *not* Positionalist; nor is it, that I can see, more *against* Positionalism than is any of the other theories which are *not* Positionalist. Perhaps the fact that Fine could find no better name for his theory than this negative one is indicative of the lack of a distinctive positive vivid quality of relational predication on this account (to provide a 'keyword' for a more positive name for the theory). Certainly here there is supposed to be *no* order, *no* direction, *no* positions, *no* Variables; but how the relation *is* supposed on this account to combine with the terms so as to produce the relational situation is less easy to describe.

Fine says (pp. 19–20): 'there are now no argument-places or orientations by which the different completions might be distinguished. A relation should now be taken to be a simple unadorned body or "magnet", to which the relata of the relation are taken to be attached by some sort of invisible bond. Different configurations may then be formed from a given body and the relata, according to how they are attached. But there will be nothing in the body itself that can be identified as the parts or areas to which the different relata are meant to attach.'

In order to distinguish the different 'completions' or values of a (non-i.s.) relation for the same arguments, Fine then invokes a notion of 'co-mannered completion', i.e. of a situation being the completion of a certain relation by certain terms in the same way as another situation is the completion of the (same) relation by certain terms; and this, he says, is to be defined in terms of *substitution*. One completion can then be 'identified' in terms of another. Fine says that relations apply 'to their relata via a network of connections' (p. 32); it seems to be impossible on his account to give a direct, 'inner' description of how a given relational situation is produced from the relation and the relata. Thus he says (pp. 29–30, my italics near end):

'Suppose the antipositionalist is asked to identify the state of Anthony's loving Cleopatra. Then he may say that it is a completion of the amatory relation by Anthony and Cleopatra and even that it is *the* completion of the relation by Anthony and Cleopatra in the same manner as another amatory state is the completion of [sic] *its* respective constituents. He can also, in a fashion, distinguish the state from its converse, the state of Cleopatra's loving Anthony, since each is obtainable from the other through the interchange of constituents. But if he is asked to identify the state absolutely, independently of any other state, then he will be stumped – for there is nothing he can say to distinguish the given state from its converse. *It must somehow be manifest from the state itself that it is the state that it is* and not the converse.'

This seems to me to be just too mysterious – almost mystical –: surely, if there are non-i.s. relations, it should be possible to describe directly the nature of a value of such a relation – how it is synthesized from the relation and the relata. On this point I am in essential agreement with the criticisms of Fine's view made in the recent paper Gaskin & Hill 2012; I particularly like their motto (p. 181) 'Substitution cannot create structure' (but only 'preserves it if it is already there'). – Fine's view seems to be part of a more general conception of constituent structure: see Fine 1989 pp. 73–6, and esp. the remark on p. 74 about the priority of substitution to 'syntactic construction'. – I myself prefer an account of constituent structure along more straightforward lines, where in particular substitution is far from fundamental and arises rather late in the series of definitions – v. Batchelor 2013. A situation for me, like any other complex object (if there are others), consists of its constituents put together in a certain way: the ultimate description of a situation is wholly 'internal' – it makes reference to nothing 'outside' the situation. It is true that it is not absurd to think that there are *some* objects the ultimate description of which must reach outside them – such as smiles and so on. But surely *situations* are perfectly ordinary substances – as 'self-standing' as anything – and so *they* should have an 'internal account'.

A further concern appears from Fine's picture of a relation as 'a simple unadorned body or "magnet", to which the relata of the relation are taken to be attached by some sort of invisible bond'. A (say) *binary* relation may be defined as a relation every completion of which has two terms (or occurrences of terms). Now, normally, if R is a binary relation, one would say that it has (unless it is i.s.) exactly two completions by two distinct objects a and b (where both are 'used'), which perhaps one will write as Rab and Rba. (And similarly for relations which are ternary etc.) But on Fine's picture it is not clear at all why there should be two completions rather than some other number. 'Different configurations may be formed from a given body and the relata, according to how they are attached.' But *why* then in this case exactly two 'configurations' can be formed? Why couldn't there be three, or four, or five, or any other number of ways of 'attaching' a and b to R? (Perhaps Fine, who likes 'bold moves', might say 'Why not indeed?'.)

A related point, made in MacBride 2007 p. 53, is that on Fine's picture it is not clear how *fixed degree* is possible: i.e. why should a certain relation always take only e.g. three arguments? (Again, one might say 'Why indeed?'.)

We have seen the similar objection already against Orderism and Directionalism; and like there, here too one might answer that it is simply the *contentual nature* of a relation that makes it have the degree it has; and now the same may be said also for the number of completions. It is always better however when we can explain why something behaves in a certain way by saying more than just that it is because it has the nature it has. *Positionalism* and *Variablism* (at least in their *Basic* version – 'one ghost, one object') have an 'anatomical' explanation both of the number of completions and of fixed degree. *Orderism* and *Directionalism* have anatomical explanations of the number of completions though not of fixed degree. *Fine's theory* has anatomical explanation for neither phenomenon.

This picture is almost exactly reversed when we come to the issue of permission of i.s. relations and multigrade relations. As Fine says, his theory permits both, whereas Positionalism permits neither. And again Orderism and Directionalism occupy a middle ground: they permit multigrade relations, but not i.s. relations. This reversal of the picture is not accidental. The anatomic richness of Positionalism gives it a high degree of explanatory power, but a low degree of permissiveness (as the relational situation is rigidly constrained by the supposed anatomy). Orderism and Directionalism have less anatomy, but still some, and accordingly permit a little more, and explain a little less. Finally, in Fine's theory, there is no anatomy at all, and thus the theory permits everything, and explains nothing.

Another difficulty with Fine's theory concerns *substitution*. As Fine says himself (p. 30, fn. 19), the 'anti-positionalist' 'can give no well-defined meaning to what it would be to substitute one neutral relation for another. Given the state of a's loving b, for example, we cannot say whether the result of substituting the relation of vertical placement for the amatory relation would be the state of a's being on b or b's being on a.' Thus Fine's theory violates what has been called above (4.3.1) the Principle of Uniqueness of Substitution. As we have said, this principle has a certain inner plausibility; and so it is at least desirable that a theory of relations should not violate it.

6.2 Russell 1913

Russell's 1913 manuscript *Theory of Knowledge* rejects the Directionalism (for primary relations) of *Principles of Mathematics*, and proposes an alternative theory of relations, where relations are taken to be (anachronistically speaking) 'neutral' in Fine's sense. (The manuscript was published posthumously in 1984. It is perhaps surprising that there has not been much discussion of this theory in the later literature [though there are exceptions, such as Hochberg 1987].) This Russell 1913 theory may be classified as a 'Rôle Theory' – i.e. it is based on the idea of the 'rôles' 'played' by the relata in the relational situation –; although it has also some affinities to Fine's theory.

A question which immediately arises for a 'Rôle-Theorist' is: *Where* are these rôles supposed to be? Are they (1) inside the relation? Or are they (2) inside the relational situation but outside the relation? Or are they perhaps (3) not even inside the relational situation but outside of it? This option (3) may sound crazy, but Russell's theory holds exactly that option.

Here it may be worth giving an extended quotation (from Pt. II, Ch. 5, pp. 145–6). (Russell's 'complexes' are only *facts*, but the theory can be straightforwardly adapted to situations in general. By 'non-permutative' Russell means basically i.s. He uses the term 'position' but obviously in the sense of rôle, not argument-place.)

'The two complexes A-before-B and A-after-B are composed of the same constituents, namely, A and B and sequence, combined in the same form. The words *before* and *after* differ, and are not names for the relation of sequence, which is unique, not twofold. What then is meant by *before* and *after*? In the complex A-before-B, which we will call γ, A has one "position" and B another; that is to say, A has one relation to γ and B has another. We may say A is *earlier in* γ, and B is *later in* γ, where *earlier in* and *later in* are heterogeneous relations, and therefore complexes in which they are the relating relations are completely determined by their constituents. Then "A-before-B" is the name of that complex (if any) in which A is earlier and B is later; while "A-after-B" is the name of that complex (if any) in which A is later and B is earlier. The distinction between *earlier* and *later* is indefinable, and must be simply perceived ...

'The proposition "There is a complex in which A is earlier and B later" is one which contains four constituents, namely A and B and earlier and later. This is a molecular proposition, whose atomic constituents are different from those of the proposition which results from interchanging A and B. Moreover, each of the atomic constituents is non-permutative. It is thus that the two words *before* and *after* become distinguishable, and that language is able to discriminate between the two complexes that may be formed of A and B and sequence.

'We may now generalize this solution, without any essential change. Let γ be a complex whose constituents are $x_1, x_2, \ldots x_n$ and a relating relation R. Then each of these constituents has a certain relation to the complex. We may omit the consideration of R, which obviously has a peculiar position. The relations of $x_1, x_2, \ldots x_n$ to γ are their "positions" in the complex; let us call them $C_1, C_2, \ldots C_n$. As soon as R is given, but not before, the n relations $C_1, C_2, \ldots C_n$ are determinate [sic], though unless R is non-permutative, it will not be determinate [sic] which of the constituents has which of the relations. If R is symmetrical with respect to two constituents, two of the C's will be identical. If R is heterogeneous with respect to two constituents, two of the C's will be incompatible. But in any case there are these relations $C_1, C_2, \ldots C_n$, and each constituent has one of these relations to γ.

'Unless the relation happens to be non-permutative, γ is not determined when we are given R and $x_1, x_2, \ldots x_n$. But it is determined when we are given also the positions of $x_1, x_2, \ldots x_n$, i.e. when we are given

$$x_1 C_1 \gamma \cdot x_2 C_2 \gamma \cdot \cdots x_n C_n \gamma.$$

Thus there is one complex at most which has the given constituents in the given positions. We may describe this as "*the* complex γ in which R is the relating relation, and $x_1 C_1 \gamma, x_2 C_2 \gamma, \ldots x_n C_n \gamma$".

If R is i.s., we can give an intrinsic description of the complex (say) Rab by saying that it is the complex where R is the 'relating relation' and a and b its relata. But if R is non-i.s., its values can only be extrinsically described, through phrases like: the situation p in which R is the 'relating relation' and which is such that $x_1 C_1 p$ and ... and $x_n C_n p$. We are forced to resort to mention of such special relations $C_1, \ldots, C_n$, extrinsic to (not constituents of) the situation we are describing.

So there is again some mystery here (we cannot describe the relational situation intrinsically, by saying what its constituents are and how they are put together), though I suppose less than in Fine's theory since here we can at least give a canonical (though extrinsic) description of each value by appealing to the Rôles.

The difficulty over Uniqueness of Substitution also applies here just as much as before. And although by allowing some of the C's to be identical Russell avoids the proliferation of situations in simple cases of symmetry (like multiple-occupancy Positionalism does), still the 'more complicated symmetries' will cause proliferations (again just like with multiple-occupancy Positionalism – v. 4.1 above). Also, the theory as formulated by Russell excludes multigrade relations (though one might of course try to modify the theory to avoid this). And no doubt here there will be 'necessary connexions amongst wholly distinct existences' (viz. the C's and the R), in violation of the Humean strictures.

It should be emphasized here that there is a great difference between the idea of *rôle* and the idea of position or argument-place. (There is also of course a great difference between the idea of rôle and the idea of Objective Variable; but this is perhaps too obvious to need emphasizing.) An argument-place is something quite *thin* and dull – a mere gap, a kind of blank. There may be lots of argument-places in the world (if there are lots of attributes), but they are not more different from one another than are different blades of grass or different grains of sand. The whole 'content' of the relation (under Positionalism, whether Nuclear or Anti-Nuclear) is in its 'constant part' (so to speak the relation apart from its gaps). A *rôle* by contrast is supposed to be something *thick* and contentual: the rôle *lover* for instance is not a mere gap but rather the locus of a good part of human misery.

Still, strictly speaking, I suppose a Positionalist *might* have a 'thick' conception of argument-places; this does not seem to be really *incoherent*. So in principle one can make a distinction between 'Thin Positionalism' where the argument-places are supposed to be 'non-contentual', and 'Thick Positionalism' where the argument-places are supposed to be 'contentual'. Indeed the names of argument-places in Fine's paper – *Lover* and *Beloved* and so on – might suggest this Thick Positionalism. Thick Positionalism would be a sort of hybrid between Positionalism and Rôle Theory; just like Armstrong's remarks (referred to above in §3) suggest a hybrid of Positionalism and Orderism. But just as I have said above that Armstrong's view would be 'theoretical over-shooting' (either positions alone or order alone being already sufficient to explain relational predication), so again the same applies to Thick Positionalism. The Positionalist does not *need* content in the positions to explain relational predication. And a Rôle-Theorist does not *need* to regard his rôles as being also some kind of 'gap'. A pure form of Rôle Theory would be like Positionalism only with the idea of position simply *replaced* by the idea of rôle (as a supposed fundamental); in the relational situation then the relata would be supposed to 'play' the rôles.

But again, if we have contentual rôles like *Lover* and *Beloved*, it is not really so clear that we need anything more – a further 'part' of the relation which 'supports' the rôles. Indeed this line of reasoning leads quite naturally to the Platonic theory to be considered in the next section.

In any case it is clear that *Russell's* so-by-him-called 'positions' are not positions in the sense of the Positionalist, i.e. argument-places. Argument-places outside the relation but only inside the relational situation would seem already a barely intelligible conception; but argument-places *outside the situation* is certainly nonsense.

It is worth adding here that Russell himself not only never published this 1913 theory of relations, but also seems to have soon afterwards reverted to Directionalism: in Russell 1919 he says (p. 49, my italics):

'We say that a relation and its converse have opposite "senses"; thus the "sense" of a relation that goes from x to y is the opposite of that of the corresponding relation from y to x. *The fact that a relation has a "sense" is fundamental*, and is part of the reason why order can be generated by suitable relations.'

7 Symmetricalism

We come finally to Symmetricalism. If non-i.s. relations are giving us so much trouble, perhaps it is because they don't exist?

Now, naively, one certainly supposes that there are non-i.s. relations (at any rate if one thinks that there are such entities as relations at all). So philosophers who have contemplated rejecting such relations have often felt a sort of duty to provide some kind of method of 'reduction' of the supposed non-i.s. relations. But I am not sure that the Symmetricalist is really under any such duty. He can give his general arguments for his view, e.g. the failure of all alternative views on the topic, and leave it at that. There need not be any general pattern or method of reduction. If the Symmetricalist is right, in the ultimate analysis of all situations only i.s. relations will appear. But surely it is too much to demand that the Symmetricalist should actually produce this ultimate analysis of all situations. – He may however give some examples to indicate how in certain cases or groups of cases i.s. relations are sufficient (and I will do that below).

Still of course if there *were* a satisfactory method of 'reduction', this would certainly boost Symmetricalism. So let us briefly consider some attempts to provide such a method of reduction. (Though, to anticipate, unfortunately none of them seems satisfactory.)

(1) *Properties of sequences*. This is the facile idea of replacing relations (at least non-i.s. ones) by monadic attributes (properties) of sequences: thus aRb becomes $R\langle a, b\rangle$. (Sequences in turn being presumably sets.)

– Objections:

(i) This goes against the classical (and plausible) conception where the immediate constituents of aRb are a, R, and b.

(ii) It violates the plausible supervenience principle given above as premiss (2) of our argument in 5.3.3.

(iii) It brings in sequences (i.e. presumably sets), which otherwise we might have hoped to dispense with. (Cf. van Inwagen 2006, §4.) It is true that the complex-attribute ontologist can easily define sets (and hence sequences) in terms of properties – e.g. $\{a, b, \ldots\}$ $=_{\mathrm{df}} \lambda x(x = a \vee x = b \vee \ldots)$ –, so that he could take aRb as a higher-order property applied to a complex property. But for the attribute-ontologist averse to complex and higher-order attributes (like myself – v. §5 above) this of course will not do. He actually *can* (given some other assumptions) give an elimination of sets (v. Batchelor 2013, p. 342, fn. 1), but *not* in the presence of an unspecified range of basic properties of sets (as would be needed here).

(2) *Repetition of terms*. One might try to turn a supposed basic binary non-i.s. predication Rab into a supposed basic ternary i.s. one $R'abb$; and similarly in other cases – i.e. simulate order by repetition of subjects. (Cf. Dorr 2004 [a paper which proposes a form of Symmetricalism], §8, *ad fin.*)

– Objections:

(i) This would go radically against the Principle of Independence of Atomic Situations: for then $R'abc$ (with a, b, c all different) would presumably be necessarily counterfactual.

(ii) It would force the Symmetricalist to complicate his account of relational predication: now he can no longer say that the subjects are a mere plurality of objects. The required modification would also have to go against what seems to be the only otherwise satisfactory method for giving a general definition of the notion of *occurrence*, viz. the one in Batchelor 2013.

(We should add that this simulation is not technically adequate if one uses, for the numbers of repetitions of subjects in relations of the various degrees, an arithmetical progression like $1, 2, 3, 4, \ldots$ [as in Dorr's paper]: for then e.g. $R'aaabbb$ would correspond ambiguously both to $Raab$ and to $Rbba$. We should take rather a *geometrical* progression, like $1, 2, 4, 8, \ldots$, for then it is guaranteed that the terms of a sum of elements of the progression are uniquely determined by the sum itself.)

(3) *Leibniz's 'et eo ipso'*. Another curious attempt at reduction is Leibniz's. 'Paris loves Helen' is of course not equivalent to 'Paris loves and Helen is loved', because *this* might be because Paris loves someone other than Helen and Helen is loved by someone other than Paris. So Leibniz suggested that 'Paris loves Helen' should be analyzed as 'Paris loves, *et eo ipso* Helen is loved', where the supposed non-i.s. relation of love is replaced by the property *loves*, the property *is loved*, and perhaps an i.s. relation (between situations) *et eo ipso*.

– Objections:

(i) 'Paris loves' surely means $\exists x$(Paris loves x) (and similarly for 'Helen is loved'), which re-introduces the love relation. Taking *loves* and *is loved* as simple, on the other hand, besides being implausible *per se*, would violate the Humean principle of 'No necessary connexions between wholly distinct existences'.

(ii) Another difficulty is the nature of the notion *et eo ipso*. It seems very implausible to think that this is a fundamental, indefinable notion. The most obvious way of trying to analyze it would be along the lines of:

> There is a fact in which Paris loves and
> in which Helen is loved.

But surely this supposed underlying fact would be none other than [Paris loves Helen], so that we are back where we started. Moreover this fact is surely not *identical* with the above existential quantification, which is what the *et eo ipso* statement became. (It is interesting to compare this line of thought with the Russell 1913 theory, discussed in 6.2 above.)

(4) *Castañeda's Plato*. According to Castañeda (1972), Plato had a 'cathartic philosophical vision' on the nature of relations, which he expounded in the *Phaedo*. Castañeda summarizes his exegesis as follows (§2, p. 471): –

'All Forms are monadic, i.e., each Form is instantiated only by one particular in each fact it is involved in: no Form is ever instantiated by pairs or other n-tuples, whether ordered or not.

'Some facts consist of a particular instantiating, or participating in, a Form: they are *single-pronged*. Other facts are *multiple-pronged*: they consist of an array of Forms each instantiated by one particular, where these instantiations do not by themselves constitute facts.

'Forms that can enter into multiple-pronged facts cannot enter into single-pronged facts. This is the *law of factual enchainment*. Forms governed by this law constitute Form-chains or *relations*.

'For an example consider the Forms Tallness and Shortness. They constitute the chain Tallness-Shortness, which is the relation taller-than. Thus, the fact that Simmias is taller than Socrates can be perspicuously represented by writing "Tallness(Simmias)-Shortness(Socrates)".'

It is unclear how much of this is Plato and how much Castañeda (as Castañeda himself admits in his 1975, p. 240); but anyway it is a curious view. Though Castañeda speaks of 'an array of Forms', and of 'Form-chains or *relations*', it is clear that this is here a mere *façon de parler* and that there are strictly speaking no relations on this view – no single entity is a relation. Castañeda says that Plato's theory 'breaks away from the dogma of the uniqueness of the relational entity'. (Castañeda sometimes [e.g. 1975 p. 241] speaks of the relation as *being* the corresponding *set* of monadic Forms. But in any case such a set would not enter into the 'relational' situation.)

There is here some affinity to Leibniz's view considered above. (See Castañeda 1982.) But here the attribution of a Form like Tallness etc. will not by itself constitute a situation. Where Leibniz has attributions of properties sort-of conjoined, Plato (or Platañeda) has a 'system' of simultaneous attributions of quasi-properties.

(Note that we have here two aspects, capable in principle of independent variation: (1) the monadic attributions may constitute bona fide situations in their own right, or not; and (2) we may have a system of simultaneous monadic attributions, or else an '*eo ipso* conjunction' of the attributions. We thus get four reductionistic views of this kind – the Leibnizian, the Platonic, and two others.)

There is also an obvious affinity with Russell 1913; but Plato does away with the R and keeps only something *like* the $C_1 \ldots C_n$; and they are now *inside* the relational situation.

– Objections:

(i) This Platañeda theory may perhaps be accused of ontic excess, since it replaces each one n-ary relation by n quasi-properties. But this seems to me to be a rather shallow objection: surely the true spirit of Ockham's razor is not to be found in such crass arithmetic.

(ii) A deeper objection to the Platañeda theory is that it violates the Humean principle of 'No necessary connexions amongst wholly distinct existences' – in particular its corollary 'No necessary connexions amongst distinct simples'. For the 'Forms' pertaining to a 'Form-chain' will be somehow essentially connected by virtue of their nature, despite being (at least in the most basic cases) ontically simple.

– So much then for attempted reductions. As I have said, I do not think any general reduction is possible, or necessary. But the following more local remarks may help lend some plausibility to Symmetricalism. (Further remarks of this kind, concerned with cases of a more empirical character, can be found in Lowe 2016.)

(1) *Logic*. It is relatively easy to found the whole of standard logic upon i.s. relations only – even with reification of logical notions.

Thus in classical propositional logic the primitives may be e.g. $\neg$ (property) and $\wedge$ (i.s. relation). The definition

$$\varphi \to \psi =_{df} \neg(\varphi \wedge \neg\psi)$$

is a paradigmatic example of construction of non-i.s. condition from basic properties and basic i.s. relations only.

Similarly for modal logic (property $\Box$), quantification theory ($\forall$ being either higher-order property, or infinitary conjunction which is i.s. relation), and identity (i.s.). The only trouble comes in set theory with $\in$: but again see Batchelor 2013, p. 342, fn. 1.

(2) *Anti-symmetric relations*. The typical *anti*-symmetric relations (in the strong sense of the term, i.e. [roughly speaking] that xRy always implies not yRx), like: before and after, above and below, smaller and larger, heavier and lighter, younger and older, etc., etc. – these can all be analyzed as disjunctions of conjunctions of suitable property-attributions. E.g. [a is younger than b] (in years) can be analyzed as the disjunction of all situations of the form [a is m years old $\wedge$ b is n years old] where $m < n$. ('<' enters into our present description of the situation but not into the situation itself.) – Here [a is m years old] might even be a relation between a and m, since it would be 'heterogeneous' in the sense of Russell 1913, and hence i.s.

Indeed, this can be extended of course to the corresponding relations of 'agreement' like *simultaneous with* etc. Thus we have here an analysis of all attributions of supposed relations 'of comparison' in the sense of Leibniz (who divides relations into such relations 'of comparison' and relations 'of connexion' like *loves* etc.).

Note that this is indeed an *analysis* – it replaces the attribution of supposed relation by an *equivalent*; this is not merely the usual remark that a 'ground' for the supposed attribution of relation can be given in terms of properties only. Moreover *this* 'ground', being a disjunct of our analysans, will indeed be a ground for it in the precise sense of Batchelor 2010.

– It should also be observed that, *were* there basic (strictly) anti-symmetric relations, this would of course go against the Principle of Independence of Atomic Situations. This alone should already cast a shadow of doubt on the naive assumption of existence of non-i.s. relations, which seems no less naive than the assumption of existence of strictly anti-symmetric relations.

(3) *Rigid attributes*. By a *rigid attribute* I mean an attribute all of whose values are rigid situations (i.e. necessarily factual or necessarily counterfactual). The following brutal method effects an elimination of all rigid attributes – so in particular all rigid *relations*, and more particularly still all rigid non-i.s. relations (which is what has to do with our present target). Suppose there *were* a (say binary) rigid relation R, which held of a, b and c, d etc. Then for any objects x, y, the situation Rxy would be strictly equivalent to

$$(x = a \wedge y = b) \vee (x = c \wedge y = d) \vee \ldots$$

But then it is enough to have *this* situation – we don't need the situation Rxy. (If the extension of the supposed R is 'too big', we can consider its restriction to a 'type' to which the given arguments belong.) E.g.: we don't need a relation Succ of immediate successor for natural numbers: what need is there for a situation e.g. Succ$(2, 3)$ when we already have

$$(2 = 0 \land 3 = 1) \lor (2 = 1 \land 3 = 2) \lor (2 = 2 \land 3 = 3) \lor \ldots?$$

The only rigid relation that might remain is identity itself. Now identity would be of course harmlessly i.s.; but I would rather eliminate it too nevertheless, for good measure, in the manner of Ramsey as explained above (5.4).

8 Concluding remark

The above examination of proposed solutions to the problem of relations reveals a number of recurrent themes and common pitfalls. This leads us to the following tentative list of *desiderata* for a theory of relational predication: –

(1) The supposed basic 'mechanism' of relational predication should be *credibly* fundamental.

(2) It should not incorporate multiple features where fewer would be explanatorily sufficient (no 'theoretical over-shooting').

(3) A *uniform* treatment (where the same 'mechanism' is supposed to operate in all relational situations) is as such preferable to a non-uniform one.

(4) One would like it to be the case that: The only (immediate) constituents of a relational situation are the relation and its relata.

(5) The theory should be compatible with existence of i.s. relations.

(6) Same for *non*-i.s. relations.

(7) Same for multigrade relations.

(8) Same for 'unigrade' relations.

(9) If there are non-i.s. relations, they should be 'neutral' in Fine's sense; in particular it should be *conceptually incoherent* to suppose that there are two distinct relations one the 'converse' of the other.

(10) Uniqueness of Substitution for relations.

(11) No facile violation of the Principle of Independence of Atomic Situations, nor of the related Humean principle of 'No necessary connexions amongst wholly distinct existences'.

(12) The relational situation should have an 'internal' account (not merely an 'external' account – let alone merely a 'holistic' one).

(13) If there are unigrade relations, their degree should have an *explanation* and not have to be taken as a brute 'fact of life'.

(14) Same for the *number of values* of a unigrade relation for given arguments.

– It can hardly be expected that a theory of relational predication should conform to *all* these desiderata (which in any case are not of course all equally important); but the extent of conformity may be regarded as a rough measure of plausibility.

The following synoptic scheme then provides a compendious recapitulation of some of the main considerations in this paper. We list the numbers corresponding to the desiderata (in the list above) which are *not* satisfied by the theory in question; and to avoid boredom we take only the *main* theories considered in the paper. (But recall that *Nuclear* Variablism is actually a hybrid theory and must be evaluated together with Orderism or Directionalism for primary relations.)

Strong Orderism: 1, 5, 9, 13.

Strong Directionalism: 1, 5, 9, 13.

Basic Positionalism ('one position, one object'): 1, 5, 7, 10, 11.

Basic Variablism: 1, 7, 10, 11.

Fine's theory: 10, 12, 13, 14.

Russell's 1913 theory: 7, 10, 11, 12.

Symmetricalism: 6, 13.

References

[Armstrong 1997] D. M. Armstrong. *A World of States of Affairs*. Cambridge University Press, 1997.

[Batchelor 2010] Roderick Batchelor. Grounds and consequences. *Grazer Philosophische Studien*, 80:65–77, 2010.

[Batchelor 2013] Roderick Batchelor. Complexes and their constituents. *Theoria (Lund)*, 79:326–52, 2013.

[Bergmann 1957] Gustav Bergmann. Elementarism. Incl. in his *Meaning and Existence*, pages 115–23. University of Wisconsin Press, 1959.

[Castañeda 1972] Hector-Neri Castañeda. Plato's *Phaedo* theory of relations. *Journal of Philosophical Logic*, 1:467–80, 1972.

REFERENCES

[Castañeda 1975] Hector-Neri Castañeda. Relations and the identity of propositions. *Philosophical Studies*, 28:237–44, 1975.

[Castañeda 1982] Hector-Neri Castañeda. Leibniz and Plato's *Phaedo* theory of relations and predication. In M. Hooker, editor, *Leibniz: Critical and Interpretive Essays*, pages 124–59. Manchester University Press, 1982.

[Dorr 2004] Cian Dorr. Non-symmetric relations. *Oxford Studies in Metaphysics*, 1: 155–92, 2004.

[Dummett 1973 / 1981] Michael Dummett. *Frege: Philosophy of Language*. London: Duckworth, 1973 (1st ed.) / 1981 (2nd ed.).

[Dummett 1981] Michael Dummett. *The Interpretation of Frege's Philosophy*. London: Duckworth, 1981.

[Fine 1985] Kit Fine. *Reasoning with Arbitrary Objects*. Oxford: Blackwell, 1985.

[Fine 1989] Kit Fine. The problem of de re modality. Incl. in his *Modality and Tense: Philosophical Papers*, pages 40–104. Oxford University Press, 2005.

[Fine 1998] Kit Fine. Cantorian abstraction: A reconstruction and defense. *Journal of Philosophy*, 95:599–634, 1998.

[Fine 2000] Kit Fine. Neutral relations. *Philosophical Review*, 109:1–33, 2000.

[Fine 2007] Kit Fine. Response to Fraser MacBride. *dialectica*, 61:57–62, 2007.

[Frege 1919] Gottlob Frege. Notes for Ludwig Darmstaedter. In M. Beaney, editor, *The Frege Reader*, pages 362–7. Oxford: Blackwell, 1997.

[Gaskin and Hill 2012] Richard Gaskin and Daniel J. Hill. On neutral relations. *dialectica*, 66:167–86, 2012.

[Gilmore 2013] Cody Gilmore. Slots in universals. *Oxford Studies in Metaphysics*, 8:187–233, 2013.

[Grossmann 1983] Reinhardt Grossmann. *The Categorial Structure of the World*. Indiana University Press, 1983.

[Grossmann 1992] Reinhardt Grossmann. *The Existence of the World: An Introduction to Ontology*. London: Routledge, 1992.

[Hansen 2016] Heine Hansen. On the road from Athens to Thebes again: Some thirteenth-century thinkers on converse relations. *British Journal for the History of Philosophy*, 24:468–89, 2016.

[Hochberg 1987] Herbert Hochberg. Russell's early analysis of relational predication and the asymmetry of the predication relation. *Philosophia*, 17:439–59, 1987.

[Leo 2008] Joop Leo. The identity of argument-places. *Review of Symbolic Logic*, 1: 335–54, 2008.

[Lowe 2016] E. J. Lowe. There are (probably) no relations. In A. Marmodoro and D. Yates, editors, *The Metaphysics of Relations*, pages 100–12. Oxford University Press, 2016.

[MacBride 2007] Fraser MacBride. Neutral relations revisited. *dialectica*, 61:25–56, 2007.

[Ramsey 1925] F. P. Ramsey. The foundations of mathematics. Incl. in his *Philosophical Papers*, pages 164–224. Cambridge University Press, 1990.

[Russell 1903] Bertrand Russell. *The Principles of Mathematics*. Cambridge University Press, 1903.

[Russell 1919] Bertrand Russell. *Introduction to Mathematical Philosophy*. London: Allen & Unwin, 1919.

[Russell 1913 / 1984] Bertrand Russell. *Theory of Knowledge: The 1913 Manuscript*. London: Allen & Unwin, 1984.

[Svenonius 1987] Lars Svenonius. Three ways to conceive of functions and relations. *Theoria* (Lund), 53:31–58, 1987.

[van Inwagen 2006] Peter van Inwagen. Names for relations. *Philosophical Perspectives*, 20:453–77, 2006.

[Weyl 1949] Hermann Weyl. *Philosophy of Mathematics and Natural Science*. Princeton University Press, 1949.

[Williamson 1985] Timothy Williamson. Converse relations. *Philosophical Review*, 94:249–62, 1985.

On formal aspects of the epistemic approach to paraconsistency *

Walter Carnielli[1], Marcelo Coniglio[1], Abilio Rodrigues[2]

[1]CLE and Department of Philosophy - State University of Campinas
[2]Department of Philosophy - Federal University of Minas Gerais

Contents

1	**Introduction**	**50**
2	**On the duality between paraconsistency and paracompleteness**	**52**
3	**Epistemic contradictions**	**53**
4	***BLE*: the Basic Logic of Evidence**	**54**
5	**A logic of evidence and truth**	**56**
6	**Valuation semantics for *BLE* and *LET$_J$***	**58**
7	**Inferential semantics for *BLE* and *LET$_J$***	**60**
8	**A calculus for factive and unfactive evidence**	**62**
9	**An algebraic approach: Fidel structures for *BLE* and *LET$_J$***	**64**
	9.1 Nelson's logic *N4* and the basic logic of evidence *BLE*: different views under equivalent formalisms	64
	9.2 Fidel-structures semantics for *N4*/*BLE*	65
	9.3 Fidel-structures semantics for *LET$_J$*	68

---

*The first and second authors acknowledge support from *FAPESP* (Fundação de Amparo à Pesquisa do Estado de São Paulo, thematic project *LogCons*), and from individual *CNPq* (Conselho Nacional de Desenvolvimento Científico e Tecnológico) research grants. The third author acknowledges support from *FAPEMIG* (Fundação de Amparo à Pesquisa do Estado de Minas Gerais, grants PEP 157-16 and 701-16).

10 Final remarks   71

Abstract

This paper reviews the central points and presents some recent developments of the epistemic approach to paraconsistency in terms of the preservation of evidence. Two formal systems are surveyed, the basic logic of evidence (*BLE*) and the logic of evidence and truth (*LET$_J$*), designed to deal, respectively, with evidence and with evidence and truth. While *BLE* is equivalent to Nelson's logic *N4*, it has been conceived for a different purpose. Adequate valuation semantics that provide decidability are given for both *BLE* and *LET$_J$*. The *meanings* of the connectives of *BLE* and *LET$_J$*, from the point of view of preservation of evidence, is explained with the aid of an *inferential semantics*. A formalization of the notion of evidence for *BLE* as proposed by M. Fitting is also reviewed here. As a novel result, the paper shows that *LET$_J$* is semantically characterized through the so-called *Fidel structures*. Some opportunities for further research are also discussed.

1 Introduction

Paraconsistency is the study of technical and philosophical aspects of formal systems in which the presence of a contradiction does not imply triviality, that is, systems with a non-explosive negation $\neg$ such that a pair of propositions A and $\neg A$ does not (always) lead to trivialization. Differently from classical (and intuitionistic) logic, in paraconsistent logics triviality is not tantamount to contradictoriness. Paraconsistent logics are able to deal with contradictory contexts of reasoning by means of the rejection of the principle of explosion, according to which anything follows from a contradiction.

From the philosophical point of view, maybe the most important question in paraconsistency addresses the nature of the contradictions allowed by paraconsistent logics. The answer to this question, of course, would better have some impacts on the formal systems. There are two basic approaches to this problem. On the one hand, the dialetheists claim that there are some *true contradictions* [?, e.g.]]dial.sta. This means that reality is contradictory in the sense that some pairs of contradictory propositions are needed in order to correctly describe reality. On the other hand, the epistemic approach to paraconsistent claims that it is much more plausible to consider that *all* contradictions that occur in real-life contexts of reasoning are epistemic in the sense that they are related to and/or originated in thought and language. The latter is the position endorsed by the authors of this text and has been already presented and defended in some papers (e.g. [Carnielli and Rodrigues(2016b)], [Carnielli and Rodrigues(2015)], [Carnielli and Rodrigues(2016d)]). Our aim here is to review the central points of the epistemic approach on paraconsistency, as well as to present some recent developments.

1 Introduction

The remainder of this text is structured as follows. In the section 2, we start by explaining the duality between paraconsistent and paracomplete (so also intuitionistic) logics. We will show that the central point is not really a duality between logics, but rather a duality between *principles of inference* that may be added to a common core, obtaining thus paracomplete or paraconsistent logics. Next, in section 3, we will present the epistemic reading of contradictions in connection with conflicting evidence. Evidence is an epistemic notion, weaker than truth, that means 'reasons for believing/accepting' a proposition as true (or false). In sections 4 and 5 we present two formal systems, the basic logic of evidence (*BLE*) and the logic of evidence and truth (LET_J). *BLE* is a natural deduction system designed to preserve evidence. It can be seen that *BLE* coincides with Nelson's paraconsistent logic *N4*; however, the motivations and interpretations of both systems are different. LET_J is a logic of formal inconsistency and undeterminateness (**LFIU**) that adds to *BLE* means to recover classical logic for formulas that have been established as true (or false). LET_J, thus, is capable of talking simultaneously about preservation of truth and preservation of evidence. Section 6 presents complete and correct valuation semantics for *BLE* and LET_J. Such semantics, however, are better understood as tools to prove technical results than semantics in the sense of providing meanings to the formal system. That the meanings of the expressions in the context of *BLE* and LET_J are given bay the inferences allowed is the topic of Section 7, where an *inferential semantics* is proposed for the logics *BLE* and LET_J. Although the notion of preservation of evidence is defined by the logic *BLE* in a precise way, the notion of evidence presented in section 3 is only intuitively explained. In section 8 we show the formalization of the notion of evidence provided by Melvin Fitting using *justification logics* [Fitting(2016)]. Fitting has shown that *BLE* has both *implicit* and *explicit* evidence interpretations in a strictly formal sense. In section 9 a semantics of Fidel structures is presented for the logics *BLE* and LET_J. In spite of the fact that the algebraizability of LET_J has not been established yet, an 'algebraic-relational semantics' like the one here presented sheds light upon the algebraic aspects of this logic. Finally, in section 10, we will point at some possible topics for further inquiry and philosophical research in the field of paraconsistency.[1]

---

[1] Some parts of this text draw on other papers by the authors. Parts of sections 2 and 3 have appeared in [Carnielli and Rodrigues(2016d)]. The formal systems presented in sections 4 and 5, as well as the valuation semantics of section 6, appear in [Carnielli and Rodrigues(2016c)]. Section 7 sums up the ideas presented in [Carnielli and Rodrigues(2016a)].

2 On the duality between paraconsistency and paracompleteness

At first sight, it seems to be an easy conclusion that paraconsistent and intuitionistic logics are 'dual', since excluded middle does not hold in the latter and some contradictions are accepted in the former.[2] Indeed, if we take a look at how Newton da Costa devises C_1, the first logic of his C_n hierarchy [?, see]p. 499]costa1974, it is not difficult to see that there is a sort of 'informal duality' between C_1 and intuitionistic logic. In the former, excluded middle and introduction of double negation hold, although in the latter non-contradiction and double elimination of negation hold.

However, in our view, this approach to paraconsistency is somewhat misleading. The central point is not that the *logics* are dual, nor that excluded middle and non-contradictions are dual *formulas*, but rather that the *inference rules* excluded middle (*PEM*) and explosion (*EXP*) are dual. This is easily seen in the framework of sequent calculus and multiple-conclusion logic.

$$\frac{}{\Gamma \Rightarrow A, \sim A, \Delta} \, PEM \qquad \frac{}{\Gamma, A, \sim A \Rightarrow \Delta} \, EXP$$

Indeed, added to the positive fragment of Gentzen's system *LK* [Gentzen(1935)], the axioms above yield classical logic. Notice that although *PEM* and *EXP* are *axioms* of sequent calculus, they express the fact that, classically, $A \vee \sim A$ follows from anything, and anything follows from $A \wedge \sim A$. From the point of view of classical logic, the invalidity of *PEM* in paracomplete (for instance, intuitionistic) logics and the invalidity of *EXP* in paraconsistent logics are like 'mirror images' of each other.

Now, to see the duality from the semantical viewpoint, let us take a look at the semantic characterization of classical negation $\sim$. A negation is classical if the following conditions hold (for classical $\vee$ and $\wedge$):

$$A \wedge \sim A \vDash, \tag{1}$$

$$\vDash A \vee \sim A. \tag{2}$$

According to condition (1), there is no model M such that $A \wedge \sim A$ holds in M. (2) expresses the fact that for every model M, $A \vee \sim A$ holds in M. A *paracomplete* negation disobeys (2), and a *paraconsistent* negation disobeys (1). Intuitionistic negation is an example of a paracomplete negation. Each one of the conditions above corresponds to half of the classical semantic clause for negation, respectively:

$$M(\sim A) = T \text{ only if } M(A) = F; \tag{3}$$

---

[2] Actually, the invalidity of the principle of non-contradiction is not an essential feature of paraconsistent logics, although the authors of this text share the opinion that both non-contradiction and explosion should be invalid in any paraconsistent logic. An example of a paraconsistent logic where explosion does not hold but non-contradiction is a valid formula is the Logic of Paradox [?, see]]priest.lp.

$$M(\sim A) = F \text{ only if } M(A) = T. \tag{4}$$

The clause (3) above forbids that both A and $\sim A$ receive *True*, and the clause (4) forbids that both receive *False*. Given the classical account of logical consequence – B follows from A iff there is no model M such that A is true but B is false in M – from the conditions above it follows that anything is a logical consequence of $A \land \sim A$ and $A \lor \sim A$ is a logical consequence of anything.

A counterexample to the principle of explosion is given by a circumstance such that a pair of propositions A and $\neg A$ hold but a proposition B does not hold ($\neg$ being a paraconsistent negation). Dually, a paracomplete logic requires a circumstance such that both A and $\neg A$ do not hold (now $\neg$ is a paracomplete negation). Notice that neither a paracomplete nor a paraconsistent negation is a contradictory-forming operator, in the sense that applied to a proposition A they do not produce a proposition $\neg A$ such that A and $\neg A$ cannot receive simultaneously the value F, nor simultaneously the value T – i.e. they do not 'invert' the semantic value of A. Besides, neither a paracomplete nor a paraconsistent negation is a 'truth-functional' operator because the semantic value of $\neg A$ is not unequivocally determined by the value of A: in a paraconsistent logic, if A receives T, the value of $\neg A$ may be T or F, and in a paracomplete logic, if A receives F, $\neg A$ may be T or F. It is important to call attention to the fact that we talk about the semantic values *True* and *False* here as a 'façon de parler'. From the epistemic viewpoint proposed here, neither paraconsistent nor paracomplete logics are talking about truth.

3 Epistemic contradictions

We have seen above the duality between the failure of explosion and excluded middle, respectively, in paraconsistent and paracomplete logics. An example of an intuitive motivation for a paracomplete negation is given by intuitionistic logic, where a circumstance such that there is no constructive proof of A nor of $\neg A$ acts as a counterexample for excluded middle. Indeed, the usual proof by cases,

$$A \to B, \neg A \to B \vdash B,$$

cannot be performed in intuitionistic logic. But what would be a justification for a paraconsistent, non-explosive negation?

There are two basic answers to this question. The dialetheist claims that there are true contradictions [Priest and Berto(2013)], what means that contradictions, so to speak, 'belong to the essence of reality'. But since it is not the case that everything holds, a paraconsistent logic is needed in order to describe reality correctly. The other answer, already mentioned here, says that a non-explosive negation should be understood from the epistemic viewpoint.

The acceptance of A and $\neg A$ in some contexts of reasoning does not need to mean, and actually does not mean, that both are true. There are a number of circumstances in which we deal with pairs of propositions A and $\neg A$ such that there are good reasons for accepting and/or believing in both. It does not mean of course that both are true, nor that we actually *believe* that both are true, although we still want to draw inferences in the presence of them. We have already argued elsewhere that a non-dialetheist position in paraconsistency ascribes a property *weaker than truth* to a pair of propositions A and $\neg A$ that 'hold' in a given context [Carnielli and Rodrigues(2016c)]. We propose the notion of *evidence*, understood as 'reasons for believing/accepting a proposition', to play the role of such a property. There may be evidence that A is true even if A is false, and conflicting evidence occurs when there are reasons for accepting A and reasons for accepting $\neg A$, both simultaneous and non-conclusive.[3]

The reading of contradictions as conflicting evidence fits well with the practices of empirical sciences. There are an extensive literature about contradictions in sciences (e.g. [da Costa and French(2003)], [Nickles(2002)]). The notion of contradictions as conflicting evidence is in line with the view that empirical theories are better seen as *tools* to solve problems, rather than *descriptions* of the world (these two approaches are discussed by [Nickles(2002)]). Of course, the occurrence of contradictions is a problem for the descriptive view of theories, since the latter requires that such a representation be correct (i.e. true). Once this non-representational view of scientific work is accepted, contradictions in the empirical sciences are better viewed as originated in limitations of our cognitive apparatus, failure of measuring instruments and/or interactions of these instruments with phenomena, stages in the development of scientific theories or even simply mistakes, to be corrected.

4 *BLE*: the Basic Logic of Evidence

In this section, we present a natural deduction system, the *Basic Logic of Evidence* (*BLE*), suited to the reading of contradictions as conflicting evidence. *BLE* ends up being equivalent to Nelson's logic *N4*, but has been conceived for a different purpose (see Section 9.1). The rules of *BLE* intend to express preservation of evidence in the following sense: supposing the availability of evidence for the truth (or falsity) of the premises, we ask whether an inference rule yields a conclusion for which evidence for its truth (or falsity) is also available. This approach has an analogy to the inference rules for intuitionistic logic, when the latter is understood epistemically as concerned with the availability of a constructive proof. Indeed, the basic idea of the Brouwer-Heyting-Kolmogorov interpretation is that an inference rule is valid if it transforms constructive proofs for one or more premises into a constructive proof of the conclu-

---

[3]The use we make here of the notion of evidence is close to how evidence in understood in epistemology – see [Kelly(2014)], [Achinstein(2010)] and also [Carnielli and Rodrigues(2016c)].

4 BLE: the Basic Logic of Evidence

sion. Natural deduction systems has been presented by [Gentzen(1935)] as formalisms capable of expressing 'natural logical reasoning'. Natural deduction fits our purpose here because we want to express how people actually, and naturally, draw inferences when the criterion is preservation of evidence.

Consider that the falsity of A is represented here by $\neg A$. 'Evidence that A is true' is understood as 'reasons for accepting/believing in A', and 'evidence that A is false' means 'reasons for accepting/believing in $\neg A$'. BLE is paraconsistent and paracomplete, neither explosion nor excluded middle hold. This is because there may be contexts with conflicting evidence as well as contexts with no evidence at all. In the former both A and $\neg A$ hold, in the latter both A and $\neg A$ do not hold.

DEFINITION 1. *The basic logic of evidence BLE*
Consider the propositional language L_1 defined in the usual way over the set of connectives $\{\wedge, \vee, \rightarrow, \neg\}$. S_1 is the set of of formulas of L_1. Roman capitals stand for meta-variables for formulas of L_1. The following natural deduction rules define the logic BLE:

$$\frac{A \wedge B}{A} \wedge E \quad \frac{A \wedge B}{B} \quad \frac{A \quad B}{A \wedge B} \wedge I$$

$$\frac{A}{A \vee B} \vee I \quad \frac{B}{A \vee B} \quad \frac{A \vee B \quad \overset{[A]}{\vdots}{C} \quad \overset{[B]}{\vdots}{C}}{C} \vee E$$

$$\frac{\overset{[A]}{\vdots}{B}}{A \rightarrow B} \rightarrow I \quad \frac{A \rightarrow B \quad A}{B} \rightarrow E$$

$$\frac{\neg A}{\neg(A \wedge B)} \neg \wedge I \quad \frac{\neg B}{\neg(A \wedge B)} \quad \frac{\neg(A \wedge B) \quad \overset{[\neg A]}{\vdots}{C} \quad \overset{[\neg B]}{\vdots}{C}}{C} \neg \wedge E$$

$$\frac{\neg A \quad \neg B}{\neg(A \vee B)} \neg \vee I \quad \frac{\neg(A \vee B)}{\neg A} \neg \vee E \quad \frac{\neg(A \vee B)}{\neg B}$$

$$\frac{A \quad \neg B}{\neg(A \rightarrow B)} \neg \rightarrow I \quad \frac{\neg(A \rightarrow B)}{A} \neg \rightarrow E \quad \frac{\neg(A \rightarrow B)}{\neg B}$$

$$\frac{A}{\neg\neg A} \quad DN \quad \frac{\neg\neg A}{A}$$

As an example, let us see how the preservation of evidence works w.r.t. the introduction rules for $\wedge$, $\vee$ and $\rightarrow$. If κ and κ' are evidence, respectively, for A and B, κ and κ' together constitute evidence for $A \wedge B$. Similarly, if κ constitutes evidence for A, then κ is also evidence for any disjunction that has A as one disjunct. For $\rightarrow I$, when the supposition that there is evidence κ for A leads to the conclusion that there is evidence κ' for B, this is evidence for $A \rightarrow B$. The implication, thus, works analogously to both classical and intuitionistic logic. It is not necessary that the contents of A and B be related.

The rules in which the conclusion is a negation of a conjunction, a disjunction or an implication cannot be obtained from the rules we already have because introduction of negation does not hold.[4] In order to obtain the negative rules we have to ask what would be sufficient conditions for having evidence for the falsity of a conclusion. So, if κ is evidence that A is false, κ constitutes evidence that $A \wedge B$ is false – *mutatis mutandis* for B. Thus, we obtain the rule $\neg \wedge I$. Analogous reasoning for disjunction and implication gives the respective introduction rules $\neg \vee I$ and $\neg \rightarrow I$.[5]

It is well-known that the elimination rules for $\wedge$, $\vee$ and $\rightarrow$ may be obtained from the introduction rules with the help of the *inversion principle*, presented by [Prawitz(1965)] as a refinement of the famous Gentzen's remarks that the introductions rules are, so to speak, 'definitions' of the connectives, and the eliminations rules are 'consequences' of these definitions [Gentzen(1935), p. 80]. Analogous reasoning works for the 'negative' elimination rules, $\neg \rightarrow E$, $\neg \wedge E$ and $\neg \vee E$. Suppose an application of the rule $\neg \rightarrow E$ that concludes A from $\neg(A \rightarrow B)$. A and $\neg B$ together are sufficient conditions for obtaining $\neg(A \rightarrow B)$. So, a derivation of the latter 'already contains' a derivation of A.[6] Notice that the negation rules exhibit a 'symmetry' with respect to the corresponding assertion rules for the dual operators.

5 A logic of evidence and truth

The logic *BLE* can express preservation of evidence. But in some contexts of reasoning we deal simultaneously with truth and evidence, that is, with propositions that are

---

[4]To see that $A \rightarrow B, A \rightarrow \neg B \vdash \neg A$ does not hold, suppose there is conflicting evidence for B and $\neg B$, but there is no evidence for $\neg A$. So, both $A \rightarrow B$ and $A \rightarrow \neg B$ hold, but $\neg A$ does not hold.

[5]The idea that natural deduction rules for concluding falsities may be obtained in a way similar to the rules for concluding truths is found e.g. in [López-Escobar(1972)] and also in [Prawitz(1965)]. Instead of asking about the conditions of assertability, the point is to ask about the conditions of *refutability*. This criterion works also for preservation of evidence.

[6]A more detailed account of the natural deduction rules of *BLE* is found in [Carnielli and Rodrigues(2016c)]. Regarding the inversion principle, see [Prawitz(1965), p. 33].

taken as conclusively established as true (or false), as well as others for which only non-conclusive evidence is available. Since preservation of truth is the criterion for a valid inference in classical logic, we get a tool for also dealing with true and false propositions if we can restore classical logic precisely for those propositions.

The *Logics of Formal Inconsistency* (from now on **LFIs**) are a family of paraconsistent logics that encompasses a great number of paraconsistent systems developed within the Brazilian tradition. **LFIs** are able to express the notion of 'consistency' of propositions inside the object language employing a unary connective: $\circ A$ means that A is consistent. Like any other paraconsistent logic, the principle of explosion does not hold in **LFIs**. But **LFIs** are so designed that some contradictions, that we call *consistent contradictions*, lead to triviality. Intuitively, one can understand the notion of a 'consistent contradiction' as a contradiction involving well-established facts, or involving propositions that have been conclusively established as true (or false) – notice that the point is precisely to prohibit consistent contradictions. A logic $\mathcal{L}$ is an **LFI** if the following holds:

For some Γ, A and B: $\Gamma, A, \neg A \nvdash B$,

For every Γ, A and B: $\Gamma, \circ A, A, \neg A \vdash B$.

LFIs start from the principle that propositions about the world can be divided into two categories: non-consistent and consistent ones. The latter are subjected to classical logic, and consequently a theory T that contains a pair of contradictory sentences $A, \neg A$ explodes only if A is taken to be a consistent proposition.[7]

The motivation of **LFIs**, restricting some logical property to some propositions, has been extended. In the *Logics of Formal Undeterminedness* (from now on **LFUs**), a class of paracomplete logics introduced in [Marcos(2005)], excluded middle can be restricted, and recovered, in a way analogous to **LFIs** restrict and recover explosion. Propositions can be divided into determined and non-determined ones, and a theory T may contain a proposition A such that neither A nor $\neg A$ hold. In an **LFU** the language is extended by a new unary connective $*$, where $*A$ means that A is (in some sense) determined. A logic $\mathcal{L}$ is an **LFU** if the following holds:

For some Γ, A and B: $\Gamma, A \vdash B, \Gamma, \neg A \vdash B$ but $\Gamma \nvdash B$,

For every Γ, A and B: if $\Gamma, A \vdash B$ and $\Gamma, \neg A \vdash B$, then $\Gamma, *A \vdash B$.

---

[7]The idea of expressing a metalogical notion within the object language is found, e.g. in the C_n hierarchy introduced by [da Costa(1963)], through the idea of 'well-behavedness' of a formula. In da Costa's hierarchy, however, this is done employing a definition: in C_1 it is expressed by A°, an abbreviation of $\neg(A \wedge \neg A)$, which makes the 'well-behavedness' of A equivalent to saying that A is non-contradictory. On the other hand, in the **LFIs**, $\circ A$ is introduced in such a way that allows $\circ A$ and $\neg(A \wedge \neg A)$ to be logically independent (non-equivalent). The family of **LFIs** incorporate a wide class of paraconsistent logics, as shown in [Carnielli et al.(2007)Carnielli, Coniglio, and Marcos] and [Carnielli and Coniglio(2016)].

An **LFI** and an **LFU** may be combined in an **LFIU** – a *Logic of Formal Inconsistency and Undeterminateness*. Explosion and excluded middle may be recovered at once with respect to a given formula A, and hence the properties of classical negation with respect to A. Since here we want to recover consistency and determinateness simultaneously, we use the symbol ∘ for both notions. The logic of evidence and truth obtained by extending *BLE*, is an **LFIU**.

DEFINITION 2. *The logic of evidence and truth LET_J.*
Consider the propositional language L_2 defined in the usual way over the set of connectives $\{\wedge, \vee, \rightarrow, \neg, \circ\}$. S_2 is the set of of formulas of L_2. The logic of formal inconsistency and undeterminedness LET_J is defined by adding to BLE the rules below:

$$\frac{\circ A \quad \begin{array}{c}[A]\\ \vdots \\ B\end{array} \quad \begin{array}{c}[\neg A]\\ \vdots \\ B\end{array}}{B} PEM^\circ \qquad \frac{\circ A \quad A \quad \neg A}{B} EXP^\circ$$

From 'outside' of the system, $\circ A$ means the truth-value of A has been conclusively established, or that there is *conclusive evidence* with respect to the truth-value of A. So, the fact that a proposition A is true is expressed as $\circ A \wedge A$, and the fact that A is false as $\circ A \wedge \neg A$.

The unary operator, ∘ may be called a *classicality* operator because when $\circ A_1, ..., \circ A_n$ hold, classical logic is recovered for all formulas that depend only on $A_1, ..., A_n$ and are formed with $\rightarrow, \wedge, \vee$ and $\neg$.[8]

6 Valuation semantics for *BLE* and *LET_J*

The valuation semantics to be presented in this section for *BLE* and *LET_J* does not intend to be a 'semantics' in the sense of a non-linguistic device that 'explains the meaning' of the corresponding deductive system – like, for example, the truth-tables for classical logic and the possible-worlds semantics for alethic modal logic. In the latter, the semantic clauses 'make sense' independently of the deductive system. On the other hand, the valuation semantics to be presented here is better seen as a mathematical tool capable of *representing* the inference rules in such a way that some technical results may be proved.

Valuation semantics have been proposed for the logics of da Costa's hierarchy C_n as a "generalization of the common semantics of the classical propositional calculus" [da Costa and Alves(1977), p. 622]. Later on, valuation semantics have been proposed also for da Costa's logic C_ω

---
[8]More details and several technical results that fit the intended intuitive interpretation of *BLE* and *LET_J* in terms of evidence and truth are to be found in [Carnielli and Rodrigues(2016c)]).

6 Valuation semantics for *BLE* and *LET$_J$*

[Loparic(1986)], intuitionistic logic [Loparic(2010)] and several Logics of Formal Inconsistency (**LFI**s) ([Carnielli et al.(2007)Carnielli, Coniglio, and Marcos] and [Carnielli and Coniglio(2016)]). Given a language L, valuations are functions from the set of formulas of L to $\{0,1\}$ in such a way that the semantic clauses are a kind of *representations* of the axioms. Roughly speaking, as we will see, assigning *1* and *0* to a formula A means, respectively, that A holds and A does not hold.

DEFINITION 3. *A semivaluation s for BLE is a function from the set S_1 of formulas to $\{0,1\}$ such that:*

(i) if $s(A) = 1$ and $s(B) = 0$, then $s(A \to B) = 0$,

(ii) if $s(B) = 1$, then $s(A \to B) = 1$,

(iii) $s(A \wedge B) = 1$ iff $s(A) = 1$ and $s(B) = 1$,

(iv) $s(A \vee B) = 1$ iff $s(A) = 1$ or $s(B) = 1$,

(v) $s(A) = 1$ iff $s(\neg\neg A) = 1$,

(vi) $s(\neg(A \wedge B)) = 1$ iff $s(\neg A) = 1$ or $s(\neg B) = 1$,

(vii) $s(\neg(A \vee B)) = 1$ iff $s(\neg A) = 1$ and $s(\neg B) = 1$,

(viii) $s(\neg(A \to B)) = 1$ iff $s(A) = 1$ and $s(\neg B) = 1$.

DEFINITION 4. *A semivaluation s for LET$_J$ is a function from the set S_2 of formulas to $\{0,1\}$ that satisfies the clauses (i)-(viii) of Definition 3 plus the following clause:*

(ix) if $s(\circ A) = 1$, then ($s(A) = 1$ if and only if $s(\neg A) = 0$).

DEFINITION 5. *A valuation for BLE/LET$_J$ is a semivaluation for which the condition below holds:*

(Val) For all formulas of the form $A_1 \to (A_2 \to ... \to (A_n \to B)...)$ with B not of the form $C \to D$:
if $s(A_1 \to (A_2 \to ... \to (A_n \to B)...)) = 0$, then there is a semivaluation s' such that for every i, $1 \leq i \leq n$, $s(A_i) = 1$ and $s(B) = 0$.

Logical consequence in *BLE* and *LET$_J$* is defined as usual: $\Gamma \vDash A$ if and only if for every valuation v, if $v(B) = 1$ for all $B \in \Gamma$, then $v(A) = 1$. The semantics above is sound and complete, and provides a decision procedure for *BLE* and *LET$_J$* by means of the *quasi-matrices* (see [Carnielli and Rodrigues(2016c)]). Below, as an example, we show how the quasi-matrices work.

EXAMPLE 6. $p \rightarrow (\neg p \rightarrow q)$ *is invalid in BLE.*

p	0							1				
$\neg p$	0			1				0			1	
q	0		1	0		1		0		1	0	1
$\neg p \rightarrow q$	0	1	1	0		1		0	1	1	0	1
$p \rightarrow (\neg p \rightarrow q)$	0	1	1	1	0	1	1	0	1	1	0	1
	s_1	s_2	s_3	s_4	s_5	s_6	s_7	s_8	s_9	s_{10}	s_{11}	s_{12}

In the example 6 above, the semi-valuation s_{11} turns out to be a valuation that acts as a counter-example. Notice that *BLE* and *LET$_J$* are not compositional, in the sense that the semantic value of a complex formula is not always functionally determined by the semantic values of its component parts.

EXAMPLE 7. $\circ p \rightarrow (p \vee \neg p)$ *is valid in LET$_J$.*

p	0				1		
$\neg p$	0		1		0		1
$\circ p$	0	0	1		0	1	0
$p \vee \neg p$	0	1	1		1	1	1
$\circ p \rightarrow (p \vee \neg p)$	0	1	1	1	1	1	1
	s_1	s_2	s_3	s_4	s_5	s_6	s_7

In the example 7 above, the semi-valuation s_1 is not a valuation, since the clause *Val* of Definition 5 is not satisfied. *BLE*, being coincident with Nelson's logic *N4*, is an extension of positive intuitionistic logic *(PIL)*. Indeed, clauses (i)-(iv) of definition 3 plus clause (Val) of definition 5 give a valuation semantics for *PIL*.[9]

7 Inferential semantics for *BLE* and *LET$_J$*

According to the standard view, syntax is concerned with the formal properties of linguistic expressions without regard to their meanings. Syntax, thus, includes formulas, axioms, rules of inference and proofs – in sum: manipulation of symbols according to certain rules. The word 'semantics' in the broad sense has to do with the *meanings* of the linguistic expressions, and such meanings are given by how the expressions are 'related to reality'. However, when a semantics is given to a deductive system, it is not always the case that the respective semantic values 'explain the meanings' of the corresponding expressions. Especially in the case of non-classical logics, it is not uncommon that the semantics, although a useful tool for providing counter-examples,

---

[9]A more detailed presentation of valuation semantics for *BLE* and *LET$_J$* may be found in [Carnielli and Rodrigues(2016c)] and [Carnielli and Rodrigues(2016a)].

decision methods and other relevant results, actually does not give any explanation of the meanings of the expressions, let alone the deductive system as a whole. An example of this situation is precisely the valuation semantics presented above for *BLE* and *LET$_J$*. However, the semantics of a deductive system, in the broad sense of an *explanation of meaning*, may be provided by the syntax, that is, by how the system is used to make inferences.

The proof-theoretic – or inferential – semantics is an approach to meaning originated in the natural deduction for intuitionistic logic.[10] Differently from the truth-conditional theory of meaning, inferential semantics provides meanings to the connectives of intuitionistic logic without the need of a semantics in the standard sense, i.e. the attribution of semantic values to formulas. The meanings are given by the deductive system itself, or more precisely, by the inference rules, that in this case do not express preservation of (a transcendent notion of) truth, but rather preservation of the availability of a constructive proof. The 'link to reality', so to speak, is given by the deductive system, more precisely, by the introduction rules. Now, since the meanings no longer depend on the semantics, but have been given by syntax, it becomes clear that valuation semantics for intuitionistic logic are nothing but mathematical representations of the formal system.

The origin of this idea is in [Gentzen(1935)], where the natural deduction system *NJ* for intuitionistic logic is presented. There we find the passage already mentioned in section 4, according to which the introduction rules are 'definitions' of the respective symbols [Gentzen(1935), p. 80]. From this perspective, the meaning of the connective $\vee$, for example, is given by how we *use* it in inferences which are not concerned with preservation of truth, but rather with preservation of availability of a (constructive) proof. The introduction rules for disjunction say that having available a proof of A (or a proof of B) is a sufficient condition for having a proof of the disjunction $A \vee B$. Intuitionistically, a disjunction cannot be obtained otherwise.

We propose to expand the basic idea of inferential semantics to the paraconsistent logics *BLE* and *LET$_J$*. On what regards *BLE*, the point is how we *use* the connectives in inferences that preserve evidence. So, the meanings of the logical connectives is also given by the inference rules, but now in a context where what is at stake is *preservation of evidence*. The same idea applies to *LET$_J$*, that is able to deal simultaneously with evidence and truth. In *LET$_J$*, classical logic holds for formulas marked with ∘. Thus, we can say that for such formulas the meaning of the connectives is classical.[11]

---

[10]Since we are going to extend the idea of proof-theoretic semantics to paraconsistent logics that are not concerned with 'truth obtained by means of a proof', but rather with 'preservation of evidence', we prefer to use the more general expression 'inferential semantics'.

[11]A more detailed analysis of the natural deduction rules of *BLE* and *LET$_J$* regarding preservation of evidence and truth is given in [Carnielli and Rodrigues(2016c)].

8 A calculus for factive and unfactive evidence

The logics *BLE* and *LET$_J$* define a notion of *preservation of evidence* in a precise way. But the corresponding notion of *evidence* is not formal. Melvin Fitting has provided in [Fitting(2016)] a formal alternative by means of the so-called *justification logics*. Fitting was able to show that *BLE* has both implicit and explicit evidence interpretations in a strictly formal sense. It is convenient to recall that *BLE* is presented through natural deduction rules, where the underlying idea is that rules should preserve evidence for an assertion, rather than its truth. A sequent calculus for the equivalent logic *N4* can be found in [Kamide and Wansing(2012)].

The plan followed in [Fitting(2016)] has a close analogy with the case of intuitionistic propositional logic, which is known since the work of [Gödel(1933)] to be embeddable into the modal logic $S4$. It was proved later (see [Artemov(2001)], [Artemov(2008)] and [Artemov and Fitting(2015)]) that $S4$ in turn embeds into the *strong justification logic LP*,[12] and the latter embeds into arithmetic. The logic *LP* provides a kind of calculus for certain *justification terms*. These terms can be regarded as representatives of proofs, and instead of $\Box A$ we may write $t : A$, where t is a justification term. The plan is the following:

1. It is first shown that *BLE* embeds into the modal logic $KX4$ explained below, a *logic of implicit uncertain evidence*, in which $\Box A$ can be interpreted as asserting that there is evidence for A, where this evidence can be partial or uncertain, and sometimes even incomplete and contradictory.

2. It is shown, furthermore, that $KX4$ in its turn embeds into the *justification logic* $JX4$, whose terms express pieces of uncertain evidence and are closed under certain operations that perform on such pieces of evidence.

The current axiomatization of the modal logic $S4$, inherited from Kurt Gödel, builds on the idea that $\Box$ has some intrinsic provability properties. The fact that intuitionistic logic embeds into $S4$ justifies the view that 'intuitionistic truth' can be understood as a version of provability *from the viewpoint of a classical mathematician*.

Provability may be considered as 'evidence of the strongest kind'. [Fitting(2016)] points out that provability coincides with the notion of evidence represented implicitly in $S4$, and explicitly in *LP*. Proofs can be seen as *factive* evidence, that is, evidence that is 'certain and never mistaken'. In contrast, the notion of evidence treated in *BLE* and represented implicitly in $KX4$, and explicitly in $JX4$, is *unfactive* in the sense of being disputable, retrievable or non-conclusive.[13]

---

[12] It should not be confused with the Logic of Paradox of Priest [Priest(1979)].
[13] We take the liberty to coin the term 'unfactive' due to its enlightening character.

8 A calculus for factive and unfactive evidence

$KX4$ is a normal modal (strict) subsystem of $S4$ obtained by dropping, precisely, the axiom of the factivity

$$\Box A \to A$$

and adding a new axiom schema called $C4$ or X for weaker or erroneous evidence:

$$\Box\Box A \to \Box A.$$

Informally, this schema expresses that evidence for the existence of evidence for A is sufficient to count as evidence for A. The other schemas for $KX4$ are the usual K:

$$\Box(A \to B) \to (\Box A \to \Box B)$$

and 4:

$$\Box A \to \Box\Box A$$

plus *modus ponens*. Obviously, in $KX4$

$$\Box\Box A \equiv \Box A$$

holds, which amounts to saying that evidence for the existence of evidence for some A is the same as evidence for A.

In this way, $KX4$ is an implicit logic of unfactive (or non-factive) evidence, in the same way $S4$ is a logic of provability (the term *implicit* refers to the fact that evidence is not explicitly shown, but just existential, as indicated by the modal operator $\Box$). As remarked in [Fitting(2016)], $KX4$ is complete with respect to frames meeting the conditions of transitivity and denseness.

The notion of 'implicit evidence against A' is also treated. Evidence is understood as something positive. The idea behind the rule $\neg \wedge I$, presented in Definition 1, is the following: if κ is evidence that A is false, κ constitutes evidence that $A \wedge B$ is false. An example given by Fitting illustrates this rule:

> We see that it is not raining, for instance, this is positive evidence that it is false that it is raining, and hence we have positive evidence that it is not both raining and cold. [Fitting(2016)]

This justifies a version of *implicit* evidence for *BLE*.

One of the two main results of [Fitting(2016)] reads:

THEOREM 8. *Theorem on Implicit Evidence for BLE:*
A is a theorem of BLE iff A^f is a theorem of $KX4$, where A^f is an inductively defined translation from the language of BLE into the language of $KX4$ using $\Box$ that reads as 'implicit evidence for A'.
Proof: see [Fitting(2016)].

An *explicit* counterpart to $KX4$, called $JX4$, can be obtained (omitting technical details) in such a way that $JX4$ serves as a justification counterpart of $KX4$ and is connected with it via a realization theorem, just as LP and $S4$ are connected in the sense of [Fitting(2015)].

The *justification formulas* of $JX4$ are built up from propositional letters using the usual propositional connectives, certain justification terms and additional justification formulas of the kind $t : A$, given by the following formation rule: if t is a justification term and A is a justification formula, then $t : A$ is a justification formula. Then it comes to the second main result of [Fitting(2016)]:

THEOREM 9. *Explicit Evidence for BLE:*
A^f *is a theorem of $KX4$ if and only if some normal realization of A^f is a theorem of $JX4$.*
Proof: see [Fitting(2016)].

The fact that the logic *BLE* embeds into the modal logic $KX4$ (via the Theorem on Implicit Evidence for *BLE*) justifies the view, in analogy with the intuitionistic case, that derivability in *BLE* can be understood as (preservation of) unfactive evidence *from the viewpoint of a classical philosopher*. The Theorem on Explicit Evidence for *BLE* grants that such evidence is rigorous and can be treated in a formal calculus. Several examples are given in [Fitting(2016)], while leaving as an open problem an investigation of LET_J in terms of formalized implicit and explicit evidence.

9 An algebraic approach: Fidel structures for *BLE* and LET_J

9.1 Nelson's logic *N4* and the basic logic of evidence *BLE*: different views under equivalent formalisms

D. Nelson introduced in [Nelson(1949)] a constructible interpretation for the first-order number theory based on intuitionistic logic. Nelson's aims was to overcome what appears to be a non-constructive feature of the intuitionistic negation $\neg$. In Nelson's logic N of 1949 some principles valid in the standard intuitionistic logic are not valid – a remarkable example is the principle of non-contradiction $\neg(A \wedge \neg A)$ – and some principles intuitionistically invalid are valid in N. In the first-order system N for number theory obtained from Nelson's interpretation, the resulting negation called *strong negation* (here denoted by $-$) satisfies all the properties of a De Morgan negation as well as the following meta-property:

$$\vdash -(A \wedge B) \text{ implies } \vdash -A \text{ or } \vdash -B.$$

Indeed, from the constructive viewpoint, it seems plausible that supposing a formula $A \wedge B$ has been proved false, either a proof of the falsity of A or a proof of the falsity of B should be available.

In 1959 Nelson introduced a system called S based on positive first-order intuitionistic logic (see [Nelson(1959)]) which turned out to be paraconsistent for secondary reasons. Together with A. Almukdad, he later proposed, in 1984, a variant of S called N^- (see [Almukdad and Nelson(1984)]). This system became the standard presentation of Nelson's paraconsistent logic. [Odintsov(2003)] rebaptized N^- as *N4*, proving that it is sound and complete with respect to a class of algebras called *N4*-lattices, as well as with respect to a variant of an algebraic-relational class of structures originally introduced by M. Fidel in [Fidel(1977b)] for da Costa's calculi C_n and afterwards for Nelson's logic N in [Fidel(1980)]. This kind of structures, called *Fidel-structures* or **F**-*structures* in [Odintsov(2003)], will be adapted here (Section 9.3 below) to give a semantical characterization for system *LET$_J$*. It is worth noting that [Odintsov(2004)] also proposed an interesting semantics for *N4* in terms of *twist-structures*, a general semantical framework which was independently proposed by [Fidel(1977a)] and by [Vakarelov(1977)].

As we have mentioned, the logic *BLE* is equivalent to *N4*. However, we must emphasize that *BLE* has been found independently of *N4*, based on a completely different motivation – namely, a logic able to express the deductive behaviour of a notion weaker than truth in order to provide an intuitive and clear interpretation for paraconsistency negation that does not depend on the simultaneous *truth* of a pair of contradictory sentences. Of course, all the technical results valid for *N4* are also valid for *BLE*, but their intended meaning are rather divergent.

9.2 Fidel-structures semantics for *N4*/*BLE*

From the contemporary perspective, the relationship between logic and algebra comes back to the ideas of A. Lindenbaum and A. Tarski of interpreting the formulas of a given logic with the aid of algebras with operations associated to the logical connectives. This approach was generalized by W. Blok and D. Pigozzi in [Blok and Pigozzi(1989)], in order to encompass a wider range of logics. Afterwards, several generalizations of Blok and Pigozzi's technique were proposed in the literature (see, for instance, [Font and Jansana(2009)] and [Font(2016)]). However, several logic systems lie outside the scope of the general methods of contemporary algebraic logic. For instance, the logics of da Costa's hierarchy C_n are not algebraizable by these methods, and the same holds for most of the **LFI**s studied in the literature (see [Carnielli and Coniglio(2016)]).

In 1977, Manuel Fidel proved, for the first time, the decidability of the calculi C_n using an original algebraic-relational class of semantical structures called C_n-*structures* [**?**, see]]fidel.1977. This kind of structure was called *Fidel-structures* or

F-*structures* in [Odintsov(2003)] (see also [Odintsov(2008)]). Briefly, a C_n-structure is a triple $\langle \mathcal{A}, \{N_a\}_{a \in \mathbf{A}}, \{N_a^{(n)}\}_{a \in \mathbf{A}} \rangle$ such that $\mathcal{A}$ is a Boolean algebra with domain **A** and each N_a and $N_a^{(n)}$ is a non-empty subset of **A**. Intuitively, $b \in N_a$ and $c \in N_a^{(n)}$ means that b and c are possible values for the paraconsistent negation $\neg a$ of a and for the 'well-behavior' (or 'consistency') a° of a, respectively. Because of the previous observations, the use of relations instead of functions for interpreting these two 'non-truth-functional' connectives seems to be appropriate.

As observed in [Odintsov(2008)], the logic *N4* lies in an intermediary stage with regards to algebraizability: the usual equivalence

$$A \leftrightarrow B \stackrel{\text{def}}{=} (A \to B) \wedge (B \to A)$$

does not define a logical congruence with respect to negation. That is, it is possible that the negations of equivalent formulas are not equivalent. For instance, given a propositional variable p, the formulas $\neg(p \to p)$ and $\neg(p \to (q \to p))$ are not equivalent in this logic, despite $(p \to p)$ and $(p \to (q \to p))$ being both valid (and so equivalent). However, it is possible to define a *strong equivalence*

$$A \Leftrightarrow B \stackrel{\text{def}}{=} (A \leftrightarrow B) \wedge (\neg A \leftrightarrow \neg B)$$

which constitutes a logical congruence in *N4*. Because of this, the following *weak replacement property* holds in *N4* (and so in *BLE*):

PROPOSITION 10. *[Odintsov(2008), Proposition 8.1.3] The logic N4 [BLE] satisfies the following weak replacement rule:*

$$\text{if } \vdash A \Leftrightarrow B \text{ then } \vdash C[p/A] \leftrightarrow C[p/B]$$

for every formula C, where $C[p/A]$ (resp., $C[p/B]$) denotes the formula obtained from C by replacing the variable p by the formula A (by the formula B, resp.).

As proved in [Odintsov(2008), Section 8.4], there exists a class of algebraic structures called *N4*-lattices associated to the logic *N4*. The class of *N4*-lattices is a *variety*, that is, it can be axiomatized by a set of equations. As Odintsov has shown, the logic *N4* (and so *BLE*) is algebraizable in the sense of [Blok and Pigozzi(1989)] by means of the variety of *N4*-lattices. Despite this algebraic characterization, Odintsov obtained another characterization of *N4* in respect of **F**-structures, by generalizing the proposal by M. Fidel in 1979 for the original Nelson's system N (see [Fidel(1980)]).

All the results mentioned above, of course, hold also for *BLE*. However, differently of *N4*/*BLE*, it is not clear whether or not the extension *LET$_J$* of *BLE* is algebraizable by Blok and Pigozzi's method. Indeed, it is possible to define, in a similar way to *N4*, the following equivalence formula:

$$A \rightleftharpoons B \stackrel{\text{def}}{=} (A \leftrightarrow B) \wedge (\neg A \leftrightarrow \neg B) \wedge (\circ A \leftrightarrow \circ B).$$

9.2 Fidel-structures semantics for N4/BLE

Clearly, it defines a logical congruence in *LET$_J$*, and so it induces a weak replacement property for *LET$_J$* analogous to that for *N4* stated in Proposition 10. It is an open problem to determine if this congruence is trivial, namely, whether or not it is the case that: if $\vdash A \rightleftharpoons B$ holds in *LET$_J$* then $A = B$, for every formulas A and B. More generally, it is a open problem to detemine if *LET$_J$* admits non-trivial logical congruences. This question justifies the present semantical approach to *LET$_J$* in terms of Fidel-structures, which expands the ones defined by Odintsov for the logic *N4*. The details of the construction will be described in Section 9.3.

Let us recall that an *implicative lattice* is an algebra $\mathcal{A} = \langle \mathbf{A}, \wedge, \vee, \rightarrow, 1 \rangle$ where $\langle \mathbf{A}, \wedge, \vee, 1 \rangle$ is a lattice with top element 1 such that there exists the supremum $\bigvee \{c \in \mathbf{A} : a \wedge c \leq b\}$ for every $a, b \in \mathbf{A}$. Here, $\leq$ denotes the partial order associated with the lattice, namely: $a \leq b$ iff $a = a \wedge b$ iff $b = a \vee b$; and $\bigvee X$ denotes the supremum of the set $X \subseteq \mathbf{A}$ w.r.t. $\leq$, whenever it exists. In addition, $\rightarrow$ is a binary operator (called *implication*) such that $a \rightarrow b \stackrel{\text{def}}{=} \bigvee \{c \in \mathbf{A} : a \wedge c \leq b\}$ for every $a, b \in \mathbf{A}$. It is well-known that, if an implicative lattice has a bottom element 0, then it is a Heyting algebra.

DEFINITION 11. *Fidel-structures for BLE (N4)*
A Fidel-structure for BLE (or an F-structure for BLE) is a pair

$$\mathcal{E} = \langle \mathcal{A}, \{N_a\}_{a \in A} \rangle$$

such that $\mathcal{A} = \langle \mathbf{A}, \wedge, \vee, \rightarrow, 1 \rangle$ *is an implicative lattice and* $\{N_a\}_{a \in A}$ *is a family of nonempty subsets of* $\mathbf{A}$ *where, for every* $a, b, c, d \in \mathbf{A}$, *the following holds:*

(1) if $c \in N_a$, *then* $a \in N_c$;

(2) if $c \in N_a$ *and* $d \in N_b$, *then* $c \wedge d \in N_{a \vee b}$;

(3) if $c \in N_a$ *and* $d \in N_b$, *then* $c \vee d \in N_{a \wedge b}$;

(4) if $d \in N_b$, *then* $a \wedge d \in N_{a \rightarrow b}$.

Intuitively, $c \in N_a$ means that c is a 'possible negation' $\neg a$ of a.

DEFINITION 12. *A valuation over an F-structure* $\mathcal{E} = \langle \mathcal{A}, \{N_a\}_{a \in A} \rangle$ *for BLE is a mapping v from the language L_1 to* $\mathbf{A}$ *satisfying the following:*

(1) $v(\neg p) \in N_{v(p)}$, *for every propositional letter p;*

(2) $v(A \# B) = v(A) \# v(B)$ *for* $\# \in \{\wedge, \vee, \rightarrow\}$;

(3) $v(\neg(A \wedge B)) = v(\neg A) \vee v(\neg B)$;

(4) $v(\neg(A \vee B)) = v(\neg A) \wedge v(\neg B)$;

(5) $v(\neg(A \to B)) = v(A) \wedge v(\neg B)$;

(6) $v(\neg\neg A) = v(A)$.

Let **P** be the set of propositional letters of L_1. A valuation is completely determined by its values over the set $\mathbf{P} \cup \{\neg p : p \in \mathbf{P}\}$. It is immediate to prove the following:

PROPOSITION 13. *Let v be a valuation over an **F**-structure $\mathcal{E}$ for BLE. Then $v(\neg A) \in N_{v(A)}$ for every formula A.*

The semantical consequence relation associated with **F**-structures is defined in a natural way:

DEFINITION 14. *Let $\Gamma \cup \{A\} \subseteq L_1$ and let $\mathcal{E}$ be a Fidel-structure for BLE. Then, A follows from Γ in $\mathcal{E}$, written as $\Gamma \models_{\mathbf{F}}^{\mathcal{E}} A$, if, for every valuation v over $\mathcal{E}$, $v(A) = 1$ whenever $v(B) = 1$ for every $B \in \Gamma$. We say that A is a semantical consequence of Γ (w.r.t. Fidel-structures for BLE, denoted by $\Gamma \models_{BLE}^{\mathbf{F}} A$, if $\Gamma \models_{\mathbf{F}}^{\mathcal{E}} A$ for every **F**-structure $\mathcal{E}$ for BLE.*

Then, the following holds (see [Odintsov(2008)]):

THEOREM 15. *Adequacy of BLE (N4) w.r.t. Fidel-structures Let $\Gamma \cup \{A\}$ be a set of formulas such that Γ is non-trivial in BLE. Then:*

$$\Gamma \vdash_{BLE} A \text{ iff } \Gamma \models_{BLE}^{\mathbf{F}} A.$$

9.3 Fidel-structures semantics for *LET_J*

Recall that the logic *LET_J* is an extension of *BLE* in the language L_2 obtained by adding the rules $PEM°$ and $EXP°$ to the latter (Definition 2). Given the adequacy of *BLE* w.r.t. **F**-structures (Theorem 15), it is natural to consider extensions of these **F**-structures, in order to capture semantically the logic *LET_J*.

DEFINITION 16. *Fidel-structures for LET_J*
*A Fidel-structure for LET_J (or an **F**-structure for LET_J) is a triple*

$$\mathcal{E} = \langle \mathcal{A}, \{N_a\}_{a \in A}, \{O_a\}_{a \in A} \rangle$$

*where $\mathcal{A} = \langle A, \wedge, \vee, \to, 0, 1 \rangle$ is a Heyting algebra, $\langle \mathcal{A}, \{N_a\}_{a \in A} \rangle$ is a Fidel-structure for BLE (N4), and $\{O_a\}_{a \in A}$ is a family of nonempty subsets of **A** such that, for every $a, b \in \mathbf{A}$, the following holds:*

(FJ) *if $b \in N_a$ then $BD_{ab} \cap BC_{ab} \neq \emptyset$, where*

$$BD_{ab} = \{c \in O_a : c \to (a \vee b) = 1\}$$

and

$$BC_{ab} = \{c \in O_a : a \wedge b \wedge c = 0\}.$$

9.3 Fidel-structures semantics for LET_J

REMARK 17. *Let $\mathcal{A}$ be a Heyting algebra, and let $\div$ be the intuitionistic negation in $\mathcal{A}$, which is defined as $\div a = a \to 0$ for every $a \in \mathbf{A}$. For each $a \in \mathbf{A}$ let $a\downarrow$ be the set $\{b \in \mathbf{A} : b \leq a\}$. Observe that $a \wedge c = 0$ if and only if $c \in (\div a)\downarrow$, and $c \to a = 1$ iff $c \in a\downarrow$, for every $a, c \in \mathbf{A}$. Then, condition (FJ) states that, if $b \in N_a$, then $O_a \cap (a \vee b)\downarrow \cap (\div(a \wedge b))\downarrow \neq \emptyset$. Equivalently, (FJ) requires that, if $b \in N_a$, then $O_a \cap ((a \vee b) \wedge \div(a \wedge b))\downarrow \neq \emptyset$.*

Intuitively, $b \in N_a$ means that b is a 'possible negation' $\neg a$ of a, while $c \in O_a$ means that c is a 'possible recovery value' $\circ a$ of a coherent with a given $b \in N_a$. This is supported by the following definition:

DEFINITION 18. *A valuation over an **F**-structure $\mathcal{E} = \langle \mathcal{A}, \{N_a\}_{a \in A}, \{O_a\}_{a \in A} \rangle$ for LET_J is a map v from L_2 to $\mathcal{A}$ satisfying the clauses (2)-(6) of Definition 12, plus the following properties, for every formula A:*

(1) $v(\neg A) \in N_{v(A)}$;

(2) $v(\circ A) \in BD_{v(A)v(\neg A)} \cap BC_{v(A)v(\neg A)}$.

REMARK 19. *Given that $B \wedge \neg B \wedge \circ B$ is a bottom (that is, trivializing) formula in LET_J, for any formula B, then $\div A$ can be represented in LET_J by $A \to (B \wedge \neg B \wedge \circ B)$. Being so, $v(\div A) = \div v(A)$ for every valuation v over an **F**-structure $\mathcal{E}$ for LET_J.*

EXAMPLE 20. *Let $\mathbb{R}$ be the set of real numbers endowed with the usual topology generated by the open intervals of the form (a, b), $(-\infty, a)$ and $(a, +\infty)$. It is well-known that the set of open subsets of $\mathbb{R}$ constitutes a Heyting algebra $\Omega(\mathbb{R})$ where $1 = \mathbb{R}$, $0 = \emptyset$ and, for every $X, Y \in \Omega(\mathbb{R})$: $X \vee Y = X \cup Y$; $X \wedge Y = X \cap Y$; and $X \to Y = Int((\mathbb{R} \setminus X) \cup Y)$, where $Int(Z)$ denotes the interior of a subset Z of $\mathbb{R}$ (that is, the greatest open contained in Z). Hence $\div X = Int(\mathbb{R} \setminus X)$. Consider an **F**-structure $\mathcal{E}$ over $\Omega(\mathbb{R})$ such that $(1, 3) \in N_{(0,2)}$. Let A, B two formulas and let v be a valuation v over $\mathcal{E}$ such that $v(A) = (0, 2)$ and $v(B) = (1, 3)$. Then $v(A \vee B) = v(A) \cup v(B) = (0, 3)$; $v(A \wedge B) = v(A) \cap v(B) = (1, 2)$; and $v(\div(A \wedge B)) = \div v(A \wedge B) = (-\infty, 1) \cup (2, +\infty)$. Thus, by Remark 17, the element $v(\circ A)$ of $O_{(0,2)}$ must be an open subset of $v(A \vee B) \cap v(\div(A \wedge B)) = (0, 1) \cup (2, 3)$.*

The next step is to prove that the proposed semantics for LET_J is adequate, that is, the logic LET_J is sound and complete w.r.t. Fidel-structures. The proof will be similar to the one obtained by Odintsov for N4 (see [Odintsov(2008)]) and the adaptation to mbC given in [Carnielli and Coniglio(2016), ch. 6].

Let Γ be a non-trivial theory in LET_J, that is, a set of formulas such that $\Gamma \nvdash_{LET_J} A$ for some formula A. Define the following relation $\equiv_\Gamma$ between the formulas of L_2:

$$A \equiv_\Gamma B \text{ iff } \Gamma \vdash_{LET_J} A \to B \text{ and } \Gamma \vdash_{LET_J} B \to A.$$

It is immediate to prove that $\equiv_\Gamma$ is an equivalence relation. Moreover, $\equiv_\Gamma$ is a congruence w.r.t. the connectives in the language of positive intuitionistic logic (*PIL*). Denote by $[A]_\Gamma$ the equivalence class of each formula A and let

$$\mathbf{A}_\Gamma \stackrel{\text{def}}{=} L_2/{\equiv_\Gamma} = \{[A]_\Gamma \ : \ A \in L_2\}$$

be the set of all the equivalence classes. From the observation above, it is possible to define the following operations:

$$[A]_\Gamma \# [B]_\Gamma \stackrel{\text{def}}{=} [A \# B]_\Gamma \qquad \text{for } \# \in \{\wedge, \vee, \to\}.$$

All these operations are well-defined, that is, they do not depend upon the representative chosen for each equivalence class. This means that

$$\mathcal{A}_\Gamma \stackrel{\text{def}}{=} \langle \mathbf{A}_\Gamma, \wedge, \vee, \to, 0_\Gamma, 1_\Gamma \rangle$$

(where $0_\Gamma \stackrel{\text{def}}{=} [p_1 \wedge \neg p_1 \wedge \circ p_1]_\Gamma$ and $1_\Gamma \stackrel{\text{def}}{=} [p_1 \to p_1]_\Gamma$) is a Heyting algebra, given that 0_Γ is a bottom element of the underlying implicative lattice. It is now possible to define from here an **F**-structure for *LET$_J$* by considering

$$N_{[A]_\Gamma} \stackrel{\text{def}}{=} \{[\neg B]_\Gamma \ : \ B \in [A]_\Gamma\}$$

and

$$O_{[A]_\Gamma} \stackrel{\text{def}}{=} \{[\circ B]_\Gamma \ : \ B \in [A]_\Gamma\}$$

for every $[A]_\Gamma \in \mathbf{A}_\Gamma$. This structure will be called *the Lindenbaum **F**-structure for LET$_J$ over* Γ. Observe that this is coherent with the intuitive reading for the sets N_a and O_a given above.

PROPOSITION 21. *Let Γ be a non-trivial theory in LET$_J$, and let $\mathcal{A}_\Gamma$ and $\mathbf{A}_\Gamma$ as above. Then, the triple*

$$\mathcal{E}_\Gamma = \langle \mathcal{A}_\Gamma, \{N_a\}_{a \in \mathbf{A}_\Gamma}, \{O_a\}_{a \in \mathbf{A}_\Gamma} \rangle$$

*is an **F**-structure for LET$_J$.*

Proof. The pair $\mathcal{E}_\Gamma = \langle \mathcal{A}_\Gamma, \{N_a\}_{a \in \mathbf{A}_\Gamma} \rangle$ is an **F**-structure for *BLE(N4)* (see [Odintsov(2008)]). It remains to prove that the family $\{O_a\}_{a \in \mathbf{A}_\Gamma}$ satisfies the requirement (FJ) of Definition 16. Thus, let $[\neg B]_\Gamma \in N_a$ (for a given $a \in \mathbf{A}_\Gamma$). Then, $B \in a$ and so $a = [B]_\Gamma$. From this, $[\circ B]_\Gamma \in O_a$ satisfies:

$$[\circ B]_\Gamma \to (a \vee [\neg B]_\Gamma) = [\circ B]_\Gamma \to ([B]_\Gamma \vee [\neg B]_\Gamma) = [\circ B \to (B \vee \neg B)]_\Gamma = 1_\Gamma$$

since $\circ B \to (B \vee \neg B) \equiv_\Gamma p_1 \to p_1$. In an analogous way it is proved that

$$a \wedge [\neg B]_\Gamma \wedge [\circ B]_\Gamma = [B]_\Gamma \wedge [\neg B]_\Gamma \wedge [\circ B]_\Gamma = [B \wedge \neg B \wedge \circ B]_\Gamma = 0_\Gamma$$

since $B \wedge \neg B \wedge \circ B \equiv_\Gamma p_1 \wedge \neg p_1 \wedge \circ p_1$. This means that condition (FJ) is satisfied. $\square$

We thus arrive at the desired result:

THEOREM 22 (Adequacy of LET_J w.r.t. Fidel-structures). *Let $\Gamma \cup \{A\}$ be a set of formulas such that Γ is non-trivial in LET_J. The following conditions are equivalent:*

(1) $\Gamma \vdash_{LET_J} A$;

(2) $\Gamma \models_{\mathbf{F}} A$;

(3) $\Gamma \models_{\mathbf{F}}^{\mathcal{E}_\Gamma} A$.

Proof. (1) ⇒ (2): This is the Soundness theorem, which can be proved in a straightforward way as usual. Indeed, it is enough to prove that all the rules of LET_J are valid w.r.t. **F**-structures.

(2) ⇒ (3): It is an immediate consequence of Definition 14.

(3) ⇒ (1): Let $v : L_2 \to L_2/_{\equiv_\Gamma}$ be the canonical mapping given by $v(B) = [B]_\Gamma$. By the very definition of $\mathcal{A}_\Gamma$, it follows that v is a valuation over $\mathcal{E}_\Gamma$ satisfying the following: $v(B) = 1_\Gamma$ iff $\Gamma \vdash_{LET_J} B$, for every formula B. Hence, $v(B) = 1_\Gamma$ for every $B \in \Gamma$, which, by hypothesis, implies that $v(A) = 1_\Gamma$. That is, $\Gamma \vdash_{LET_J} A$. □

10 Final remarks

This paper reviewed the main points of the approach to paraconsistent with reference to preservation of evidence. The ideas presented also suggests a promising approach to the issue of logical pluralism. The difference between classical, intuitionistic and paraconsistent logics, the last two understood from the epistemic point of view, is what is being preserved – respectively, truth, availability of a constructive proof and availability of evidence. Notice that there is a kind of informal duality in this reading of these three logics, since proof is a notion stronger (and evidence weaker) than truth. This helps to understand that the pluralist perspective is perfectly coherent, and in principle nothing prevents these three logics to be combined in some kind of 'general approach to rationality'.

It is also worth noting that the formalization of the notion of evidence provided by M. Fitting, as surveyed in Section 8, is yet another indication that we have taken the correct path basing the epistemic approach on the (formal and informal) duality between paraconsistency and paracompleteness. Actually, there are several 'convergences' in our approach. As it has been mentioned, the logic *BLE* has been conceived independently of Nelson's *N4*, although they are equivalent. The 'evidence interpretation' of *BLE* is endorsed by the fact that *BLE* is related to justification logics, as Fitting has shown. The paper also proves in Section 9 that both *BLE* and LET_J are semantically characterized through **F**-structures, a kind of algebraic-relational semantic structures. There are, however, several points yet to be developed and investigated. The first is to check how much **F**-structures can help to solve the algebraizability problem for LET_J, an open problem by now (*BLE*, being equivalent to *N4*, is algebraizable in the sense of Blok and Pigozzi). The second problem was raised by Fitting: how to formalize the notions of implicit and explicit evidence for LET_J (as it was done for BLE in [Fitting(2016)]). The third, philosophically more ambitious, is how to frame the classical, intuitionistic and paraconsistent paradigms in terms of preservation of levels of evidence. This would be a leap towards a better understanding of logical pluralism.

References

[Achinstein(2010)] P. Achinstein. Concepts of evidence. In *Evidence, Explanation, and Realism*. Oxford University Press, 2010.

[Almukdad and Nelson(1984)] A. Almukdad and D. Nelson. Constructible falsity and inexact predicates. *The Journal of Symbolic Logic*, 49(1), 1984.

[Artemov(2001)] S. Artemov. Explicit provability and constructive semantics. *Bulletin of Symbolic Logic*, 7:1–36, 2001.

[Artemov(2008)] S. Artemov. The logic of justification. *The Review of Symbolic Logic*, 4:477–513, 2008.

[Artemov and Fitting(2015)] S. Artemov and M. Fitting. Justification logic. *Stanford Encyclopedia of Philosophy*, 2015. URL http://plato.stanford.edu/entries/logic-justification/.

[Blok and Pigozzi(1989)] W. J. Blok and D. Pigozzi. Algebraizable logics. *Memoirs of the American Mathematical Society*, 77, 1989.

[Carnielli and Coniglio(2016)] W. Carnielli and M. E. Coniglio. *Paraconsistent Logic: Consistency, Contradiction and Negation*, volume 40 of Logic, Epistemology, and the Unity of Science series. Springer, 2016.

[Carnielli and Rodrigues(2015)] W. Carnielli and A. Rodrigues. Towards a philosophical understanding of the logics of formal inconsistency. *Manuscrito*, 38: 155–184, 2015.

[Carnielli and Rodrigues(2016a)] W. Carnielli and A. Rodrigues. Inferential semantics, paraconsistency and preservation of evidence. *Submitted*, 2016a.

[Carnielli and Rodrigues(2016b)] W. Carnielli and A. Rodrigues. On the philosophy and mathematics of the Logics of Formal Inconsistency. In J.-Y. Beziau, M. Chakraborty, and S. Dutta, editors, *New Directions in Paraconsistent Logic*, pages 57–88. Springer India, 2016b.

[Carnielli and Rodrigues(2016c)] W. Carnielli and A. Rodrigues. An epistemic approach to paraconsistency: a logic of evidence and truth. *Submitted*, 2016c.

[Carnielli and Rodrigues(2016d)] W. Carnielli and A. Rodrigues. Paraconsistency and duality: between ontological and epistemological views. In *The Logica Yearbook 2015*. College Publications, 2016d.

[Carnielli et al.(2007)Carnielli, Coniglio, and Marcos] W. Carnielli, M. E. Coniglio, and J. Marcos. Logics of Formal Inconsistency. In D. M. Gabbay and F. Guenthner, editors, *Handbook of Philosophical Logic (2nd. edition)*, volume 14, pages 1–93. Springer, 2007.

[da Costa(1963)] N. C. A. da Costa. Sistemas formais inconsistentes (Inconsistent formal systems, in Portuguese). Habilitation thesis, Universidade Federal do Paraná, Curitiba, Brazil, 1963. Republished by Editora UFPR, Curitiba, Brazil,1993.

[da Costa and Alves(1977)] N. C. A. da Costa and E. H. Alves. A semantical analysis of the calculi C_n. *Notre Dame Journal of Formal Logic*, 18:621–630, 1977.

[da Costa and French(2003)] N. C. A. da Costa and S. French. *Science and Partial Truth: A Unitary Approach to Models and Scientific Reasoning*. Oxford: Oxford University Press, 2003.

[Fidel(1977a)] M. Fidel. An algebraic study of a propositional system of Nelson. In A. I. Arruda, N. C. A. da Costa, and R. Chuaqui, editors, *Mathematical Logic. Proceedings of the First Brazilian Conference on Mathematical Logic, Campinas 1977 (volume 39 of Lecture Notes in Pure and Applied Mathematics)*. Marcel Dekker, 1977a.

[Fidel(1977b)] M. Fidel. The decidability of the calculi C_n. *Reports on Mathematical Logic*, 8:31–40, 1977b.

[Fidel(1980)] M. Fidel. An algebraic study of logic with constructive negation. In A. I. Arruda, N. C. A. da Costa, and A. M. A. Sette, editors, *Proceedings of the Third Brazilian Conference on Mathematical Logic, Recife, 1979*. Sociedade Brasileira de Logica, Campinas, 1980.

[Fitting(2015)] M. Fitting. Modal logics, justification logics, and realization. *Submitted, 2015*, 2015.

[Fitting(2016)] M. Fitting. Paraconsistent logic, evidence, and justification. *Studia Logica (to appear)*, 2016. Unpublished.

[Font(2016)] J.M. Font. *Abstract Algebraic Logic: An Introductory Textbook*, volume 60. Mathematical Logic and Foundations series, College Publications, London, 2016.

[Font and Jansana(2009)] J.M. Font and R. Jansana. *A General Algebraic Semantics for Sentential Logics*. Volume 7 of Lecture Notes in Logic, Association for Symbolic Logic, Ithaca, NY, USA, 2009.

[Gentzen(1935)] G. Gentzen. Investigations into logical deduction. In *The Collected Papers of Gerhard Gentzen (ed. M.E. Szabo)*. North-Holland Publishing Company (1969), 1935.

[Gödel(1933)] K. Gödel. An interpretation of the intuitionistic propositional calculus. In *Collected Works, vol. 1 (1986), Feferman, S., et. al. (ed.)*. Oxford: Oxford University Press, 1933.

[Kamide and Wansing(2012)] N. Kamide and H. Wansing. Proof theory of Nelson's paraconsistent logic: A uniform perspective. *Theoretical Computer Science*, 415:1–38, 2012.

[Kelly(2014)] T. Kelly. Evidence. *The Stanford Encyclopedia of Philosophy (Fall 2014, ed. E. Zalta)*, 2014. URL http://plato.stanford.edu/archives/fall2014/entries/evidence.

[Loparic(1986)] A. Loparic. A semantical study of some propositional calculi. *The Journal of Non-Classical Logic*, 3(1):73–95, 1986.

[Loparic(2010)] A. Loparic. Valuation semantics for intuitionistic propositional calculus and some of its subcalculi. *Principia*, 14(1):125–133, 2010.

[López-Escobar(1972)] E.G.K. López-Escobar. Refutability and elementary number theory. *Indagationes Mathematicae*, 34:362–374, 1972.

[Marcos(2005)] J. Marcos. Nearly every normal modal logic is paranormal. *Logique et Analyse*, 48:279–300, 2005.

[Nelson(1949)] D. Nelson. Constructible falsity. *The Journal of Symbolic Logic*, 14, 1949.

[Nelson(1959)] D. Nelson. Negation and separation of concepts in constructive systems. In A. Heyting, editor, *Constructivity in Mathematics - Proceedings of the colloquium held at Amsterdam*. North-Holland, 1959.

[Nickles(2002)] T. Nickles. From Copernicus to Ptolemy: inconsistency and method. In *Inconsistency in Science (Ed. J. Meheus)*. Dordrecht: Springer, 2002.

REFERENCES

[Odintsov(2003)] S. Odintsov. Algebraic semantics for paraconsistent Nelson's logic. *Journal of Logic and Computation*, 13(4):453–468, 2003.

[Odintsov(2004)] S. Odintsov. On representation of N4-lattices. *Studia Logica*, 76 (3), 2004.

[Odintsov(2008)] S. Odintsov. *Constructive Negations and Paraconsistency*. Springer, 2008.

[Prawitz(1965)] D. Prawitz. *Natural Deduction: A Proof-Theoretical Study*. Dover Publications (2006), 1965.

[Priest(1979)] G. Priest. The logic of paradox. *Journal of Philosophical Logic*, 8(1): 219–241, 1979.

[Priest and Berto(2013)] G. Priest and F. Berto. Dialetheism. *Stanford Encyclopedia of Philosophy*, 2013. URL http://plato.stanford.edu/archives/sum2013/entries/dialetheism/.

[Vakarelov(1977)] D. Vakarelov. Notes on N-lattices and constructive logic with strong negation. *Studia Logica*, 36(1/2):109–125, 1977.

Limitaciones expresivas y fundamentación: la teoría ingenua de conjuntos basada en LP

Luis Estrada González
Instituto de Investigaciones Filosóficas, UNAM
loisayaxsegrob@gmail.com

28 de noviembre de 2018

Índice

1. **Preliminares conjuntistas**   78
2. **Preliminares lógicos: LP**   79
3. **$LP - NS$**   81
4. **Los resultados de Thomas**   82
5. **Evaluación de los resultados de Thomas**   83

Introducción

Este trabajo es parte de un proyecto más amplio en el que investigo la matematicidad. Como veo el problema, parte de entender qué es la matemática requiere entender las conexiones entre la matemática clásica y la matemática no clásica. Sin embargo, aquí no hablaré de la matematicidad en general y sólo hablaré de las conexiones entre la matemática clásica y la no clásica tangencialmente para concentrarme en una pregunta un tanto menos ambiciosa: dado que parece que la teoría de conjuntos es central para la matemática tanto clásica como no clásica, ¿cuáles son las consecuencias conceptuales y filosóficas para una $\mathcal{L}$-matemática de resultados limitativos para la $\mathcal{L}$-teoría de conjuntos, donde $\mathcal{L}$ es una lógica dada, especialmente una no clásica?

Como puede notarse, esta pregunta es acerca de las conexiones entre la teoría de conjuntos basada en $\mathcal{L}$ y el resto de la $\mathcal{L}$-matemática.[1] Por ejemplo, ¿qué pasaría con la matemática constructiva en caso de que hubiera una teoría de conjuntos constructiva distinguida y ésta tuviera serias limitaciones como, digamos, no tener recursos expresivos suficientes para definir alguna noción conjuntista que pareciera indispensable? ¿Dicha limitación se trasladaría *ipso facto* de la teoría constructiva de conjuntos a la matemática constructiva en general? Argumentaré que cualquier respuesta a la pregunta principal de este trabajo involucra diferentes posturas acerca del rol de una teoría de conjuntos dentro de la matemática.

Para que la discusión sea un tanto menos abstracta, me concentraré en un caso particular: la evaluación filosófica de los resultados limitativos probados por Morgan (antes Nick) Thomas para la teoría ingenua de conjuntos basada en la lógica LP y en otras lógicas relacionadas. Estos resultados son importantes porque, a primera vista, parecen golpes duros contra los prospectos de hacer matemática inconsistente o, por lo menos, matemática basada en LP y lógicas relacionadas.

Asumo que quien lea esto conoce la lógica clásica de primer orden y los rudimentos de teoría clásica de conjuntos, por ejemplo, ZF, así como ciertas convenciones notacionales asociadas (como que '$\neg(x = y)$' y '$(x \neq y)$' son intercambiables). El plan del artículo es como sigue. En la primera sección presento algunos preliminares conjuntistas necesarios para el resto del trabajo; básicamente, se trata de una rápida presentación de los rasgos centrales de una teoría ingenua de conjuntos. En la segunda sección presento los preliminares lógicos: algunas propiedades importantes de la lógica LP y de las LP mínimamente inconsistentes. En la tercera sección introduzco algunas características básicas de la teoría ingenua de conjuntos basada en LP. En la cuarta sección presento los resultados de Thomas para, en la quinta, evaluarlos lógica y matemáticamente, pero sobre todo filosóficamente.

1. Preliminares conjuntistas

Una *teoría de conjuntos ingenua* (basada en una lógica $\mathcal{L}$, lo que denotaré con '$\mathcal{L} - NS$') es una teoría de conjuntos en la que valen los siguientes dos principios:
Extensionalidad $\quad \forall x \forall y (\forall z ((z \in x) \equiv (z \in y)) \supset (x = y))$
Comprehensión $\quad \exists x \forall y ((y \in x) \equiv A)$ (con x no libre en A)

Hay dos hechos más o menos bien conocidos acerca de $\mathcal{L} - NS$. El primero es el teorema de Russell, a saber, que si $\mathcal{L}$ es la lógica clásica, $\mathcal{L} - NS$ es inconsistente y, con ello, trivial. El segundo es que no basta cualquier lógica no clásica para evitar la inconsistencia o la trivialidad de $\mathcal{L} - NS$.

---

[1] Para simplificar la discusión, estoy suponiendo que, como en la matemática clásica, en cualquier $\mathcal{L}$-matemática hay una teoría de conjuntos distinguida, por las razones que fuere, y me refiero a ella como 'la $\mathcal{L}$-teoría de conjuntos', pero esto no debe leerse como implicando la existencia de una sola $\mathcal{L}$-teoría de conjuntos.

Una idea común es que, puesto que lo que trivializa a $\mathcal{L} - NS$ es una contradicción –la existencia de un conjunto de Russell– y la lógica clásica es explosiva, esto es, que permite inferir cualquier proposición a partir de una contradicción, una lógica paraconsistente podría evitar la trivialización. Sin embargo, el teorema de Curry dice que si $\mathcal{L}$ satisface Transitividad, Separación (para el condicional, esto es, *Modus Ponens*), Prueba Condicional o Contracción, entonces $\mathcal{L} - NS$ es trivial. Pero estos principios son centrales a muchas lógicas no clásicas, incluidas muchas paraconsistentes –notablemente las lógicas Cn de da Costa y las lógicas de la relevancia E y R–, por lo que las esperanzas de desarrollar una teoría ingenua de conjuntos no trivial se reducen bastante. Sin embargo, la lógica LP, propuesta primero por González Asenjo (1966) y redescubierta y ampliamente aplicada por Priest (1979, 2006), al no satisfacer *Modus Ponens*, parece un candidato viable para hacer teoría de conjuntos ingenua no trivial.

2. Preliminares lógicos: LP

Hay varias maneras de presentar la lógica LP; aquí usaré una presentación modelo-teórica bivalente pero no veritativo-funcional. Usaré un lenguaje de primer orden típico, con colecciones de variables individuales (Var), constantes individuales ($Cons$), símbolos de funciones n-arias ($Func_n$) y símbolos de relaciones n-arias (Rel_n), las últimas dos para todo n. Las colecciones de términos (*Térm*) y fórmulas (*Form*) se forman de la manera usual. Una interpretación para este lenguaje es un par $M = \langle D, d \rangle$, donde D es un dominio no vacío de interpretación y d una función de denotación tal que

para toda $c \in Cons$, $d(c) \in D$;
para toda $f \in Func_n$, $d(f) : D^n \to D$
para toda $R \in Rel_n$, $d(R) = \langle E^+, E^- \rangle$ donde $E^+ \cup E^- = D^n$
Escribiré 'E^+' y 'E^-' como '$d^+(R)$' y '$d^-(R)$' y los llamaré la *extensión y la antiextensión* de R, respectivamente. Intuitivamente, son las colecciones de cosas que satisfacen R, en un caso, y su negación, en el otro. Los valores de verdad son, como en el caso clásico, sólo dos: $V = \{\bot < \top\}$, interpretados como "falso" y "verdadero", respectivamente.

Dada una interpretación M y una función $s : Var \to D$ que especifica la denotación de cada variable, se define la denotación de cada término t, $den(t)$, de la manera usual:
si $t \in Var$, $den(t) = s(t)$;
si $t \in Cons$, $den(t) = d(t)$;
si $f \in Func_n$ y $t_1, \ldots, t_n \in$ *Térm*, $den(f(t_1 \ldots t_n)) = d(f(den(t_1) \ldots den(t_n)))$

Una evaluación, v, es un subconjunto de $(F \times D^{Var} \times \{\bot, \top\}$ definida por las siguientes cláusulas recursivas, donde $t_1, \ldots, t_n \in$ *Térm* y $R \in Rel_n$:

$v((Rt_1 \ldots t_n, s), \top)$ si y sólo si $\langle den(t_1) \ldots den(t_n)\rangle \in d^+(R)$
$v((Rt_1 \ldots t_n, s), \top)$ si y sólo si $\langle den(t_1) \ldots den(t_n)\rangle \in d^-(R)$
$v((\neg A, s), \top)$ si y sólo si $v((A, s), \bot)$
$v((\neg A, s), \bot)$ si y sólo si $v((A, s), \top)$
$v((A \wedge B, s), \top)$ si y sólo si $v((A, s), \top)$ y $v((B, s), \top)$
$v((A \wedge B, s), \bot)$ si y sólo si $v((A, s), \bot)$ o $v((B, s), \bot)$
$v((A \vee B, s), \top)$ si y sólo si $v((A, s), \top)$ o $v((B, s), \top)$
$v((A \vee B, s), \bot)$ si y sólo si $v((A, s), \bot)$ y $v((B, s), \bot)$
$v((A \supset B, s), \top)$ si y sólo si $v((A, s), \bot)$ o $v((B, s), \top)$
$v((A \supset B, s), \bot)$ si y sólo si $v((A, s), \top)$ y $v((B, s), \bot)$
$v((\forall x A, s), \top)$ si y sólo si para todo $a \in D$, $v((A, s(x/a)), \top)$
$v((\forall x A, s), \bot)$ si y sólo si para algún $a \in D$, $v((A, s(x/a)), \bot)$
$v((\exists x A, s), \top)$ si y sólo si para algún $a \in D$, $v((A, s(x/a)), \top)$
$v((\exists x A, s), \bot)$ si y sólo si para todo $a \in D$, $v((A, s(x/a)), \bot)$
donde $x \in Var$ y $s(x/a)$ es como s excepto que su valor en x es a. Finalmente, '=' es una relación binaria particular tal que
$d^+(=) = \{\langle x, x \rangle | x \in D\}$
y $d^-(=)$ debe ser tal que $d^+(=) \cup d^-(=) = D^2$.

Usaré $\Gamma \models \mathcal{L}C$' para denotar que el argumento con premisas Γ y conclusión C es válido en $\mathcal{L}$. La definición de 'validez' es la usual:
Validez: $\Gamma \models_{LP} C$ si y sólo si para cualquier v, si $\top \in v(P)$ para toda $P \in \Gamma$, $\top \in v(C)$.

La única diferencia con la lógica clásica, y por lo cual las segundas cláusulas en las condiciones de satisfacibilidad anteriores no son redundantes, es que las interpretaciones no tienen que ser funciones, sino que pueden ser relaciones.

LP tiene las siguientes características:
$\models LPC$ si y sólo si $\models_{LC} C$
Si $\Gamma \models_{LP} C$ entonces $\Gamma \models_{LC} C$

Es decir, LP tiene exactamente las mismas verdades lógicas que la lógica clásica y, si un argumento es válido en LP es válido, también lo es en lógica clásica. Sin embargo, la recíproca de lo anterior no es cierta, pues en LP se tiene que
Paraconsistencia: x $A, \neg A \not\models_{LP} B$; esto es, falla el *Principio de Explosión*
No separabilidad de $\supset$: $A, A \supset B \not\models_{LP} B$; esto es, falla *Modus Ponens*

Como contraejemplo tanto al *Principio de Explosión* como a *Modus Ponens*, considérese la evaluación $v((A, s), \{\bot, \top\})$ y $v((B, s), \bot)$. En ambos casos, $\top$ pertenece a la evaluación de todas las premisas y, sin embargo, la conclusión es sólo falsa ($\{\bot\}$).

En una LP *mínimamente inconsistente*, se miden cuán inconsistentes son diferentes modelos de LP el uno respecto al otro (según cierta medida de inconsistencia), y se restringen los modelos de una teoría a los modelos mínimamente inconsistentes.

Aunque interesante, presentar las construcciones precisas de las medidas de inconsistencia, de las comparaciones entre modelos y de las restricciones a los modelos mínimamente inconsistentes (según la medida establecida) no es necesario para los propósitos del trabajo. Para lo que resta del texto, basta con que el lector conozca la idea general de las LP mínimamente inconsistentes tal como la he presentado; los detalles pueden consultarse en Priest (1991) y Crabbé (2011).

3. $LP - NS$

$LP - NS$, la teoría de conjuntos ingenua basada en LP, tiene algunas propiedades interesantes. Entre los teoremas notables de $LP - NS$ están los axiomas

Pares $\quad\quad\quad\quad \forall w \forall z \exists x \forall y ((y \in x) \equiv ((y = w) \vee (y = z)))$
Potencia $\quad\quad\quad \forall w \exists x \forall y ((y \in x) \equiv \forall z ((z \in x) \supset (z \in w)))$
Unión $\quad\quad\quad\quad \forall w \exists x \forall y ((y \in x) \equiv \exists z ((z \in w) \wedge (y \in z)))$
Separación $\quad\quad \forall w \exists x \forall y ((y \in x) \equiv ((y \in w) \wedge Py))$
Reemplazo $\quad\quad \forall w \exists x \forall y ((y \in x) \equiv \exists z ((z \in w) \wedge ((Pz) = y)))$
Conjunto vacío $\quad \exists x \neg \exists y (y \in x)$
Infinito $\quad\quad\quad\quad \exists x ((y \in x) \wedge \forall z ((z \in x) \supset \exists w ((z \in w) \wedge (w \in x))))$

La prueba de que los primeros cinco son teoremas de $LP - NS$ es bastante sencilla: todos ellos son instancias de *Comprehensión* poniendo alguna condición en lugar de A y generalizando universalmente. Por ejemplo, *Pares* resulta de sustituir 'A' por '$((y = w) \vee (y = z))$'; *Potencia*, de sustituir 'A' por $\forall z ((z \in x) \supset (z = w))$', etcétera.

$LP - NS$ difiere de ZF (clásica) en que la primera cuenta con conjunto universal y en que no valida el axioma de Regularidad, $\forall x ((x \neq \emptyset) \supset \exists y ((y \in x) \wedge \forall z (z \in x \supset \neg (z \in y))))$. Esto último constituye una prueba de la no trivialidad de $LP-NS$: hay por lo menos una fórmula que no es teorema.[2]

Pero validar buena parte de los axiomas usuales junto con *Comprehensión* tiene un costo. Uno de ellos es que ciertas fórmulas parecen significar algo diferente a lo que significan sus contrapartes tipográficamente idénticas en ZF. Por ejemplo, en $LP-NS$ es un teorema que $\exists y \exists x (y \neq x)$. Sin embargo, la prueba no deja determinar si hay (al menos) dos objetos distintos en $LP - NS$. En $LP - NS$ hay un conjunto de Russell, esto es, $\exists r ((r \in r) \equiv (r \in r))$; por tanto, $\exists r \neg ((r \in r) \equiv (r \in r))$. Así, $\neg \forall z ((r \in z) \equiv (r \in z))$, y $\exists x \forall z ((r \in z) \equiv (r \in z))$. Pero de eso se sigue que

---

[2] Véase Restall (1992) para una prueba de la no trivialidad de $LP - NS$. Como no podría ser de otro modo, *Elección* es problemático en este contexto. Puede añadirse como axioma junto a *Extensionalidad* y *Comprehensión*. Sin embargo, puede recuperarse a partir de *Comprehensión* si se añade maquinaria lógica a LP, como el cálculo ϵ de Hilbert (véase Leisenring 1969: cap. 4), o si se retira el requisito de que la variable del cuantificador particular en Comprehensión no aparezca libre en A; véase Routley (1980). Sin embargo, la interpretación del axioma es complicada, del mismo modo que lo son otras fórmulas, como las que se discutirán más abajo. Para una discusión de *Elección* en este contexto, véase Weber (2013).

$\exists x \neg (x = x)$, y de esto $\exists y \exists \neg x (y = x)$. Sin embargo, nótese que el teorema se sigue de que hay un objeto distinto de sí mismo. Pero, puesto que todo teorema clásico es un teorema de LP, en $LP - NS$ vale $\forall x (x = x)$. Pero todo esto es compatible con que haya sólo un conjunto distinto de sí mismo. En cualquier teoría basada en LP, A y $\neg A$ se verifican por diferentes procedimientos: si se ha establecido la verdad de $y \neq x$, determinar el valor de $y = x$ es un asunto adicional, pues ambas pueden ser verdaderas.

La siguiente instancia de *Comprehensión* representa un problema similar: $\forall z \exists x \forall y ((y \in x) \equiv (x = z))$. Usualmente, ésta sería la expresión de que hay conjuntos unitarios. Sin embargo, no está claro que x en $LP - NS$ sea un conjunto unitario: la fórmula $\forall y ((y \in x) \equiv (x = z))$ es satisfecha por x si éste es tal que $\forall z ((z \in x) \wedge \neg (z \in x))$.

4. Los resultados de Thomas

Éste es precisamente el punto de partida de los resultados que me ocupan en este trabajo. Thomas investiga fórmulas ϕ que satisfagan ciertos grupos de condiciones con cláusulas de las formas $NS \models_{LP} (c)\phi$ y $NS, \phi, \Gamma \models_{LP} \psi$ -donde '$(c)\phi$' es una cuantificación, ya sea universal o particular, sobre ϕ que liga todas sus variables- para que definan nociones conjuntistas. Por ejemplo, '$\phi(x, y)$' con dos variables libres es una *definición de unitario para* $LP - NS$ si y sólo si las siguientes cláusulas se satisfacen:
$DU1. NS \models_{LP} \forall a \exists b \phi(a, b)$
$DU2. NS, \phi(a, b) \models_{LP} a \in b$
$DU3. NS, \phi, c \in b \models_{LP} c = a$
Estas condiciones están formuladas en términos de consecuencia lógica porque, como vimos, las fórmulas podrían no significar lo que uno desearía. La fórmula '$\phi(a, b)$' se lee "b es el unitario $\{a\}$". Las cláusulas requeridas son que $LP - NS$ pruebe que tal conjunto existe para todo a; que cualquier b que satisfaga $\phi(a, b)$ contiene a a; y que cualquier elemento de tal b es igual a a.

Por último, sea A una fórmula que contiene los términos $\tau_1, \ldots, \tau_n$ y el símbolo de relación n-aria R. Diremos que una teoría T es *casi trivial* si y sólo si, para alguna de sus relaciones n-arias R y todo τ_i, tanto φ como $\neg \varphi$ valen en T.

Los resultados de Thomas son como siguen. Sean las distintas variantes de LP las variantes mínimamente inconsistentes estudiadas en Crabbé (2001) y $LP° \in \{LP, LP_=, LP_\supseteq\}$ y $LP° \in \{LP_m, LP_\subseteq\}$, en tanto que '$LP^* - NS$' denota una *teoría de conjuntos ingenua basada en cualquiera de esas lógicas*. Entonces

-$LP° - NS$ no puede probar la existencia de conjuntos unitarios, pares Cartesianos u órdenes infinitamente ascendentes, y

-$LP^\bullet - NS$ es casi trivial, ya que prueba $\forall x \forall y ((x \in y) \wedge (x \in y))$, y su único modelo es el modelo con un solo elemento, pertenencia inconsistente e igualdad consistente.

5. Evaluación de los resultados de Thomas

A partir de ahora, usaré indistintamente las expresiones 'X recupera a Y', 'X reproduce a Y' y 'X reconstruye a Y', donde X y Y son teorías matemáticas, y significarán que X prueba teoremas tipográficamente idénticos a los de Y y que valen en modelos tan parecidos como sea posible a los modelos pretendidos de Y.

Thomas (2014: 341s) dice que con sus resultados pretende "(...) sugerir que $LP-NS$ no puede reproducir ningún fragmento interesante de la matemática clásica (...)", en particular, que "(...) la carencia de órdenes lineales infinitamente ascendentes sugiere que $LP-NS$ no puede construir los números naturales." La implicatura aquí es que $LP-NS$ no puede recuperar la aritmética (clásica). Pero dado que las LP mínimamente inconsistentes también tienen limitaciones expresivas o son casi triviales, "(...) NS en LP mínimamente inconsistente no es "mucho mejor" que $[LP-NS]$ como base para la matemática."

Estos parecen golpes duros contra los prospectos de hacer matemática inconsistente, o al menos LP^*-matemática. Sin embargo, el veredicto no es tan claro, pues hay varias razones para no aceptar la solidez de las pruebas de Thomas. La primera es que las pruebas tiene supuestos controvertidos desde un punto de vista puramente lógico o matemático. Thomas asume, por ejemplo, que la igualdad es primitiva, como lo hace Priest en su versión de $LP-NS$, cuando bien podría haberla considerada como definida a la manera usual, como en Restall (1992).[3] Por otro lado, buena parte de su teoría es más clásica de lo que debería. Por ejemplo, la noción de igualdad usada en los modelos es consistente y las pruebas de Thomas proceden por reducción al absurdo, sin haber asegurado que el uso de este método de prueba es paraconsistentemente aceptable en este contexto. Finalmente, las extensiones de LP que podrían servir como matemática subyacente para una teoría de conjuntos ingenua no se reducen a las LP mínimamente inconsistentes. En Verdée (2013) y Omori (2015) se presentan teorías de conjuntos ingenuas basadas en extensiones de LP que no están en el alcance de los resultados de Thomas.

No obstante, incluso concediéndole a Thomas toda la maquinaria formal requerida para sus pruebas e ignorando teorías de conjuntos como las de Verdée y Omori, uno no tiene por qué concluir que la LP-matemática, y mucho menos la matemática inconsistente, está igual de limitada que LP^*-NS. La opinión contraria parece estar basada en un argumento del siguiente tipo:

---

[3] Aunque la ganancia podría no ser mucha. La igualdad definida a la manera usual suele conducir, en NS, a cierto tipo de trivialidad, pues es posible probar $\forall x \forall y (x = y)$.

(1) Si una teoría de conjuntos S basada en una lógica $\mathcal{L}$, $\mathcal{L} - S$, no logra recuperar (al menos una parte significativa de) cierta teoría matemática o construcción clásica, $LC - M$, o es suficientemente trivial, entonces $\mathcal{L} - S$ no es digna de ser considerada una teoría matemática seria.

(2) Si $\mathcal{L} - S$ no es digna de ser considerada una teoría matemática seria, la $\mathcal{L}$-matemática tampoco es digna de ser considerada una teoría matemática seria.

(3) $LP^* - NS$ o bien falla en recuperar los números naturales (y otros objetos clásicos) o es suficientemente trivial.

(4) $LP^* - NS$ no es digna de ser considerada una teoría matemática seria.

(5) Por tanto, La LP^*-matemática tampoco es digna de ser considerada una teoría matemática seria.

La premisa (1) tiene cierta aceptación entre los teóricos de la paraconsistencia. Priest dice, por ejemplo: "Una condición mínima de adecuación para una teoría paraconsistente de conjuntos parecería ser que podamos obtener de ella por lo menos una parte decente de la teoría estándar, ortodoxa, de conjuntos." (Priest 2006: 248) Nótese que hay una premisa similar a (1) subyacente al trabajo de Thomas: puesto que $LP^* - NS$ no puede reproducir "ningún fragmento interesante de la matemática clásica", $LP^* - NS$ no es una buena base para la matemática. ¿Cómo evitar la pendiente resbaladiza que llevaría de las limitaciones de $LP^* - NS$ a desechar toda la LP^*-matemática?

La primera opción sería rechazar la idea de que una teoría matemática no clásica $NC\mathcal{L} - T$ tenga que recuperar, reconstruir o reproducir partes significativas de su contraparte clásica, $LC - T$, pues la teoría no clásica puede o bien generalizar la teoría clásica o tratar de un dominio incomparable al de ésta. En el caso de teoría de conjuntos, la teoría paraconsistente, al considerar objetos como los conjuntos de Russell, podría o bien estar ampliando el dominio de investigación o tratando de un dominio diferente. En cualquier caso, no es claro que la teoría paraconsistente tenga que recuperar teoremas de las teorías clásicas de conjuntos, y ya no digamos "una parte decente" de éstos, como pide Priest. Recuperar la matemática clásica tiene, por supuesto, el atractivo de la conservatividad, pero la apelación a la conservatividad, si no es una petición de principio, sólo desplaza la cuestión: ¿Por qué una teoría tendría que ser conservativa con respecto a otra teoría a la cual generaliza o de cuyo dominio de investigación se separa?

Quizá una $\mathcal{L}$-teoría de conjuntos no clásica no necesariamente tenga que recuperar, reconstruir o reproducir partes significativas de la matemática clásica, pero quizá sí de otras teorías matemáticas basadas en $\mathcal{L}$. El argumento de antes, con la premisa (1) modificada

(1*) Si una teoría de conjuntos S basada en una lógica $\mathcal{L}$, $\mathcal{L} - S$, no logra recuperar (al menos una parte significativa de) otras $\mathcal{L}$-teorías matemáticas, $\mathcal{L} - M$, o es suficientemente trivial, entonces $\mathcal{L} - S$ no es digna de ser considerada.

Funcionaría prácticamente igual. ¿Qué podría objetarse ahora, si ya se eliminó la conexión problemática entre una teoría de conjuntos no clásica y la matemática clásica? Sigue habiendo una conexión problemática que, sin embargo, lógicos clásicos y no clásicos por igual dan por sentada: una $\mathcal{L}$-teoría de conjuntos ocupa un lugar privilegiado en la $\mathcal{L}$-matemática. Así, por ejemplo, dice Verdée (2013: 108):

> Asumo que idealmente uno preferiría una teoría de conjuntos (...) con fuerza matemática (...) [tal que] uno debería ser capaz de traducir la mayoría de la matemática a la teoría de conjuntos. Esto significa, por ejemplo, que uno debería ser capaz de expresar todos los teoremas interesantes de la teoría de números complejos dentro de la teoría de conjuntos. Esto puede sonar poco realista, pero es generalmente aceptado que la mayoría de la matemática clásica puede reducirse a la teoría de conjuntos ZFC clásica.

Y entonces resulta que la premisa más problemática es en realidad (2):

(2) Si $\mathcal{L} - S$ no es digna de ser considerada una teoría matemática seria, la $\mathcal{L}$-matemática tampoco es digna de ser considerada una teoría matemática seria.

Sin embargo, no está claro que una teoría de conjuntos ocupe un lugar privilegiado en la matemática en general de manera tal que un problema en la $\mathcal{L}$-teoría de conjuntos constituya necesariamente un problema para la $\mathcal{L}$-matemática. Pero incluso suponiendo que lo tenga, faltaría aclarar en qué consistiría dicho carácter especial. Mientras esto no se haga, no es claro que resultados limitativos como los de Thomas para $LP^* - NS$ se traduzcan *ipso facto* en limitaciones para la LP^*-matemática y mucho menos para la matemática inconsistente en general.

Así, la valoración del impacto de resultados limitativos para una teoría de conjuntos sobre la matemática en general depende en buena medida de una postura acerca del lugar que aquella tenga en ésta. Sin pretender ser exhaustivo ni atribuirle ninguna de las posturas a individuos en particular para no distraerme con cuestiones exegéticas, presento aquí algunas posturas:

$(S - M1)$ Toda la $\mathcal{L}$-matemática debería ser reducida a (alguna) $\mathcal{L}$-teoría de conjuntos. De acuerdo con posturas como ésta, todos los objetos matemáticos son en última instancia conjuntos.

$(S - M2)$ Una $\mathcal{L}$-teoría de conjuntos es una entre varias $\mathcal{L}$-teorías matemáticas, pero debería ser capaz de reconstruir toda la $\mathcal{L}$-matemática. Según esta postura, no es necesario que todos los objetos matemáticos sean conjuntos, pero la(s) teoría(s) de conjuntos son *sui generis* de manera tal que ella(s) sirve(n) para expresar teorías acerca de otros objetos matemáticos.

$(S - M3)$ Una $\mathcal{L}$-teoría de conjuntos debería ser capaz de reducir o reconstruir toda la $\mathcal{L}$-matemática elemental -donde 'matemática elemental' se refiere, digamos, a la matemática que se enseña en la primera mitad de una licenciatura en matemática. Hay que hacer dos anotaciones con respecto a esta postura. Primero, hay dos versiones de la misma, que son como $(S - M1)$ y $(S - M2)$ pero restringidas a la matemática elemental: según la primera, que es reduccionista, todos los objetos de la matemática elemental son conjuntos; de acuerdo con la segunda, la(s) teoría(s) de conjuntos son sólo *sui generis* con respecto a la matemática elemental. Nótese que esta postura es fuertemente contextual, pues incluso suponiendo que la enseñanza superior de la matemática es más o menos uniforme alrededor del mundo en un momento dado, qué cuenta como "matemática elemental" sí que cambia diacrónicamente.

$(S - M4)$ En principio, una $\mathcal{L}$-teoría de conjuntos es una entre varias $\mathcal{L}$-teorías matemáticas, pero si tuviera capacidades expresivas o inferenciales especiales sobre el resto de las $\mathcal{L}$-teorías matemáticas, deberían ser usados. Aquí no es necesario ni que todos los objetos matemáticos sean conjuntos ni que la teoría de conjuntos sea sui generis pero, en caso de que lo fuera, esto debería ser aprovechado para unificación conceptual, por ejemplo.

$(S - M5)$ Una $\mathcal{L}$-teoría de conjuntos es sólo una entre varias $\mathcal{L}$-teorías matemáticas. Según esta postura no hay mayor razón para pensar que una teoría de conjuntos tenga un rol fundacional especial sobre, digamos, la geometría, la aritmética o cualquier otra rama de la matemática. De hecho, el proyecto mismo de fundamentación, incluso en las versiones más débiles de unificación conceptual, no tendría mayor importancia matemática.

Este recuento es útil para notar que no todos los resultados limitativos para una $\mathcal{L}$-teoría de conjuntos se traducen en limitaciones para la $\mathcal{L}$-matemática. En el escenario $(S - M1)$, un resultado limitativo para su $\mathcal{L}$-teoría de conjuntos representativa sí significaría una limitación igual para el resto de la $\mathcal{L}$-matemática, pues todos los objetos matemáticos son en última instancia conjuntos. Aunque el impacto no es tan devastador en $(S - M2)$ y $(S - M3)$, sí es más allá de lo deseable: la $\mathcal{L}$-matemática sería deficiente a causa de las limitaciones de su teoría matemática principal. Pero la situación es diferente para $(S - M4)$ y $(S - M5)$. En estos escenarios, un resultado limitativo para una $\mathcal{L}$-teoría de conjuntos sería un resultado entre otros.

Por supuesto, cada una de las opciones requiere ser argumentada. En particular, en opciones como $(S - M4)$ y $(S - M5)$ hay que ofrecer una explicación convincente de por qué la casi trivialidad de una teoría de conjuntos sería un resultado entre muchos otros, cuando tal casi trivialidad está muy lejos de las motivaciones ingenuas que motivaron ese otro tipo de teorías en primer lugar. Aquí sólo he querido mostrar que los resultados de Thomas están lejos de haber mostrado una limitación para toda la LP^*-matemática.

Conclusiones

En este trabajo abordé la cuestión de cuáles serían las consecuencias conceptuales y filosóficas para una $\mathcal{L}$-matemática de resultados limitativos para una $\mathcal{L}$-teoría de conjuntos, donde $\mathcal{L}$ es una lógica dada, especialmente una no clásica. Por mor de la concreción, me concentré en la evaluación filosófica de los resultados limitativos probados por Morgan Thomas para la teoría ingenua de conjuntos basada en la lógica LP y en otras lógicas relacionadas. A primera vista, dichos resultados parecerían golpes duros contra los prospectos de hacer matemática inconsistente o, por lo menos, matemática basada en LP y lógicas relacionadas.

Después de presentar los preliminares lógicos -algunas propiedades importantes de la lógica LP y de las LP mínimamente inconsistentes- y de la teoría ingenua de conjuntos necesarios para el resto del trabajo, presenté los resultados de Thomas para luego evaluarlos lógica y matemáticamente, pero sobre todo filosóficamente. Presenté algunas posturas acerca del lugar de una teoría de conjuntos en la matemática; según algunas, la teoría de conjuntos es la teoría matemática más importante, mientras que para otras es una más entre las teorías matemáticas. Dependiendo de cuál que sea el lugar que uno concibe para la teoría de conjuntos en la matemática en general, resultados limitativos para aquélla tendrán diferentes implicaciones para ésta.

Referencias

[Crabbé(2011)] Marcel Crabbé. Reassurance for the logic of paradox. *Review of Symbolic Logic*, 4(3):479–485, 2011.

[González Asenjo(1966)] Florencio González Asenjo. A calculus of antinomies. *Notre Dame Journal of Formal Logic*, 7(1):103–105, 1966.

[Leisenring(1969)] Albert C. Leisenring. *Mathematical Logic and Hilbert's ε Symbol*, Londres. MacDonald, 1969.

[Omori(2015)] Hitoshi Omori. Remarks on naïve set theory based on LP. *Review of Symbolic Logic*, 8(2):279–295, 2015.

[Priest(1979)] Graham Priest. The logic of paradox. *Journal of Philosophical Logic*, 8:219–241, 1979.

[Priest(1991)] Graham Priest. Minimally inconsistent LP. *Studia Logica*, 50(2): 321–331, 1991.

[Priest(2006)] Graham Priest. *In Contradiction: A Study of the Transconsistent*. Oxford: Oxford University Press, 2 edition, 2006.

[Restall(1992)] Greg Restall. A note on naïve set theory in LP. *Notre Dame Journal of Formal Logic*, 33(3):422–432, 1992.

[Routley(1980)] Richard Routley. *Exploring Meinong's Jungle and Beyond. An Investigation of Noneism and the Theory of Items*, Canberra: Research School of Social Sciences. Australian National University, 1980.

[Thomas(2014)] Nick Thomas. Expressive limitations of naïve set theory based in LP and minimally inconsistent LP. *Review of Symbolic Logic*, 7(2):341–350, 2014.

[Verdée(2013)] Peter Verdée. Strong, universal and provably non-trivial set theory by means of adaptive logic. *Logic Journal of the IGPL*, 21(1):108–125, 2013.

[Weber(2013)] Zach Weber. Notes on inconsistent set theory. Koji Tanaka et al., eds., *Paraconsistency: Logic and Applications, Dordrecht: Springer*, 315–328, 2013.

Este trabajo fue escrito gracias al financiamiento de los proyectos PAPIIT IA401015 "Tras las consecuencias. Una visión universalista de la lógica (I) y Conacyt CCB 2011 166502 "Aspectos filosóficos de la modalidad". Versiones previas de este trabajo fueron presentadas en el Taller "20 Years of Inconsistent Mathematics" (Estambul, junio de 2015) y en la Cuarta Escuela de Lógica y Conjuntos (Ciudad de México, diciembre de 2015). Agradezco a los organizadores de ambos eventos por sus invitaciones y a la audiencia por las discusiones, especialmente a Diderik Batens, David Meza Alcántara, Maarten McKubre-Jordens, Favio Miranda y Hitoshi Omori, y a Elisángela Ramírez Cámara por la corrección de varias infelicidades lingüísticas.

Sobre la función de la teoría de modelos y el significado de la teoría de conjuntos[*]

Mario Gómez-Torrente
Instituto de Investigaciones Filosóficas,
Universidad Nacional Autónoma de México (UNAM)
México DF 04510, México
mariogt@unam.mx

Una de las funciones fundamentales de la teoría de modelos es proporcionar definiciones de las nociones de consecuencia lógica y verdad lógica y ofrecer un marco teórico en el que poder utilizar de manera fructífera esas definiciones[1]. Pero las definiciones habituales de consecuencia lógica y verdad lógica en la teoría de modelos se han visto sometidas a importantes críticas en los últimos años. Las críticas más radicales buscan argumentar que las definiciones son extensionalmente incorrectas para algunos o todos los lenguajes estudiados por la teoría de modelos estándar[2], pero parece justo decir que estas críticas no han obtenido una gran aceptación[3]. Otras críticas aparentemente menos radicales no cuestionan abiertamente la corrección extensional de las definiciones, pero son quizá insidiosas, pues cuestionan los fundamentos existentes para aceptarlas. Una de esas críticas se basa en la observación de que las definiciones habituales de consecuencia lógica y verdad lógica en la teoría de modelos no proporcionan un análisis conceptual de las nociones preteóricas de consecuencia lógica y verdad lógica[4]. Si esta observación es correcta –y sospecho que una mayoría

---

[*]Investigación apoyada por el CONACyT de México (CCB 2011 166502), por el proyecto PAPIIT-UNAM IA 401015, y por el MINECO de España (proyecto FFI2011-25626).

[1]Véase Gómez-Torrente (2000) para una introducción a las definiciones habituales de verdad lógica y consecuencia lógica en la teoría de modelos estándar. Como es bien sabido, la forma de implementar matemáticamente estas definiciones se debe a Tarski (1936).

[2]Véanse por ejemplo Etchemendy (1990) y McGee (1992).

[3]Algunas reacciones a este tipo de críticas pueden verse en Gómez-Torrente (1998/9). En Gómez-Torrente (2008b) he sugerido que cuando uno aplica las definiciones modelistas de verdad y consecuencia lógicas a ciertos lenguajes extensionales a los que Tarski pretendía aplicarlas, pero que no caen dentro del ámbito de la teoría de modelos estándar, sí resultan problemas de inadecuación extensional. En este trabajo nos concentraremos exclusivamente en lenguajes estudiados por la teoría de modelos estándar, y de hecho sólo en algunos de esos lenguajes, a saber en los lenguajes cuantificacionales clásicos de órdenes primero y superiores.

[4]Véase de nuevo Etchemendy (1990).

de filósofos de la lógica aceptaría sin mayor reparo que lo es– un posible fundamento para aceptar las definiciones queda invalidado, a saber una justificación analítica de su adecuación. Pero de nuevo parece justo decir que una mayoría de filósofos de la lógica no piensa que la aceptación de las definiciones se base de hecho o deba siquiera basarse en una justificación conceptual. Otra crítica potencialmente más insidiosa, sin embargo, se basa en la idea de que simplemente no tenemos un fundamento lo suficientemente fuerte (analítico o no) de que las definiciones sean extensionalmente correctas (incluso si resulta que lo son)–donde por "fundamento lo suficientemente fuerte" se entiende generalmente un fundamento que muestre que la corrección extensional de las definiciones es, en algún sentido menos exigente que el de un análisis conceptual, especialmente confiable o sólido. No tenemos, señala esta crítica, una justificación lo suficientemente fuerte de que no exista alguna oración de un lenguaje formal estudiado por la teoría de modelos estándar que sea verdadera en todas las estructuras conjuntistas que la puedan interpretar pero que al mismo tiempo sea falsa en la interpretación deseada de la teoría de conjuntos (o, de forma equivalente, en alguna otra interpretación con una clase propia como dominio de cuantificación). Y sólo una justificación suficientemente fuerte de que no existen oraciones así –continúa la crítica– podría otorgar a la definición habitual de verdad lógica en la teoría de modelos un estatus de legitimidad suficientemente satisfactorio[5]. (La misma crítica se dirige *mutatis mutandis* contra la definición habitual de consecuencia lógica: si una oración con esas características existe, según esa definición será consecuencia lógica de todo conjunto de premisas, pero preteóricamente no será consecuencia lógica del conjunto vacío de premisas. En lo que sigue centraremos nuestra discusión en la noción de verdad lógica; lo que digamos al respecto podrá decirse también, *mutatis mutandis*, de la noción de consecuencia lógica.) Llamamos a esta crítica "la crítica sobre la fuerza del fundamento" de la definición modelista de verdad lógica.

En un trabajo anterior[6] he desarrollado la observación condicional de que, si es posible obtener una justificación suficientemente fuerte de la tesis de que la teoría de conjuntos es (en parte) una teoría de las interpretaciones pero no tiene ella misma una interpretación perfectamente determinada, entonces existirá un fundamento suficientemente fuerte para rechazar la crítica sobre la fuerza del fundamento e, indirectamente, para aceptar (*ceteris paribus*) la definición usual de verdad lógica. Pero, como también señalé en el mencionado trabajo, quizá no es posible obtener una justificación lo bastante fuerte del antecedente de este condicional. En el presente trabajo pretendo desarrollar brevemente otra observación que en este momento me parece más efectiva con vistas a disolver la crítica que nos ocupa. Actualmente poseemos una justificación muy confiable (según concederían todas las partes de la discusión) para aceptar la definición de verdad lógica en la teoría de modelos, una justificación que se basa en los

---

[5]Ejemplos de este tipo de preocupación pueden verse en Boolos (1985), McGee (1992), Rayo y Uzquiano (1999) y Williamson (2003).

[6]Gómez-Torrente (2008a).

"principios de reflexión" propuestos para la teoría de conjuntos. La observación que desarrollaré es que la crítica sobre la fuerza del fundamento no puede verse como un elemento que socave la confiabilidad de esa justificación, porque parte de un supuesto menos justificado y más controvertible que los principios de reflexión.

I

Como es bien sabido, un razonamiento inspirado en Kreisel[7] proporciona una justificación tan confiable como pudiéramos desear para la tesis de que no existen oraciones de primer orden verdaderas en todas las estructuras conjuntistas pero falsas en la interpretación deseada de la teoría de conjuntos (si ésta existe). Supongamos que una oración O de un lenguaje de primer orden es verdadera en todas las estructuras conjuntistas. Entonces, por el teorema de compleción de Gödel, O es derivable sin premisas en un cálculo deductivo correcto de los habituales para los lenguajes de primer orden. Y es claro por inspección que los axiomas de esos cálculos son verdaderos en cualquier interpretación y las reglas de esos cálculos preservan la verdad en cualquier interpretación, conjuntista o no. Por tanto, O es verdadera en todas las interpretaciones.

En lenguajes de segundo orden y órdenes superiores no podemos aplicar el razonamiento de Kreisel, pues en estos casos no hay cálculos deductivos que sean al mismo tiempo completos y correctos. Sin embargo, los teóricos de conjuntos aceptan comúnmente ciertos principios, llamados "principios de reflexión", de los que se siguen las conclusiones deseadas. Un principio de reflexión típico busca capturar la idea de que si una oración de un cierto lenguaje clásico de primer orden o de un orden superior finito es verdadera, entonces esa oración debe ser también verdadera en una estructura conjuntista[8]. Esta idea implica que si una oración de un lenguaje de orden superior es verdadera en todas las estructuras conjuntistas, no puede ser falsa en la interpretación deseada de la teoría de conjuntos (ni en ninguna otra interpretación con una clase propia como dominio de cuantificación): si fuera falsa, entonces su negación sería verdadera y, por reflexión, verdadera en alguna estructura conjuntista, y por tanto la oración original sería falsa en esa misma estructura conjuntista, contra el supuesto inicial.

La crítica sobre la fuerza del fundamento debe verse por tanto como un cuestionamiento de la fuerza justificatoria de los principios de reflexión, y en particular como un cuestionamiento de su confiabilidad en vista de la *posibilidad* (epistémica, cuando menos) de la existencia de oraciones de órdenes superiores verdaderas en todas las estructuras conjuntistas pero falsas en la interpretación deseada de la teoría de conjuntos

---

[7]En Kreisel (1965, 1967); también en Kreisel y Krivine (1967), p. 190.

[8]Sobre principios de reflexión pueden verse a un nivel relativamente accesible Gloede (1976), Shapiro (1987, 1991).

(o de forma equivalente, en alguna otra interpretación con una clase propia como dominio de cuantificación). Esta posibilidad debe socavar efectivamente nuestra confianza en los principios de reflexión. ¿Son los principios de reflexión lo suficientemente débiles como para ser socavados por esta posibilidad? La crítica que nos ocupa debe sostener que sí, y para ello puede apoyarse en la idea de que esos principios no forman parte del núcleo de principios aceptados canónicamente en la teoría de conjuntos, por mucho que sean comúnmente aceptados. Quizá, al menos a primera vista, esta característica de los principios de reflexión los hace lo suficientemente vulnerables como para ser socavados por la observación de que existe la posibilidad mencionada: dicha posibilidad sería, después de todo, una *posibilidad* (epistémica) porque nuestro grado de confianza en los principios de reflexión no es lo bastante fuerte como para excluirla. (Por contraste, el teorema de compleción, que se sigue de los axiomas canónicos de la teoría de conjuntos, excluye junto con otros principios obvios la posibilidad de que haya oraciones verdaderas en todas las estructuras conjuntistas pero falsas en la interpretación deseada de la teoría de conjuntos. Un principio de reflexión para fórmulas de primer orden, que puede verse en esencia como una versión del teorema de compleción, es ya un teorema de la teoría canónica de conjuntos.)

Una reacción posible ante la crítica sobre la fuerza del fundamento es simplemente aceptarla. Si hacemos esto, quizá podemos continuar aceptando las definiciones usuales de verdad lógica y consecuencia lógica renunciando a obtener una justificación suficientemente fuerte para esa aceptación; pero en ese caso nos encontraremos en una situación epistémicamente incómoda. Por otro lado, si aceptamos la crítica podemos también rechazar las definiciones usuales de verdad lógica y consecuencia lógica, buscando definir alguna nueva noción técnica de verdad lógica y consecuencia lógica en términos de una nueva colección de interpretaciones. ¿Es posible hacer esto, de tal manera que sea también posible justificar de manera suficientemente fuerte la tesis resultante de que ninguna verdad lógica en el nuevo sentido técnico es falsa en alguna interpretación sugerida por nuestra misma consideración de la teoría necesaria para hablar de la nueva colección de interpretaciones?

Una posibilidad es extender nuestra teoría de conjuntos a una teoría que hable también de clases propias, estructuras cuyo universo es una clase propia, y otras "hiperclases": clases de clases, clases de clases de clases, etc. Aquí la restricción a niveles finitos aparecerá sin duda como arbitraria, lo cual nos llevará a postular hiperclases con estratos transfinitos en el proceso de generación, y presumiblemente a imaginar un universo similar al generado por la jerarquía acumulativa habitual, pero ¡con la jerarquía misma en su estrato inicial! En una teoría apropiada de este tipo, sería posible definir una nueva noción de verdad lógica como verdad en todas las estructuras cuyo universo es una clase propia de las que hable la teoría. Claramente tendríamos entonces una justificación tan fuerte como pudiéramos desear de la afirmación de que todas las verdades lógicas en este sentido tienen la propiedad de no ser invalidadas por ninguna de las estructuras de las que habla la teoría. Sin embargo, un problema para un proyecto de este tipo es que parece únicamente posponer el problema, de una forma relativamente clara: si el universo de la teoría de conjuntos y otras clases propias pasan a ser objetos de nuestra teoría de las interpretaciones, seguiremos sin una garantía suficientemente fuerte de que toda oración verdadera en todas las interpretaciones postuladas en la teoría sea también verdadera en la interpretación deseada de *esa* teoría.

Otra propuesta en esencia similar niega sin embargo que las interpretaciones sean objetos, postulando que son simplemente valores (plurales) de las variables de orden superior (de lenguajes de orden superior para la teoría de conjuntos). La propuesta es entonces definir la noción de verdad lógica para un lenguaje de un orden dado como una oración verdadera en todos los valores apropiados de las variables de un lenguaje de un orden inmediatamente superior (para la teoría de conjuntos). Pero aunque esta propuesta no se basa en ampliar el ámbito de los objetos postulados como interpretaciones, no puede evitar implicar la existencia o concebibilidad de una jerarquía de nuevos ámbitos de pluralidades o valores plurales: a los valores plurales de las variables de segundo orden de la teoría de conjuntos habrá que añadir los valores plurales de las variables de tercer orden, los valores plurales de las variables de cuarto orden, y así sucesivamente. Y sin duda la restricción a niveles finitos habrá de parecer arbitraria, lo cual nos llevará a postular valores plurales para variables de órdenes transfinitos. Pero de nuevo me parece relativamente claro que esta construcción nos deja sin la garantía que momentáneamente creíamos haber alcanzado: si hay un ámbito de pluralidades que son los valores de las variables en, digamos, el lenguaje de tercer orden para la teoría de conjuntos, una definición de verdad lógica para un lenguaje de segundo orden como verdad en todos esos valores de tercer orden parecerá en seguida insuficientemente incluyente; pues los valores de las variables de cuarto orden proporcionarán intuitivamente nuevas interpretaciones posibles para las constantes no lógicas de segundo orden. Recordemos que en el marco de la teoría de tipos los conjuntos de tercer orden (conjuntos de conjuntos de individuos) pueden verse como interpretaciones posibles de las constantes de segundo orden, simplemente si tomamos como "individuos" a los conjuntos de segundo orden y por tanto como "conjuntos de segundo orden" a los conjuntos de tercer orden; y entonces ciertos conjuntos de cuarto orden, esto es conjuntos de conjuntos de tercer orden, pueden verse como estructuras para la interpretación de un lenguaje o conjunto de constantes de segundo orden. Evitar considerar a estos conjuntos de cuarto orden como posibles estructuras para interpretar un lenguaje de segundo orden parece claramente injustificado. Y naturalmente, las preocupaciones de este tipo sólo pueden agravarse si consideramos iteraciones ulteriores en la jerarquía transfinita de ámbitos de valores plurales imaginada hace unas líneas. Dada la analogía profunda entre la concepción "pluralista" de la teoría de tipos y la concepción tradicional "conjuntista", estas preocupaciones imposibilitan una garantía suficientemente fuerte de que la concepción pluralista excluya el tipo de problema que llevaba a la crítica sobre la fuerza del fundamento en el caso modelista.

Estas dos propuestas, por tanto, no pueden proporcionar después de todo un alivio para quien vea en la crítica sobre la fuerza del fundamento una razón de peso para dudar de la aceptabilidad de las definiciones habituales de verdad lógica y consecuencia lógica en la teoría de modelos: el mismo tipo de preocupación que ponía en cuestión esa aceptabilidad puede generarse para las nuevas definiciones de esas dos propuestas.

La debilidad de estas propuestas me llevó (aunque por medio de consideraciones diferentes[9]) a la mencionada conjetura (compartida con muchos otros autores que la han hecho en contextos diferentes) de que la teoría de conjuntos es (en parte) una teoría de las interpretaciones pero no tiene ella misma una interpretación perfectamente determinada –conjetura que de ser correcta validaría las definiciones habituales de verdad lógica y consecuencia lógica en la teoría de modelos estándar. Pero, como también mencioné, por todo lo que hemos dicho hasta ahora, si hemos de encontrar por este camino un alivio contra la crítica que nos ocupa, entonces necesitaremos una justificación suficientemente fuerte de la conjetura. Y, si bien en mi mencionado trabajo anterior considero algunas razones que por su peso acumulado nos podrían llevar a ver la conjetura con buenos ojos (la misma aceptación común de los principios de reflexión, la idea de que la teoría de conjuntos debe ser "máximamente comprensiva", la concepción frecuente de las clases propias como ficciones), debe reconocerse que, por sí mismas, estas razones no proporcionan seguramente un fundamento lo bastante fuerte como para aceptar la conjetura y por ende para aceptar las definiciones habituales de verdad lógica y consecuencia lógica en la teoría de modelos estándar.

Sin embargo, mi opinión actual es que ofrecer una justificación fuerte de la conjetura en cuestión no es realmente necesario para disolver la crítica sobre la fuerza del fundamento. Como ya adelanté, la razón es que esta crítica presupone una tesis menos justificada, y más controvertible, que los principios de reflexión que proporcionan nuestra ya existente y altamente confiable justificación para aceptar las definiciones de verdad y consecuencia lógica habituales en la teoría de modelos.

II

Antes de entrar en una discusión de esta afirmación, vale la pena considerar una posible forma de rechazar la idea de que deba haber una justificación especialmente fuerte de la adecuación de las definiciones modelistas habituales de verdad lógica y consecuencia lógica. Podría observarse, en particular, que en casos similares no siempre la hay: compárese el caso de la definición (o definiciones) de función computable en teoría de la recursión, donde si bien en tiempos recientes se han propuesto ciertas pruebas informales de la tesis de Church-Turing de que las funciones recursivas coin-

---

[9] En el trabajo mencionado, Gómez-Torrente (2008a).

ciden con las computables en el sentido preteórico relevante[10], parece justo decir que dichas pruebas no son (al menos todavía) mayoritariamente aceptadas y la tesis se ve aún como una conjetura no demostrada (e indemostrable en la matemática clásica). Eso no impide que los matemáticos acepten la tesis de Church-Turing y aun la usen en algunas demostraciones. Además, en el caso de la definición modelista de verdad lógica –podría continuar esta línea argumental– una justificación *prima facie* confiable la proporcionan los principios de reflexión, que aunque no sean aceptados como canónicos, son aceptados de hecho por la práctica totalidad de los teóricos de conjuntos. De todos modos, un problema para quien quisiera oponerse de esta manera a la crítica sobre la fuerza del fundamento es que cabría razonablemente objetar que los casos de la teoría de la recursión y la teoría de modelos no son realmente análogos, pues en el caso modelista la definición en juego deja abiertamente sin considerar la interpretación deseada de la teoría de conjuntos, mientras que en la definición en teoría de la recursión no se da una situación que se pueda considerar como análoga.

¿Cuál es, entonces, el presupuesto menos confiable y más controvertible que los principios de reflexión en el que se apoya la crítica que nos ocupa? Simplemente, el presupuesto de que existe una interpretación en el sentido clásico que podamos invocar meramente pronunciado la descripción "la interpretación deseada de la teoría de conjuntos".[11]

Como ya he argumentado en otro lugar[12], hay muchas razones para desconfiar de este presupuesto. En realidad, uno de los teoremas de la propia teoría de conjuntos parecería ser (leído "inocentemente") que su "interpretación deseada" no existe: no existe un conjunto de todos los conjuntos, y por tanto no existe una estructura conjuntista que tenga a todos los conjuntos en su dominio de cuantificación. ¿Por qué habríamos de negarnos a aceptar aquí la guía de la propia teoría, si la seguimos en tantos otros casos? Y si aceptáramos lo que la misma teoría parece sugerir, tendríamos además una justificación absolutamente transparente del aspecto de adecuación de la definición modelista de verdad lógica que nos interesa: puesto que no habría interpretaciones no conjuntistas (o que no fueran representables por medio de estructuras

---

[10]Véanse por ejemplo Dershowitz y Gurevich (2008) y Kripke (2013).

[11]No distinguiré en lo que sigue entre el presupuesto de que existe una interpretación deseada de la teoría de conjuntos y las tesis de que tal interpretación es metafísicamente posible o epistémicamente posible. Alguien podría quizá argumentar que la crítica sobre la fuerza del fundamento sólo requiere una de estas tesis, que son más débiles que la de la existencia de una interpretación deseada de la teoría de conjuntos. Pero bajo una concepción habitual de los objetos matemáticos, la existencia y la posibilidad metafísica de una interpretación deseada de la teoría de conjuntos son equivalentes. Por otro lado, la tesis de que una tal interpretación es epistémicamente posible seguirá siendo más controvertible que los principios de reflexión, a la luz de los argumentos ofrecidos más abajo para dudar de la existencia de esa interpretación y para aceptar los principios de reflexión.

[12]Me refiero una vez más al trabajo Gómez-Torrente (2008a), en el contexto del desarrollo de la observación condicional mencionada, según la cual si es posible obtener una justificación suficientemente fuerte de la tesis de que la teoría de conjuntos no tiene una interpretación deseada, entonces hay un fundamento suficientemente fuerte para rechazar la crítica que nos ocupa.

conjuntistas), no habría ninguna oración verdadera en todas las estructuras conjuntistas que fuera también falsa en alguna otra interpretación.

Bajo la hipótesis de que no existe su interpretación deseada, la teoría de conjuntos no la describe. Pero entonces surge inmediatamente la pregunta de qué es lo que hace realmente la teoría (si esa hipótesis es correcta), y cómo lo hace. Superficialmente su lenguaje es un lenguaje cuantificacional clásico, un lenguaje de los que clásicamente se supone que describen una estructura matemática, como la estructura de los números naturales, o la de los números reales. Pero es muy razonable pensar que esta apariencia es efectivamente sólo superficial. Aunque quizá el propósito original de los creadores de la axiomatización usual de la teoría de conjuntos, como Zermelo, era describir una estructura, paulatinamente se hizo evidente para los teóricos, incluido Zermelo en sus últimos trabajos sobre teoría de conjuntos[13], que parecería deseable postular ciertos principios de existencia de grandes cardinales independientes de los axiomas usuales, y que esos principios no parecían agotarse: la postulación de uno sugería siempre maneras de postular principios más poderosos. Esta parece ser básicamente la situación aún hoy en día. Pero si esta es la situación, como mínimo se confirma la seria duda de que con la teoría de conjuntos se lleve a cabo el propósito de describir una estructura bien determinada, y se sugiere más bien la idea de que la teoría en último término codifica principios que valen para una estructura inicialmente vislumbrada y que se pretende que se preserven bajo cualesquiera ampliaciones razonables de esa estructura. Esas ampliaciones teóricas razonables describen entonces sucesivas ampliaciones "ontológicas" correspondientes de la estructura inicial.

Naturalmente, si se concede que la existencia de una interpretación deseada de la teoría de conjuntos es controvertible, la idea de su existencia no puede usarse para cuestionar una definición cuya adecuación se sigue de ideas menos controvertibles. Y, en mi opinión, los principios de reflexión son de hecho principios menos controvertibles, y ciertamente menos controvertidos, que la idea de la existencia de una interpretación deseada de la teoría de conjuntos. Los principios de reflexión, al fin y al cabo, recogen una de las muchas ideas acerca de cómo obtener nuevos conjuntos a partir de conjuntos ya postulados o cuya existencia ya ha sido demostrada, y son en ese sentido menos controvertibles que la idea de que todos esos postulados buscan describir una estructura acabada y bien determinada. (Los principios de reflexión son de hecho en esencia principios de existencia de conjuntos grandes).

Podría objetarse que la crítica sobre la fuerza del fundamento presupone la existencia de la interpretación deseada de la teoría de conjuntos sólo en la misma medida en que los principios de reflexión presuponen la existencia de esa misma interpretación. Pues, ¿acaso no dice un principio de reflexión que si una oración (de un cierto tipo) es *verdadera en la interpretación deseada de la teoría de conjuntos* entonces es verdadera en alguna estructura conjuntista? Pero la respuesta es que, estrictamente hablando, no es así. Es cierto que la idea intuitiva que subyace en los principios de

---

[13] Véase en particular Zermelo (1930).

reflexión se enuncia a menudo así. Sin embargo, la verdad es más sutil, y al apreciarla bien queda claro que los principios de reflexión no presuponen la existencia de una interpretación deseada de la teoría de conjuntos.

Los principios de reflexión se formulan habitualmente como esquemas de fórmulas de un lenguaje (de primer orden, de orden superior, y de otros tipos) para la teoría de conjuntos. Así, por ejemplo, si φ^C es la relativización en el sentido habitual de una fórmula φ del lenguaje de segundo orden de la teoría de conjuntos a un conjunto C, un tipo de principio de reflexión nos dirá que para toda oración σ de ese lenguaje, la oración

$$\sigma \supset \exists C \sigma^C \tag{1}$$

es un axioma de la teoría. No hay aquí mención alguna de la existencia de la interpretación deseada de la teoría de conjuntos. En particular, es posible aceptar el principio de reflexión mencionado leyendo intuitivamente una oración de la forma de (1) como afirmando que, si σ, entonces σ es verdadera en algún conjunto C. No es necesario mencionar en ningún momento en nuestra lectura intuitiva de las oraciones de la forma de (1) la predicación "σ es verdadera en la interpretación deseada de la teoría de conjuntos" (o "σ es verdadera").

Así pues, los principios de reflexión no mencionan la interpretación deseada de la teoría de conjuntos. Pero además es también claro que no la presuponen en algún sentido menos evidente. La clave es que los principios de reflexión buscan codificar una idea abstracta acerca de la teoría de conjuntos que es intuitivamente correcta tanto si se acepta la concepción de que la teoría de conjuntos no describe una realidad acabada como si se prefiere la concepción opuesta. De hecho, cuando Lévy propone originalmente sus principios de reflexión[14], su concepción filosófica subyacente es claramente que un sistema axiomático para la teoría de conjuntos describe (de forma incompleta) un *estadio* en el "desarrollo" de un ente que nunca llega a alcanzar una forma acabada[15]. Al añadir un principio de reflexión a un sistema axiomático, el teórico busca "afirmar la existencia de modelos (...) que reflejan en algún sentido el *estado* del universo[16]". La idea es que una oración de la forma de (1) (y de otras formas asociadas a otros tipos de principios de reflexión), cuando es añadida a un sistema axiomático para la teoría de conjuntos, asegura que si la oración σ correspondiente es verdadera en una cierta estructura asociada de forma privilegiada a ese sistema, entonces σ es verdadera en una estructura conjuntista. Lévy tiene primordialmente en mente la idea de que la estructura asociada de forma privilegiada a un sistema representa efectivamente un estadio de un ente que nunca llega a alcanzar una forma acabada, un estadio o estado en el proceso de "desarrollo" del universo. Pero la idea en cuestión es general, y subsume el caso (si es que en efecto es un caso genuino) en que la estructura

---

[14] En su tesis doctoral de 1958 y en Lévy (1960).
[15] Véase también Fraenkel, Bar-Hillel y Lévy (1973), 118, para comentarios en la misma dirección, presumiblemente debidos al propio Lévy.
[16] Lévy (1960), 1. (La cursiva es mía.)

asociada de forma privilegiada a un sistema es "la interpretación deseada de la teoría de conjuntos"; en este caso, de forma análoga a los demás, una oración de la forma de (1), cuando es añadida a un sistema axiomático, asegura que si la oración σ correspondiente es verdadera en la interpretación deseada de la teoría de conjuntos, entonces σ es verdadera ya en un conjunto. Así, quien acepte (aunque sólo sea en aras de algún argumento), que existe la interpretación deseada de la teoría de conjuntos, y acepte al mismo tiempo el mencionado principio de reflexión, rechazará que exista una fórmula de segundo orden verdadera en todas las estructuras conjuntistas y falsa en la interpretación deseada de la teoría de conjuntos. (Pues quien acepte que una cierta oración σ_0 es falsa en la interpretación deseada de la teoría de conjuntos aceptará la negación de σ_0, y si acepta la fórmula de tipo (1) para la negación de σ_0, aceptará también que σ_0 es falsa en alguna estructura conjuntista.) Queda pues claro que los principios de reflexión dan cuerpo a una intuición muy general de los teóricos de conjuntos que es válida tanto desde la perspectiva de quienes no ven la teoría de conjuntos como una descripción de una estructura determinada como desde la perspectiva de quienes sí la ven así. En este sentido, los principios de reflexión no presuponen la existencia de una estructura que se pueda considerar como "la interpretación deseada de la teoría de conjuntos[17]".

La crítica sobre la fuerza del fundamento, por el contrario, debe presuponer la existencia de una interpretación deseada de la teoría de conjuntos (o de otras interpretaciones no conjuntistas). La crítica en sí no está constituida por una oración o un conjunto de oraciones del lenguaje de la teoría de conjuntos, abierta por tanto a lecturas que no presuponen la existencia de "la interpretación deseada" de la teoría. Está constituida por un enunciado o conjunto de enunciados esencialmente semánticos y metateóricos, en el que la interpretación deseada de la teoría de conjuntos (u otras interpretaciones no conjuntistas) es mencionada de manera fundamental. En este sentido, la crítica sobre la fuerza del fundamento de la definición modelista de verdad lógica tiene un presupuesto controvertido que los principios de reflexión, teóricamente más básicos, no tienen. Por tanto, la crítica en cuestión no puede socavar de manera efectiva la aceptación de la definición modelista de verdad lógica. La crítica sobre la

---

[17]Hay formalizaciones de la teoría de conjuntos y esquemas de traducción de su lenguaje que buscan explicitar completamente, ya en las fórmulas mismas de esas formalizaciones, que la teoría no presupone ni describe una realidad acabada, sino sólo realidades en algún sentido meramente "creables" o potenciales. Por ejemplo, hay axiomatizaciones donde los cuantificadores siguen una lógica aproximadamente intuicionista, con lo que se busca reflejar la idea de que el teórico de conjuntos no cuantifica sobre una totalidad acabada, sino que al generalizar quiere enunciar principios que valgan para los conjuntos que se puedan "crear" o postular. De forma más pertinente en nuestro contexto presente, hay traducciones del lenguaje de la teoría de conjuntos donde los cuantificadores se transforman en locuciones cuantificacionales modales, que en último término cuantifican precisamente sobre conjuntos potenciales. (Véanse por ejemplo Putnam 1967, Parsons 1983, Hellman 1989, Linnebo 2013.) Bajo estas traducciones los axiomas canónicos de la teoría de conjuntos se convierten en proposiciones plausibles acerca de conjuntos potenciales. Y bajo ninguna de estas lecturas se presupone que el lenguaje de la teoría de conjuntos tenga una interpretación deseada en el sentido clásico.

fuerza del fundamento presupone un principio menos transparente y más controvertible que los principios de reflexión: el principio de que la teoría de conjuntos describe una estructura bien determinada. Puede que esa estructura exista y puede que no exista. Pero la idea de su existencia no puede usarse para cuestionar una definición cuya adecuación se sigue de ideas más confiables.

Referencias

[Boolos(1988)] G. Boolos. Nominalist platonism. En *Logic, Logic, and Logic*, pp. 73–87. Cambridge (Mass.): Harvard University Press, 1988.

[Dershowitz y Gurevich(2008)] N. Dershowitz y Y. Gurevich. A natural axiomatization of computability y proof of Church's thesis. *Bulletin of Symbolic Logic*, 14: 299–350, 2008.

[Etchemendy(1990)] J. Etchemendy. *The Concept of Logical Consequence*. Cambridge (Mass.): Harvard University Press, 1990.

[Fraenkel et al.(1973)Fraenkel, Bar-Hillel, y Lévy] A. A. Fraenkel, Y. Bar-Hillel, y A. Lévy. *Foundations of Set Theory*. Amsterdam: North-Holland, 2nd revised edition, 1973.

[García de la Sienra(2008)] Adolfo García de la Sienra. *Reflexiones sobre la Paradoja de Orayen*. México, D.F.: Instituto de Investigaciones Filosóficas-U.N.A.M., 2008.

[Gloede(1976)] K. Gloede. Reflection principles and indescribability. En *Sets and Classes. On the Work by Paul Bernays*, pp. 277–323. Amsterdam: North-Holland, 1976.

[Gómez-Torrente(1998/9)] M. Gómez-Torrente. Logical Truth and Tarskian Logical Truth. *Synthèse*, 117:375–408, 1998/9.

[Gómez-Torrente(2000)] M. Gómez-Torrente. *Forma y Modalidad*. Buenos Aires: Eudeba, 2000.

[Gómez-Torrente(2008a)] M. Gómez-Torrente. Interpretaciones y conjuntos. En García de la Sienra, 2008.

[Gómez-Torrente(2008b)] M. Gómez-Torrente. Are There Model-Theoretic Logical Truths that Are not Logically True? En *New Essays on Tarski y Philosophy*, pp. 340–368. Oxford: Oxford University Press, 2008b.

[Hellman(1989)] G. Hellman. *Mathematics without Numbers*. Oxford: Clarendon, 1989.

[Kreisel(1965)] G. Kreisel. Mathematical Logic. En *Lectures on Modern Mathematics, vol. III*, pp. 95–195. Nueva York: John Wiley & Sons, 1965.

[Kreisel(1967)] G. Kreisel. Informal Rigour and Completeness Proofs. En *Problems in the Philosophy of Mathematics*, pp. 138–171. Amsterdam: North-Holland, 1967.

[Kreisel y Krivine(1967)] G. Kreisel y J.L. Krivine. *Elements of Mathematical Logic (Model Theory)*. Amsterdam: North-Holland, 1967.

[Kripke(2013)] S. Kripke. The Church-Turing 'Thesis' as a Special Corollary of Gödel's Completeness Theorem. En *Computability. Turing, Gödel, Church, y Beyond*, pp. 77–104. Cambridge (Mass.): MIT Press, 2013.

[Lévy(1960)] A. Lévy. Principles of Reflection in Axiomatic Set Theory. *Fundamenta Mathematicae*, 49:1–10, 1960.

[Linnebo(2013)] Ø. Linnebo. The potential hierarchy of sets. *Review of Symbolic Logic*, 6:205–228, 2013.

[McGee(1992)] V. McGee. Two Problems with Tarski's Theory of Consequence. *Proceedings of the Aristotelian Society, n.s.*, 92:273–292, 1992.

[Parsons(1983)] C. Parsons. Sets and Modality. En su *Mathematics in Philosophy*, Ithaca (N.Y.): Cornell University Press, pp. 298–341, 1983.

[Putnam(1967)] H. Putnam. Mathematics without Foundations. *Journal of Philosophy*, 64:5–22, 1967.

[Rayo y Uzquiano(1999)] A. Rayo y G. Uzquiano. Toward a Theory of Second-Order Consequence. *Notre Dame Journal of Formal Logic*, 40:315–25, 1999.

[Shapiro(1987)] S. Shapiro. Principles of Reflection and Second-Order Logic. *Journal of Philosophical Logic*, 16:309–333, 1987.

[Shapiro(1991)] S. Shapiro. *Foundations without Foundationalism: a Case for Second-Order Logic*. Oxford: Clarendon Press, 1991.

[Tarski(1936)] A. Tarski. On the concept of logical consequence. en Tarski (1983), pp. 409-420. Traducción de "Über den Begriff der logischen Folgerung", en Actes du Congrès International de Philosophie Scientifique, fasc. 7 (Actualités Scientifiques et Industrielles, vol. 394). *París: Hermann et Cie.*, pages 1–11, 1936.

[Tarski(1983)] A. Tarski. *Logic, Semantics, Metamathematics*. Indianapolis: Hackett, 2a edición, 1983.

[Williamson(2003)] T. Williamson. Everything. En *Philosophical Perspectives 17: Language and Philosophical Linguistics*, pp. 415–465. Oxford: Blackwell, 2003.

[Zermelo(1930)] E. Zermelo. On Boundary Numbers and Domains of Sets. New Investigations in the Foundations of Set Theory. En Zermelo (2010), pp. 401-431. Traducción de "Über Grenzzahlen und Mengenbereiche. Neue Untersuchungen über die Grundlagen der Mengenlehre". *Fundamenta Mathematicae*, 16:29–47, 1930.

[Zermelo(2010)] E. Zermelo. *Collected Works - Gesammelte Werke*, volumen I. Berlín: Springer, 2010.

//
Which is the least complex explanation? Abduction and complexity

Fernando Soler-Toscano

Grupo de Lógica, Lenguaje e Información
University of Seville
fsoler@us.es

Contents

1	How an abductive problem arises?	103
2	How an abductive problem is solved?	105
3	Which is the least complex explanation?	106
4	A proposal using epistemic logic	111
5	Discussion	117

1 How an abductive problem arises?

What is abductive reasoning and when is it used? First, it is a kind of inference. In a broad sense, a «logical inference» is any operation that, starting with some information, allows us to obtain some other information. People are continually doing inference. For example, I cannot remember which of the two keys in my pocket opens the door of my house. I try with the first one but it does not open. So I conclude that it should be the other. In this case, the inference starts with some data (premises): one of the two keys opens the door, but the first I tried did not open. I reach some new information (conclusion): it is the second key.

Inference (or reasoning) does not always follow the same way. In the example, if I know that one of the keys opens my house and I cannot open with the first key, it is *necessary* that the second opens. When this is the case (that is, the conclusion follows *necessarily* from the premises), then we are facing a *deductive inference*. Given that the conclusion is a necessary consequence of the premises, there is no doubt that the conclusion is true, whenever premises are all true: the door must be opened with the second key.

But there are many contexts in which we cannot apply deductive reasoning. Sometimes, we use it but we become surprised by the outcome. What happens if finally the second key does not open the door? This kind of *surprise* was studied by the philosopher Charles S. Peirce as the starting point of abductive reasoning:

> The surprising fact, C, is observed;
> But if A were true, C would be a matter of course,
> Hence, there is reason to suspect that A is true. (*CP 5.189, 1903*).

Peirce mentions a surprising fact, C, that in our example is that none of the keys opens the door, despite we strongly believed that one of them was the right one. This is an abductive problem: a surprising fact that we cannot explain with our current knowledge. Then, we search for a solution, an explanation A that would stop C from being surprising. To *discover* this A we put into play our knowledge about how things usually happen. For example, we may realise that maybe someone locked the door from the inside. Also, if someone usually locks the door from the inside, then the explanation becomes stronger and, as Peirce says, «there is reason to suspect» that it is true.

Not all abductive problems are identical. A common distinction is between *novel* and *anomalous* abductive problems [Aliseda(2006)]. A novel abductive problem is produced when the surprise produced by C is coherent with our previous information. Contrary, in an anomalous abductive problem we previously thought that C could not be the case, and the surprise contradicts our previous belief. The example of the key that does not open the door is a case of anomalous abductive problem.

To clarify the notions, we now offer some informal definitions of the concepts that are commonly used in the logical study of abductive reasoning [Soler-Toscano(2012)]. Suppose that the symbol $\vdash$ represents our reasoning ability, so that $A, B \vdash C$ means that from premises A and B it is possible to infer C by a *necessary* inference (deduction, as explained above). The negated symbol, as in $A, B \nvdash C$, means that the conclusion C cannot be obtained from premises A and B. Also, consider that Θ is a set of sentences (logical propositions) representing our knowledge (that I have two keys, one of them is the right one, etc.) and φ is the surprising fact (none of the keys opens the door). Then, in a novel abductive problem (Θ, φ) the following holds:

1. $\Theta \nvdash \varphi$

2. $\Theta \nvdash \neg\varphi$

The first condition is necessary for φ to be surprising: it does not follow from our previous knowledge. The second condition is specific for a novel abductive problem: the negation (the opposite) of φ, represented by $\neg\varphi$, does not follow from our knowledge Θ. So, in a novel abductive problem our previous knowledge was not useful to predict either the surprising fact φ or the contrary $\neg\varphi$.

In an anomalous abductive problem (Θ, φ), the conditions that are satisfied are the following:

1. $\Theta \nvdash \varphi$

2. $\Theta \vdash \neg\varphi$

Now, although the first condition is the same, the second is different: our previous knowledge Θ predicted $\neg\varphi$, the negation of the surprising fact φ.

2 How an abductive problem is solved?

Logicians say that in deductive reasoning the conclusion is contained in the premises. This means that the information given by the conclusion is implied by the information in the premises. For example, the information that one of my keys open the door but the first does not open contains the information that the second key will open. But this does not happen in abductive reasoning: the information that the door is locked from the inside is not implied by the information of my keys not opening the door. So, abductive reasoning raises conclusions that introduce new information not present in the premises. Because of this, abduction requires a dose of creativity to propose the solutions. Moreover, there are frequently several different solutions, and the ability to select the best of them is required. We will return later to this issue.

We have distinguished two kinds of abductive problems. Now we will comment the kinds of abductive solutions that are usually considered. We denoted above by (Θ, φ) an abductive problem that arises when our knowledge is represented by Θ and the surprising fact is φ. The solution to this problem is given by some information α such that, together with the previous knowledge we had, allows us to infer φ, logically represented by

$$\Theta, \alpha \vdash \varphi$$

This is the minimal condition for an abductive solution α to solve the problem (Θ, φ). Atocha Aliseda [Aliseda(2006)] calls *plain* to those abductive solutions satisfying this requirement.

There are other very interesting kinds of abductive solutions. For example, *consistent* solutions satisfy the additional condition of being coherent with our previous knowledge. It is formally represented by

$$\Theta, \alpha \nvdash \bot,$$

where the symbol $\perp$ represents any contradiction. It is important that our abductive solutions are consistent. Possibly, we will not know whether the abductive solution α is true, but usually, if it is inconsistent with our previous knowledge, we have reason to discard it.

Finally, *explanatory* abductive solutions are those satisfying

$$\alpha \not\vdash \varphi,$$

that is, the surprising fact φ cannot be inferred with α alone without using the knowledge given by Θ. This is to avoid self-contained explanations: the key idea behind this criterion is that a good abductive explanation offers the missing piece to solve a certain puzzle, but all the other pieces were previously given.

When an abductive solution satisfies the three conditions above, we call it a *consistent explanatory solution*. To avoid useless or trivial solutions (the key does not open the door because it does not open it), it is frequent to focus on consistent explanatory abduction.

The logical study of abductive reasoning has been receiving a notable attention for several years, and many calculi have been proposed for abduction in different logical systems [Mayer and Pirri(1993), Soler-Toscano(2012), Soler-Toscano et al.(2009)Soler-Toscano, Nepomuceno-Fernández, and Aliseda-Llera]. Now, we are not interested in offering a specific calculus for a particular logic, but in looking to an old problem in abductive reasoning: the selection of the best hypothesis. Which is the best abductive solution? First, we will proceed conceptually, by introducing some notions from information theory. It will be in Section 4 when, as an example, we will apply the introduced idea in the context of epistemic logic.

3 Which is the least complex explanation?

It may happen that for a certain abductive problem there are several possible explanations, not all of them mutually compatible. For example, to explain why the key does not open the door, we have proposed that someone locked it from the inside. But it could also happen that the key or the door lock are broken, or that someone changed the lock while we were outside, or made a joke, etc. It is necessary to select one of the many possible explanations, because it cannot be that case that all of them happened, it is enough just one of them to explain that we cannot open the door with our key. What explanation is selected and which criteria are used to select it? This is the well-known problem of the selection of abductive hypotheses [Soler-Toscano and Velázquez-Quesada(2014)].

Moreover, different to deductive reasoning, abductive conclusions (selected solutions) are not necessary true. It is easy to observe that, despite we think that someone locked the door from the inside, it may have not been the case, and that in fact the lock is broken. So, we usually have to replace an explanation with another one, when we come to know that the originally chosen is false.

Several criteria have been proposed to solve the problem of the selection of abductive hypotheses. A common one is *minimality*, that prefers explanations assuming fewer pieces of new information. So, if I can solve a certain abductive problem both assuming α_1 or α_2, it is possible that I can also solve it by simultaneously assuming α_1 and α_2, or maybe α_1 and a certain β, but we will usually discard those options because they are not the simplest possible ones. In logical terms, if $A \vdash B$ and both A and B can solve a certain abductive problem, we prefer B, given that A is at least equally strong than B, and maybe stronger, in the sense of assuming more information.

Frequently, the minimality criterion is not enough to select the best explanation. Which one is simpler: to think that someone locked the door from the inside, or that they spent a joke by changing the door lock?

Are there criteria that can help us to select the simplest explanation in a broad spectrum of abductive problems? To give an (affirmative) answer to this question we will move to a field in theoretical computer science: Algorithmic Information Theory (AIT), which is due to the works of Ray Solomonoff [Solomonoff(1964a), Solomonoff(1964b)], Andréi Kolmogórov [Kolmogorov(1965)], Leonid Levin [Levin(1973)] and Gregory Chaitin [Chaitin(July 1975)].

A central notion in AIT is the measure known as *Kolmogórov complexity*, or *algorithmic complexity*. To understand it, let us compare these two sequences of 0s and 1s:

$$0101010101010101010101010101010101010101$$
$$0001101000100110111101010010111011100100$$

If we were asked which one of them is simpler, we will answer that the first one. Why? It is built up from 20 repetitions of the pattern «01». The second sequence is a random string. The difference between the regularity of the first sequence and the randomness of the second one is related with one property: the first sequence has a much shorter description than the second. The first sequence can be described as 'twenty repetitions of «01»', while the second one can be hardly described with a description shorter than itself.

The idea behind the notion of Kolmogórov complexity is that if some object O can be fully described with n bits (*bit*: *b*inary dig*it*, information unit), then O does not contain more information. So the shorter description of the object O indicates how much information is contained in O. We would like to measure in this way the complexity of abductive solutions, and introduce an informational minimality criterion: we select the least complex explanation, that is, the least informative one. But we will look at how this complexity measure is quantified.

Kolmogórov uses the concept of universal Turing machine [Turing(1936)]. An universal Turing machine (UTM) M is a programmable device capable of implementing any algorithm. The important point for us now is that the machine M, similar to our computers, takes a program p, runs it and eventually (if the computation stops) produces a certain output o. To indicate that o is the output produced by UTM M with program p we write $M(p) = o$. Then, for a certain string of characters s, we define its Kolmogórov complexity, $K_M(s)$, as

$$K_M(s) = \min \{l(p) \mid M(p) = s,\ p \text{ is a program}\}$$

where $l(p)$ is the length in bits of the program p. That is, $K_M(s)$ is equal to the size of the shorter program producing s in the UTM M. The subindex M in $K_M(s)$ means that its value depends on the choice of UTM, because not all of them interpret the programs in the same way, despite all having the same computational power. If we choose another machine M' instead of M, it can happen that $K_{M'}(s)$ is pretty different to $K_M(s)$. However, these bad news are only relative, given that the Invariance Theorem guarantees that the difference between $K_M(s)$ and $K_{M'}(s)$ is always lower than a certain constant not depending on s, but on M and M'. So, as we face more and more complex strings, it is less relevant the choice of UTM. Then, we can simply write $K(s)$ to denote the Kolmogórov complexity of s.

The use of Turing machines and programs allows us to set an encoding to describe any computable (that is, that can be produced by some algorithm) object O. Then, for the first binary sequence above, the shortest description will not be 'twenty repetitions of «01»' (26 characters) but the shortest program producing that string.

To approach the relation between algorithmic complexity and abductive reasoning, we can look at the work of Ray Solomonoff, that conceives algorithmic complexity as a tool to create a model that explains *all the regularities in the observed universe* (see [Solomonoff(1964a)], Section 3.2). The idea of Solomonoff is ambitious, but it is in line with the common postulates of the inference to the best explanation [Lipton(1991)]. A theory can be conceived as a set of laws (axioms, hypotheses, etc., depending on the kind of theory) trying to give account of a set of observations in a given context. The laws in the theory try to explain the regularities in those observations. So, what is the best theory? From Solomonoff's point of view, the best theory is the most compact one, that describing the highest number of observations (the most general one) with the fewest number of postulates (the most elegant from a logical point of view). It is the condition for the lowest algorithmic complexity. The best theory is then conceived as the shortest program *generating* the observations that we want to explain. Such *generation* consist of the inferential mechanism underlying the postulates of the theory. If it were a set of logical rules, the *execution* of the program given by the theory is equivalent to what logicians denote by the *deductive closure* of the theory: the set of all consequences that can be deduced from the theory axioms. Such execution does not always finish in a finite number of steps, because logical closures frequently (always in classical logic) are infinite sets, but usually there are procedures (in decidable logical systems) that, in a finite number of steps, check whether a certain formula belongs to such closure.

As we can see, the algorithmic complexity measure $K(s)$ can be used to determine which is the best theory within those explaining a set of observations. However, the problem with $K(s)$ is its uncomputability: there is no algorithm such that, given the object s (a binary string or a set of observations) returns, in a finite number of steps, the value $K(s)$ (the size of the shortest program producing s, or the smallest theory explaining our observations). Therefore, if we can not generally know the value of $K(s)$, we cannot know which is the shortest program (or theory) generating the string s (explaining our observations). The most interesting consequence of the above explanation is that, in general, the problem of determining which is the *best explanation* for a given set of observations is uncomputable (if we understand *the best* as *the most compact*).

However, despite the uncomputability of $K(s)$, there are good approximations that allow to measure the algorithmic complexity of an object. One of the most used approximations is based on lossless compression algorithms. These algorithms are frequently used in our computers to compress documents. A very common compression algorithm is Lempel-Ziv, on which the ZIP compression format is based. It allows to define a computable complexity measure that approximates $K(s)$ [Lempel and Ziv(1976)]. If we have some file s and the output of the compression algorithm is $c(s)$ (it is important to use a lossless compression algorithm so that when decompressed it produces exactly the original file s), we can understand $c(s)$ as a program that, when is run in certain computer (the decompressor program) produces s. So, the length of the compressed file $c(s)$ is an approximation to $K(s)$. It is not necessary that $c(s)$ is the shortest possible description of s, as we can consider it an approximation. In fact, many applications based on $c(s)$ to measure complexity are used in different disciplines like physics, cryptography or medicine [Li and Vitányi(2008)].

How can we approximate $K(s)$ to compare the complexity of several abductive explanations and choose the best one? Compression-based approximations to $K(s)$ are frequently good when s is a character string. It also happens with other approximations based on the notion of *algorithmic probability* [Soler-Toscano et al.(05 2014)Soler-Toscano, Zenil, Delahaye, and Gauvrit]. However, abductive explanations are usually produced in the context of theories with a structure that can be missed when treated as character strings. However, we can use several tricks to reproduce some aspects of the structure of the theories into the structure of the strings. For example, in classical propositional logic, both sets of formulas $A = \{p \to q\}$ and $B = \{\neg q \to \neg p, \neg p \lor q\}$ are equivalent, but if we understand A and B as character sequences and we compress them, B will probably seem more complex than A. We can avoid this problem by converting both sets into a normal form, for example the minimal clausal form—sets (conjunctions) of sets (disjunctions) of literals (propositional variables or their negations)—which in both cases is $\{\{\neg p, q\}\}$.

Another important point to be considered is that the complexity of an abductive explanation α should be measured related to the context in which it is proposed: the theory Θ. Hence, $K(\alpha)$ may not be a good approximation to the complexity of α as an abductive solution to a certain abductive problem (Θ, φ). Because of that, in certain cases it is more reasonable to use the notion of *conditional algorithmic complexity* $K(s \mid x)$ measuring the length of the shortest program that produces s with input x. So, $K(\alpha \mid \Theta)$ would be a better approximation to the complexity of the abductive solution α (within the theory Θ) than just $K(\alpha)$. Using lossless compression, if $c(f)$ represents the compression of f and $|c(f)|$ is the length in bits of $c(f)$, a common approximation to $K(s \mid x)$ is given by $|c(xs)| - |c(x)|$, where xs represents the concatenation of x and s. It can be observed that, in general, this approximation gives different values for $K(y \mid x)$ and $K(x \mid y)$, and the value of $K(x \mid x)$ approaches 0 for an ideal compressor, given that the size of the compression of xx is almost equal to the compression of x, only one instruction to repeat all the output has to be included.

As we can see, the notions of algorithmic complexity make sense to approach the problem of the complexity of abductive solutions and to tackle with computational tools the problem of the selection of the best explanation. However, good choices have to be made about the way to represent the theories (for example, in clausal form) and which approach to $K(s)$ o $K(s \mid x)$ is to be used. We presented above a very simple example on propositional logic where clausal form can be fine. But other options are also possible. For example, Kripke frames can be used to represent relations between theories [Soler-Toscano et al.(2012)]. That way, each world w represents a possible theory Θ_w, and the accessibility relation indicates which modifications can be done to the theories. Then, if world w can access to u, then Θ_w can be modified to become Θ_u. An abductive problem appears when we are in a certain world w and there is a certain formula φ which does not follow from theory Θ_w. Then, we solve the abductive problem by moving to another accessible world u (we modify our theory Θ_w to get Θ_u) such that φ is a consequence of Θ_u. This way we give account of modifications in theories that go beyond adding new formulas. That is, if does not necessary happen $\Theta_w \subset \Theta_u$, because the change of theory can entail deeper modifications, for example in the structural properties of the logical consequence relation. Then, it may be possible to pass, for example, from Θ_w with a monotonous reasoning system, to a non-monotonic reasoning in Θ_u. However, within all accessible theories from w that explain φ, the problem of determining the least complex explanation still remains. The complexity measure that should be used here is $K(\Theta_u \mid \Theta_w)$ and, among all accessible theories from w explaining φ, the one which minimises this complexity measure should be chosen.

Despite offering resources to compare different abductive solutions and to choose the simplest one, algorithmic complexity notions have two problems: (1) to determine a good representation for theories (or formulas) and (2) to choose a computable approximation to $K(s)$ or $K(s \mid x)$, through lossless compression or by other means. In the next section we present an example, based on epistemic logic, illustrating how we can do this in a specific case.

4 A proposal using epistemic logic

In this section we introduce an application of $K(s)$ to the selection of the best abductive explanation, in the context of dynamic epistemic logic (DEL). The presentation is based on previous papers where we use the same logical tools [Soler-Toscano(2014), Soler-Toscano and Velázquez-Quesada(2014), Nepomuceno-Fernández et al.(2013)], but the selection criteria are now different. Here, an approximation to $K(s)$ is applied to choose among several abductive explanations.

One of the possible ways to model the knowledge and belief of an agent is offered by *plausibility models* [Baltag and Smets(2008)]. We start by presenting the semantic notions that will be later used to propose and solve abductive problems.

Definition 1 (Language $\mathcal{L}$) *Given a set of atomic propositions* P, *formulas φ of the language $\mathcal{L}$ are given by*

$$\varphi ::= p \mid \neg\varphi \mid \varphi \vee \varphi \mid \langle \leq \rangle \varphi \mid \langle \sim \rangle \varphi$$

where $p \in$ P. *Formulas of the form $\langle \leq \rangle \varphi$ are read as "there is a world at least as plausible as the current one where φ holds", and those of the form $\langle \sim \rangle \varphi$ are read as "there is a world epistemically indistinguishable from the current one where φ holds". Other Boolean connectives ($\wedge$, $\rightarrow$, $\leftrightarrow$) as well as the universal modalities, $[\leq]$ and $[\sim]$, are defined as usual ($[\leq] \varphi := \neg \langle \leq \rangle \neg \varphi$ and $[\sim] \varphi := \neg \langle \sim \rangle \neg \varphi$ for the latter).*

It can be observed that the language $\mathcal{L}$ is like propositional logic with two new modal connectives, $\langle \leq \rangle$ and $\langle \sim \rangle$, that will allow to define the notions of belief and knowledge. These notions will depend on a *plausibility order* that the agent sets among the worlds in the model. We now see how these models are built.

Definition 2 (Plausibility model) *Let* P *be a set of atomic propositions. A* plausibility model *is a tuple $M = \langle W, \leq, V \rangle$, where:*

- *W is a non-empty set of* possible worlds

- *$\leq \,\subseteq (W \times W)$ is a locally connected and conversely well-founded preorder*[1], *the agent's* plausibility relation, *representing the plausibility order of the worlds from her point of view ($w \leq u$ is read as "u is at least as plausible as w")*

---

[1] A relation $R \subseteq (W \times W)$ is *locally connected* when every two elements that are R-comparable to a third are also R-comparable. It is *conversely well-founded* when there is no infinite $\overline{R}$-ascending chain of elements in W, where $\overline{R}$, the *strict* version of R, is defined as $\overline{R}wu$ iff Rwu and not Ruw. Finally, it is a *preorder* when it is reflexive and transitive.

- $V: W \to \wp(\mathrm{P})$ *is an* atomic valuation function, *indicating the atoms in* P *that are true at each possible world.*

A *pointed plausibility model* (M, w) *is a plausibility model with a distinguished world* $w \in W$.

The key idea behind plausibility models is that an agent's beliefs can be defined as what is true *in the most plausible worlds from the agent's perspective*, and modalities for the plausibility relation $\leq$ will allow this definition to be formed. In order to define the agent's knowledge, the approach is to assume that two worlds are epistemically indistinguishable for the agent if and only if she considers one of them at least as plausible as the other (i.e., if and only if they are comparable via $\leq$). The *epistemic indistinguishability relation* $\sim$ can therefore be defined as the union of $\leq$ and its converse, that is, as $\sim \, := \, \leq \cup \geq$. Thus, $\sim$ is the symmetric closure of $\leq$ and hence $\leq \, \subseteq \, \sim$. Moreover, since $\leq$ is reflexive and transitive, $\sim$ is an *equivalence* relation. This epistemic indistinguishability relation $\sim$ should not be confused with the *equal plausibility* relation, denoted by $\simeq$, and defined as the *intersection* of $\leq$ and $\geq$, that is, $\simeq \, := \, \leq \cap \geq$. For further details and discussion on these models, their requirements and their properties, the reader is referred to [Baltag and Smets(2008), Velázquez-Quesada(2014)].

Now we can see how a formula is evaluated in a plausibility model. Modalities $\langle \leq \rangle$ and $\langle \sim \rangle$ are interpreted in the standard way, using their respective relations.

Definition 3 (Semantic interpretations) *Let* (M, w) *be a plausibility model* $M = \langle W, \leq, V \rangle$ *with distinguished world* $w \in W$. *By* $(M, w) \Vdash \psi$ *we indicate that the formula* $\psi \in \mathcal{L}$ *is true in the world* $w \in W$. *Formally,*

$(M, w) \Vdash p$ *iff* $p \in V(w)$, *for every* $p \in \mathrm{P}$
$(M, w) \Vdash \neg \varphi$ *iff* $(M, w) \not\Vdash \varphi$
$(M, w) \Vdash \varphi \wedge \psi$ *iff* $(M, w) \Vdash \varphi$ *and* $(M, w) \Vdash \psi$
$(M, w) \Vdash \langle \leq \rangle \varphi$ *iff*   *there exists* $u \in W$ *such that* $w \leq u$ *and* $(M, u) \Vdash \varphi$
$(M, w) \Vdash \langle \sim \rangle \varphi$ *iff*   *there exists* $u \in W$ *such that* $w \sim u$ *and* $(M, u) \Vdash \varphi$

In plausibility models, knowledge is defined using the indistinguishability relation. So, an agent knows φ in some world w iff φ is true in all worlds that cannot be distinguished from w by her, that is, all worlds considered epistemically possible for her. However, within those worlds there is a plausibility order, not all of them are equally plausible for the agent. This is relevant for the notion of belief: agent believes φ in a certain world w iff φ is true in *the most plausible worlds* that are reachable from w. Due to the properties of the plausibility relation, φ is true in the most plausible worlds iff by following the plausibility order, from some stage we only reach φ-worlds [Baltag and Smets(2008)]. We can express this idea with modalities $\langle \leq \rangle$ and $[\leq]$. Formally[2],

---

[2] Operator K is commonly used for knowledge. It should not be confused with Kolmogórov complexity $K(s)$ also usually represented by K.

4 A proposal using epistemic logic

Agent *knows* φ $\quad K\varphi := [\sim]\varphi$
Agent *believes* φ $\quad B\varphi := \langle\leq\rangle[\leq]\varphi$

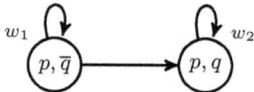

Figure 1: Example of a plausibility model

Fig. 1 shows a plausibility model example M. Plausibility relation $\leq$ is represented by arrows between worlds. For the agent, world w_2 is more plausible than w_1. In this case, p is true in both worlds, while q is true only in w_2 ($\bar{q}$ represents that q is false). So, agent knows p in w_1 but does not know q, that is, $(M, w_1) \Vdash Kp \wedge \neg Kq$. However, agent believes q, $(M, w_1) \Vdash Bq$. Indeed, she also believes p, $(M, w_1) \Vdash Bp$.

If a formula φ is true in all the states of a certain model M, then φ is valid in M, represented as $M \Vdash \varphi$. In the example, $M \Vdash Kp \wedge \neg Kq \wedge Bq$.

In the logical literature about abduction and belief revision in general [Alchourrón et al.(1985)Alchourrón, Gärdenfors, and Makinson], logical operations adding or removing information of the theory are frequently considered. In the same way, in the context of plausibility models, agents can perform epistemic actions modifying the agent's information. We now present two of the main actions that agents in plausibility models can perform. On of the actions modifies the knowledge and the other the belief. For more details about the properties of these actions, see [Baltag and Smets(2008)].

The first operation, *observation*, modifies the agent's knowledge. It is defined in a very natural way: it consists of removing all worlds where the observed formula is not satisfied, so that the domain of the model is reduced.

Definition 4 (Observation) *Let* $M = \langle W, \leq, V \rangle$ *be a plausibility model. The observation of* ψ *produces the model* $M_{\psi!} = \langle W', \leq', V' \rangle$ *where* $W' := \{w \in W \mid (M, w) \Vdash \psi\}$, $\leq' := \leq \cap (W' \times W')$ *and, for each* $w \in W'$, $V'(w) := V(w)$.

This operation removes worlds of W, keeping only those that satisfy (before the observation) the observed ψ. The plausibility relation is restricted to the conserved worlds.

Another operation that agents can do is to modify just the plausibility relation. It can be done in several ways. The operation we call *conjecture* is also known as *radical upgrade* in the literature.

Definition 5 (Conjecture) *Let $M = \langle W, \leq, V \rangle$ be a plausibility model and ψ a formula. The* conjecture *of ψ produces the model $M_{\psi\Uparrow} = \langle W, \leq', V \rangle$, that differs from M only in the plausibility relation, which is now,*

$$\leq' := \{(w, u) \mid w \leq u \text{ and } (M, u) \Vdash \psi\} \cup$$
$$\{(w, u) \mid w \leq u \text{ and } (M, w) \Vdash \neg\psi\} \cup$$
$$\{(w, u) \mid w \sim u \text{ and } (M, w) \Vdash \neg\psi \text{ and } (M, u) \Vdash \psi\}$$

The new plausibility relation indicates that, after the conjecture of ψ, all ψ-worlds (before the conjecture) are more plausible than all $\neg\psi$-worlds. The previous order between ψ-worlds or between $\neg\psi$-worlds does not change [van Benthem(2007)]. This operation preserves the properties of the plausibility relation, as shown in [Velázquez-Quesada(2014)].

We now discuss how an abductive problem can appear and be solved within the plausibility models formalism. In the classical definition of abductive problem, formula φ is an abductive problem because it is not entailed by the theory Θ. But, where does φ come from? For Peirce, φ is an observation, that is, it comes from an agent's epistemic action. As we have seen in Def. 4, the action of observing φ can be modelled in DEL. What does it mean, then, that φ is an abductive problem? After observing φ, if it is a propositional formula, the agent knows φ, so we cannot affirm that an abductive problem arises when the agent does not know φ. However, we can go back to the moment before observing φ; if the agent did not know φ, then after the observation it becomes an abductive problem. Formally,

$$\varphi \text{ is an abductive problem in } (M_{\varphi!}, w) \text{ iff } (M, w) \not\Vdash K\varphi \qquad (1)$$

This definition of abductive problem within plausibility models is in line with Peirce's idea that an abductive problem appears when the agent observes φ.

The notion of abductive problem in (1) has been defined in terms of knowledge. If could have been defined in terms of belief too, considering that φ is an abductive problem in $(M_{\varphi!}, w)$ iff $(M, w) \not\Vdash B\varphi$. Then, the condition for φ to be an abductive problem becomes stronger than in (1), because $\neg B\varphi$ implies $\neg K\varphi$.

Now the notion of abductive solution can also be interpreted in the plausibility models semantics. According to Peirce's idea, the agent knows that if ψ were true, then the truth of the surprising fact φ would be obvious. It is expressed in DEL by requiring that the agent knows $\psi \to \varphi$, that is, $K(\psi \to \varphi)$. Then, when the agent faces an abductive problem φ and knows $\psi \to \varphi$, how does she solve it? Again, Peirce says that *there is reason to suspect that ψ is true*. We now discuss what does 'to suspect' ψ mean in DEL and how can it be modelled as an epistemic action.

Something not usually considered within logical approaches to abductive reasoning is how to integrate the solution. Maybe because in classical logic there is no way to *suspect* a formula. But in epistemic logic there are beliefs. It is an information kind

4 A proposal using epistemic logic

weaker than knowledge, as we have seen. So, belief seems the most natural candidate to model the suspicion. In line with Peirce, we then distinguish the agent's knowledge of $\psi \to \varphi$ from her belief in ψ.

But for a reasonable suspicion, as Peirce requires, it is necessary that the agent knows $\psi \to \varphi$. Joining all the presented ideas, given abductive problem φ in $(M_{\varphi!}, w)$ (see (1)), formula ψ is a solution for it iff

$$(M, w) \Vdash K(\psi \to \varphi) \tag{2}$$

Condition $(M_{\varphi!}, w) \Vdash K(\psi \to \varphi)$ cannot be required because it is trivially verified in all cases in which φ is a propositional formula, as it becomes known after being observed, so the agent knows also $\psi \to \varphi$ for every ψ.

What does the agent do to suspect ψ? The most adequate abductive action to integrate ψ into the agent's information is to *conjecture* ψ (def. 5). In this way, ψ is integrated into the agent's information as a belief.

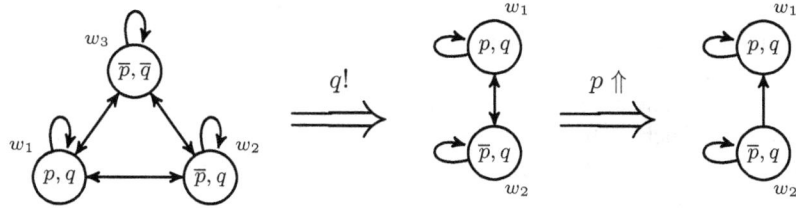

Figure 2: Solving an abductive problem

Fig. 2 shows an example of the whole explained process. In the model of the left, $K(p \to q)$ is verified, but also $\neg Kq$. In the central model, after observing q agent knows q, and of course she continues knowing $p \to q$, that is, $Kq \wedge K(p \to q)$. Then, q is an abductive problem, given that the agent knows it after the observation but not before. A possible solution is then p. After the abductive action of conjecturing p, the model on the right shows that the agent believes p, Bp, given that p is true in the most plausible world.

Briefly, (1) shows the condition for φ being an abductive problem and (2) for ψ being a solution for it. As we have proposed, is it reasonable to conjecture the abductive solution as a belief (def. 5).

We are now ready to use Kolmogórov complexity as a selection criterion within solutions of an abductive problem. First, observe that a binary relation R over a set $\{l_1, l_2, \ldots, l_n\}$ can be represented as a binary matrix A with dimension $n \times n$. In such matrix, each cell $a_{i,j}$ is equal to 1 iff $(l_i, l_j) \in R$ and is 0 otherwise. For example, consider relation $R = \{(1,1), (1,2), (2,1), (2,3), (3,1)\}$ over the set $\{1, 2, 3\}$. The matrix representing R is

$$\begin{pmatrix} 1 & 1 & 0 \\ 1 & 0 & 1 \\ 1 & 0 & 0 \end{pmatrix}$$

Given a matrix A, its Kolmogórov complexity[3] $K(A)$ can be approximated for example using lossless compression [4]. The least complex relations are equally the empty relation (all the matrix is filled with 0s) and the total relation (all filled with 1s). Those are the most compressible matrices. Other very regular relations, as that containing only pairs (e, e) for each element e (diagonal matrix), are also quite compressible. The more random a relation is, the more random is the obtained matrix and the less compressible it is.

In addition to approximate the value of $K(A)$ for a given matrix A, compression allows us to approximate $K(B \mid A)$, given matrices A and B. To do that, the concatenation of A and B (in this order) is first compressed and then A alone is also compressed. The difference of the size of both files approximates $K(B \mid A)$. This approximation to $K(B \mid A)$ can be used to determine which is the best abductive solution in the context of plausibility models.

Consider an abductive problem φ in $(M_{\varphi!}, w)$, and two competing explanations ψ_1 and ψ_2 for it. The argument below can be extended to any number of competing explanations. Which of both explanations will be chosen? It was explained below that the way to integrate an abductive explanation ψ_i is to conjecture ψ_i in the model $(M_{\varphi!}, w)$ (def. 5). The effect of conjecturing ψ_i is only to modify the plausibility relation of $M_{\varphi!}$. Then, which is the best explanation? The answer offered by the notion of algorithmic complexity if that the best explanation is the one modifying the model (the plausibility relation) in the least complex way: this is not the one making the smallest modification, but the modification with the shortest description given the initial model. If $A_{\varphi!}$ is the matrix representing the plausibility relation of $M_{\varphi!}$ and A_{ψ_i} the matrix for the plausibility relation of $(M_{\varphi!})_{\psi_i \Uparrow}$, then the best explanation is the one minimising the value of

$$K(A_{\psi_i} \mid A_{\varphi!}) \qquad (3)$$

The explanation minimising (3) is the one having the shortest description[5] starting at

---

[3]Now K is used for algorithmic complexity and not for knowledge as in previous paragraphs.

[4]Lossless compression approximations are not good for small matrices as the one in the example, because habitual compressors cannot detect regularities in very short binary sequences. For small matrices, it is more convenient to use the tools presented in [Soler-Toscano et al.(05 2014)Soler-Toscano, Zenil, Delahaye, and Gauvrit, Zenil et al.(2014)Zenil, Soler-Toscano, Dingle, and Louis] also available at The Online Algorithmic Complexity Calculator: http://www.complexitycalculator.com/.

[5]Formally, the chosen solution ψ is the one, among all possible solutions, that minimises the length of a program that, when given to an universal Turing machine together with the encoding of $M_{\varphi!}$, produces $(M_{\varphi!})_{\psi_i \Uparrow}$.

the plausibility relation for the agent after observing φ.

This methodology cannot be applied to small epistemic models, as those in the previous examples, because compression is not a good approximation to the complexity of small matrices. But it can be applied to models of medium and large size appearing in applications of DEL to multi-agent systems. For small models there are other methods that can be applied [Zenil et al.(2014)Zenil, Soler-Toscano, Dingle, and Louis], only changing in the way to approximate $K(s)$.

5 Discussion

Ray Solomonoff, one of the drivers of algorithmic information theory, was convinced that the notion of algorithmic complexity can be used to define a system explaining all the observed regularities in the Universe. But there are strong limitations that make impossible to build such a system, mainly the uncomputability of $K(s)$. However, by using computable approximations, though not being able to build a system explaining all the regularities in the Universe, it is possible, in a specific context (plausibility models for us), to establish criteria based on $K(s)$ (and its conditional version) that allow to select the best explanation among the possible ones.

There are still many issues to explore. For example, it would be interesting to study the relevance of the notion of *facticity* introduced by Pieter Adriaans [Adriaans(2012)]. He considers $K(s)$ the sum of two terms, one is the structural information of s and the other the *ad hoc* information in s. Then the best explanation could be selected by specially looking at the contained structural information. Also, algorithmic complexity measures can be combined with other common selection criteria that avoid triviality. Ideally, the least complex solution should be selected among all possible consistent and explanatory ones.

References

[Adriaans(2012)] Pieter Adriaans. Facticity as the amount of self-descriptive information in a data set. *CoRR*, abs/1203.2245, 2012.

[Alchourrón et al.(1985)Alchourrón, Gärdenfors, and Makinson] C.E. Alchourrón, P. Gärdenfors, and D. Makinson. On the logic of theory change: partial meet contraction and revision functions. *Journal of Symbolic Logic*, 50(2):510–530, 1985.

[Aliseda(2006)] Atocha Aliseda. *Abductive Reasoning: Logical Investigations into Discovery and Explanation*, volume 330 of *Synthese Library*. Springer, 2006.

[Baltag and Smets(2008)] Alexandru Baltag and Sonja Smets. A qualitative theory of dynamic interactive belief revision. In *Logic and the Foundations of Game and Decision Theory (LOFT7)*, volume 3 of *Texts in Logic and Games*, pages 13–60. Amsterdam University Press, 2008.

[Chaitin(July 1975)] Gregory J. Chaitin. A theory of program size formally identical to information theory. *J. ACM*, 22(3):329–340, July 1975.

[Kolmogorov(1965)] A. N. Kolmogorov. Three approaches to the quantitative definition of information. *Problems of Information Transmission*, 1(1):1–7, 1965.

[Lempel and Ziv(1976)] Abraham Lempel and Jacob Ziv. On the complexity of finite sequences. *IEEE Transactions on Information Theory*, 22(1):75–81, 1976.

[Levin(1973)] Leonid A Levin. On the notion of a random sequence. *Soviet Math. Dokl*, 14(5):1413–1416, 1973.

[Li and Vitányi(2008)] Ming Li and Paul M. B. Vitányi. *An Introduction to Kolmogorov Complexity and Its Applications*. Texts in Computer Science. Springer, third edition, 2008.

[Lipton(1991)] Peter Lipton. *Inference to the Best Explanation*. Routledge, New York, 1991.

[Mayer and Pirri(1993)] Marta Cialdea Mayer and Fiora Pirri. First order abduction via tableau and sequent calculi. *Bulletin of the IGPL*, 1:99–117, 1993.

[Nepomuceno-Fernández et al.(2013)] Ángel Nepomuceno-Fernández, Fernando Soler-Toscano, and Fernando R. Velázquez-Quesada. An epistemic and dynamic approach to abductive reasoning: selecting the best explanation. *Logic Journal of IGPL*, 21(6):943–961, 2013.

[Soler-Toscano(2012)] Fernando Soler-Toscano. *Razonamiento abductivo en lógica clásica*. College Publications, 2012.

[Soler-Toscano(2014)] Fernando Soler-Toscano. El giro dinámico en la epistemología formal: el caso del razonamiento explicativo. *THEORIA*, 29(2):181–199, 2014.

[Soler-Toscano and Velázquez-Quesada(2014)] Fernando Soler-Toscano and Fernando R. Velázquez-Quesada. Generation and selection of abductive explanations for non-omniscient agents. *Journal of Logic, Language and Information*, 23(2):141–168, 2014.

REFERENCES

[Soler-Toscano et al.(05 2014)Soler-Toscano, Zenil, Delahaye, and Gauvrit]
Fernando Soler-Toscano, Hector Zenil, Jean-Paul Delahaye, and Nicolas Gauvrit. Calculating Kolmogorov complexity from the output frequency distributions of small turing machines. *PLoS ONE*, 9(5):e96223, 05 2014.

[Soler-Toscano et al.(2009)Soler-Toscano, Nepomuceno-Fernández, and Aliseda-Llera]
Fernando Soler-Toscano, Ángel Nepomuceno-Fernández, and Atocha Aliseda-Llera. Abduction via c-tableaux and δ-resolution. *Journal of Applied Non-Classical Logics*, 19(2):211–225, 2009.

[Soler-Toscano et al.(2012)] Fernando Soler-Toscano, David Fernández-Duque, and Ángel Nepomuceno-Fernández. A modal framework for modelling abductive reasoning. *Logic Journal of IGPL*, 20(2):438–444, 2012.

[Solomonoff(1964a)] R.J. Solomonoff. A formal theory of inductive inference. part i. *Information and Control*, 7(1):1–22, 1964a.

[Solomonoff(1964b)] R.J. Solomonoff. A formal theory of inductive inference. part {II}. *Information and Control*, 7(2):224–254, 1964b.

[Turing(1936)] Alan M. Turing. On computable numbers, with an application to the Entscheidungsproblem. *Proceedings of the London Mathematical Society*, 2(42): 230–265, 1936.

[van Benthem(2007)] Johan van Benthem. Dynamic logic for belief revision. *Journal of Applied Non-Classical Logics*, 17(2):129–155, 2007.

[Velázquez-Quesada(2014)] Fernando R. Velázquez-Quesada. Dynamic epistemic logic for implicit and explicit beliefs. *Journal of Logic, Language and Information*, 23(2):107–140, 2014.

[Zenil et al.(2014)Zenil, Soler-Toscano, Dingle, and Louis] Hector Zenil, Fernando Soler-Toscano, Kamaludin Dingle, and Ard A. Louis. Correlation of automorphism group size and topological properties with program-size complexity evaluations of graphs and complex networks. *Physica A: Statistical Mechanics and its Applications*, 404(0):341 – 358, 2014.

On the naturalness of new axioms in set theory

Giorgio Venturi

Univ. Est. de Campinas

Contents

1 **On the justification of axioms**   123

2 **Is naturalness intrinsic or extrinsic?**   126

3 **First dogma**   130
   3.1 Theoretical difficulties 130
   3.2 Practical difficulties 132

4 **Second dogma**   135

5 **Naturalness revisited**   138

Abstract

In this article we analyze the notion of natural axiom in set theory. To this aim we review the intrinsic-extrinsic dichotomy, finding both theoretical and practical difficulties in its use. We will describe and discuss a theoretical framework, that we will call conceptual realism, where the standard justification strategy is usually placed. In outlining our view, we suggest that the extensive use of naturalness calls for a revision of the standard strategy, in favor of a justification that takes into account also the historical process that lead to the formalization of set theory. Specifically we will argue that an axiom can be considered natural when it helps the clarification of the notion of arbitrary set.

Introduction

The foundations of set theory is one of the most exciting areas of research in the field of formal sciences that combines both challenging mathematical problems and a deep philosophical reflection. Since the discovery of the phenomenon of incompleteness by

Gödel – and even more after the invention of forcing by Cohen – it became clear that if set theory was to be considered the right foundations for mathematics the widespread presence of independence results had to be contained. As a consequence, in his famous article on the Continuum Problem ([Gödel(1983)]) Gödel suggested what is now called *Gödel's program* for new axioms in set theory, that was later refined by Woodin in his programmatic paper on the Continuum Hypothesis ([Woodin(2001)]). The background motivation of both programs consists in interpreting the limits of ZFC not as intrinsic to set theory, but only as a defect of the formal presentation that Zermelo and Fraenkel – among others – gave to Cantor's theory of sets. Hence Gödel's program proposes to supplement ZFC with new axioms able to give a definitive solutions to independent problems and, so, to restore the foundational role of set theory.

> [T]hese axioms [i.e. large cardinal axioms] show clearly, not only that the axiomatic system of set theory as known today is incomplete, but also that it can be supplemented without arbitrariness by new axioms which are only the natural continuation of those set up so far[1].

Philosophically, this move had the consequence of opening a major discussion on the foundations of set theory and the justification of new axioms extending ZFC.

In what follows we will try to understand what 'without arbitrariness' means and how the notion of naturalness acts in the clarification of this expression. It is indeed a matter of fact that the reference to natural components of mathematics, in general, and more specifically the attribution of naturalness to good axiom candidates has become fairly common. Our aim consists in unveiling the philosophical background of mathematical reasons. Indeed we believe that, due to the technical character of this subject, philosophical ideas are often obscured by mathematical results. In doing so we will critically discuss two major difficulties of the justification of new axioms: first the belief that the use of the intrinsic-extrinsic dichotomy helps in a philosophical elucidation of the problem, secondly the assumption that it is with respect to a stable and well defined concept of set that we should develop Gödel's, or Woodin's, program: a step by step solution of all set-theoretical problems.

The article is structured as follows. In the section 1 we clear the philosophical background where to pursuit a meaningful discussion of the justification of new axioms. Under the name conceptual realism we will then identify the theoretical assumptions implicit in what we consider to be the standard[2] strategy of justification of new set-theoretical principles. In the attempt of clarifying the notion of natural axiom, in section 2, we review the distinction between intrinsic and extrinsic reasons, and in section 3 we describe both theoretical and practical difficulties in the use of this dichotomy. Then, in section 4 we discuss the problem, connected with conceptual

---

[1] [Gödel(1983)], p. 182, in [Gödel(1990)].

[2] The term 'standard' refers here to what is common among set-theorists working in the context of Woodin's program. In the next sections we will clarify more precisely the theoretical meaning of standard.

realism, of a global perspective in the justification of axioms meant to pursue Gödel's program. Finally, in section 5 we will suggest a different strategy of justification: one that takes into account the theoretical and philosophical reasons that motivate the axiomatization of a theory. We will then explain in which sense we believe that an axiom should be considered natural and we will find in the notion of arbitrary set the idea with respect to which is possible to argue in favor of the naturalness of new axioms of set theory. In discussing the reason for the attribution of naturalness we will describe a new theoretical framework that will try to overcome both the limits of the intrinsic-extrinsic dichotomy and the difficulties that conceptual realism encounters in the justification of new local axioms of set theory.

1 On the justification of axioms

One of the major imports of modern axiomatics is a radical change in perspective about truth and meaning in mathematics. We assisted at a progressive detachment from the ancient practice of considering the basic principles of a theory as sentences expressing self-evident propositions, towards a more abstract conception that views axioms as legitimate components of mathematical enquire.

The old notion rested on a conception of truth by reference that considered the direct link with the subject matter of a theory as a secure ground for both validity and justification. This point of view is exemplified by the classical idea that the truth of the axioms rests ultimately on their ability to capture essential properties of the objects described by a theory. Against this attitude we find a modern perspective that ceases to consider the truth as an evident property of axiom, but instead as a notion that depends on a complex mixture of internal or external mathematical reasons. As an extreme example of this perspective we may find Hilbert's idea according to which truth is considered an internal property of an axiomatic system and the ultimate criterion for truth is thus consistency.

> As long as I have been thinking, writing and lecturing on these things, I have been saying the exact reverse: if the arbitrarily given axioms do not contradict each other with all their consequences, then they are true and the things defined by them exist. This is for me the criterion of truth and existence[3].

Although Hilbert thought to have eliminated the problem of justification by means of a philosophy of mathematics based on the extensive use of implicit definitions, nevertheless the widespread presence of independence in set theory urged a choice

---
[3]Letter from Hilbert to Frege December 29th, 1899; in [Frege(1980)], pp. 39-40.

between incompatible axiom candidates and consequently the justification of our preferences. However, how to conciliate a modern perspective on axioms and the need to extend ZFC without arbitrariness?

Of course truth remains the main concern of mathematical research, but it is clear that it is not anymore an intuitive or evident property. Moreover, the idea that new advances in set theory comes from the mutual interaction between conceptual analysis and the confrontation of our expectations with the outcomes of our discoveries resembles closely a scientific method applied to pure mathematics. Therefore, in order to avoid the dependency of the justification of new axioms on an alleged theory of truth, we might assume a minimal conception of truth, viewing it as a limit notion that, although playing the role of a regulative idea, cannot be considered attained at a given moment of our scientific progress. In other terms, we maintain a sharp separation between the positive outcome of a justification and the truth of an axiom.

Consequently, in dealing with the problem of justification we do not intend to deal directly with the matter of truth in mathematics. Indeed we believe that justification, alone, cannot give the certainty able to ground the truth of an axiom. As part of a dynamic process, the reasons proposed should only be seen as suggestions that point in the direction where to seek truth. Indeed, the main fact that the justification process happens to be revised called for a qualification of the role of truth in this process.

Therefore, it is important to distinguish between the fact that an axiom is true and the reasons – connected with issues of meaning – in virtue of which it is true. Although, from a contemporary perspective, the truth of an axioms is deeply intertwined with mathematical reasons, more philosophical reasons are needed for its justification, precisely because the notion of meaning is involved.

Since the center of our interest is the justification of new axioms, we set the stage of our discussion choosing a non formalist philosophical perspective. As a matter of fact either our problem is philosophically trivial – in a formalist setting it is indeed solved vacuously – or we should then be allowed to assume the presence of a correspondence between syntax and semantics, able to inform the criteria of justification. There are many reasons for such a choice: in first place the difficulties of reductionist formalist positions – like eliminative structuralism[4] – in dealing with foundational theories like set theory, but even more importantly because we believe that on this particular matter a general formalist position should be more substantial than the simple rejection of the problem of justification.

Thus, it is within the context of a discussion on meaning and reference that we can tackle the problem of the justification of new axioms of set theory. But once a correspondentist theory has been deprived of its content of truth, what is left is the link between syntax and semantics given by the meaning of mathematical propositions.

---

[4]We intend here the position that does not admit the existence of neither mathematical object, nor structures and that proposes to reduce the latter to some more fundamental notion like for example set theory ([Linnebo(2008)]) or modality ([Hellman(1989)]).

Indeed, at a very general level, we may have justified reasons to accept an axiom when we are in the position to argue that it accords with the basic ideas or principles formalized by the theory to which the axiom belongs. The difficult philosophical task here is thus to determine the theoretical status of these ideas or principles. A possible easy solution consists in substituting objects with objective concepts, however, as we will discuss later, this move is problematic in many respects.

Let us now come back to set theory. The problem of justification of new axioms has been shaped by Gödel's program: a step by step supplementation of the axioms of ZFC aimed at finding a solution to concrete mathematical questions like the Continuum Problem i.e., the problem of determining the cardinality of the set of real numbers. In more recent years this program has been taken up by Woodin who proposed a step by step program for the completeness of the first initial segments of the universe of all sets.

Without committing ourselves to a general philosophical description of Gödel's or of Woodin's position, we would like here to give a rough description of the conceptual framework that underlines both programs. It is a common understanding of the problem that meaning points at a – not universally understood – concept of set, able to legitimate directly the axioms of set theory or to play the role of an abstract idea with respect to a more concrete set-theoretical reality. This general picture is supplemented by two general features that shape the related justification strategy.

In first place we find the belief in the existence of a clear and stable notion of set (e. g. the iterative conception of set: the notion of "set of") able to justify new set-theoretical principles. In other words, the belief that it is possible to give meaning to new axioms in terms of fundamental properties of *the* concept of set. Second, the conviction that the notion of set is sufficiently well shaped that the solution of a mathematical problem depends on the recognition of a particular property of this concept. For example, we may believe that it is the notion of "set of" that needs to be analyzed – directly, in a more philosophical way or indirectly in a more mathematical way – in search for a solution to the Continuum Problem.

A consequence of these two ideas is a weak form of realism, since, in order to justify a statement, it is assumed that there is something in virtue of what the justification process can be satisfactory conducted. Since it suggests a correspondence between axioms and concepts, we may call this attitude conceptual realism[5]. Furthermore, a consequence of this attitude is a global point of view on set theory that links the solution of specific problems to the general concept of set (whatever this may be): only its clarification – via a conceptual analysis or via the understanding of mathematical results – is able to solve a set theoretical problem. Even if the problem is local as the Continuum Problem. Thus the uniformity of the universe of set witnessed by the reflection principles is extended to the conceptual level: it is the general concept of set

---

[5]This notion is weaker than Gödel's conceptual realism as described in [Martin(2005)], but as we will see in the discussion of the analytic and synthetic distinction it is connected.

that determines the solution of (even local) set theoretical questions.

It is with respect to these two ideas that we will try to understand what the use of naturalness suggests in the process of justification. Before, let us offer another interesting quote, again by Gödel, that connects the idea of extending ZFC without arbitrariness and the notion of natural axiom. We stress that although this notion has been used in the literature after Gödel, for example in [Bagaria(2004)] or in [Friedman(2006)], however it has not been subject of a sufficient theoretical clarification.

> The proposition [V = L]... added as a new axiom, seems to give a *natural* completion of the axioms of set theory, in so far as it determines the vague notion of an arbitrary infinite set in a definite way.[6]

As a first attempt in the clarification of the notion of naturalness we propose to understand its place within the classification offered by the intrinsic-extrinsic dichotomy. Indeed, the latter is considered having both a mutually exclusive and a jointly exhaustive character. Although many different criteria for accepting new axioms have been proposed since (at least) fifty years, we choose to discuss intrinsic and extrinsic reasons because these criteria are not specific – as for example maximality[7] or fairness[8] or stability[9] – but are *forms* of justification. For this reason the considerations one could make about them should hold in general without been affected by the specific context of their application. Our aim is to unveil the philosophical ideas behind the standard strategy of justification of new axioms sketched above and not to move a criticism towards specific instances of it. Indeed, in the literature it is possible to find good arguments – often mathematically well informed – that sometimes lack an appropriate elucidation of the philosophical ideas that motivate them. When dealing with the reasons in favor of new axioms it is important to recall that philosophy in not only welcome, but necessary due to the main status of *axioms*, whose validity cannot be ascertained with rigorous mathematical reasons based on previously accepted principles, if we want them to be *new*.

2 Is naturalness intrinsic or extrinsic?

First of all it is useful to recall the distinction between intrinsic and extrinsic justifications. This dichotomy has been proposed and discussed in Gödel's work and in Maddy's papers. Both authors share a realist position[10] that consists in a clear and

---

[6][Gödel(1938)], p. 557.
[7]See, among others, [Maddy(1997)].
[8]See [Bagaria(2004)] on this respect.
[9]See [Friedman(2006)] on this respect.
[10]Or at least they shared. In the latest years Maddy's position evolved toward an attempt to overcome the antithesis between realism and anti-realism. See [Maddy(2011)].

proud platonism in the case of Gödel, while a more many-sides position in the case of Maddy.[11]

- *intrinsic reasons*: the justification of an axiom originates from the concept of set itself. Axioms are deduced - in a Kantian sense - by *the* concept of set, that is supposed to be knowable and describable in an axiomatic setting. The rational arguments used in this act of justification borrow their legitimacy from the subject matter of the theory. The objectivity of such an argument rests on the stability of a well definite concept of set.

- *extrinsic reasons*: the burden of the justification of an axiom rests on the success of its use as a set-theoretical principle. An axiom is extrinsically justified if its validity is confirmed by many mathematical facts and is able to give new interesting mathematical results. The prediction-confirmation model, typical of the empirical science, finds its place in a purely mathematical context. A more inductive reasoning is used in this act of justification. The objectivity of an argument rests on the possibility to give an explanation of a mathematical phenomenon by means of a mathematical law.

An aspect that is implicitly sustained by both intrinsic and extrinsic reasons is the descriptive character of a justification strategy. Indeed, both intrinsic and extrinsic justifications apply when the principles we intend to justify are able to describe appropriately aspects of the concept of set or, respectively, of set-theoretical reality. For this reason this dichotomy perfectly fits with the context of conceptual realism we described before: the realist component is expressed here by the referential component of the descriptive character, while either the presence of a nature of the concept, or the fact that mathematical reality is describable by its consequences, presupposes a stable conceptual level.

More concretely, intrinsic reasons presuppose the stability of a concept whose objectivity is autonomous from our formal presentation, but that is able to inform, alone, the criteria for accepting new axioms. Indeed, intrinsic reasons are legitimated both by the independent existence of a concept and by the possibility to describe its essential features by means of formal tools.

Moreover, and more interestingly, we believe that also extrinsic reasons presuppose a form of realism: a realism that assimilates mathematics to natural sciences. As a matter of fact the form of arguments that are named extrinsic resemble closely inferences to the best explanation in a purely mathematical setting. The fact that set-theoretic principles have so many desired consequences that force us to accept them,

---

[11] Indeed, she started her reflections maintaining a very strong form of existence with respect to sets (even in space and time, as argued in [Maddy(1990)]), for later developing an anti-(first)philosophical naturalism aimed at justifying Woodin's realist view (see [Maddy(1997)] in this respect).

due to their success, can be seen as confirmations of hypotheses by means of experiments. Axioms are confirmed by theorems and not the opposite. Inductive more than deductive arguments are used in the act of an extrinsic justification. Axioms are meant to describe such a stable mathematical reality and their truth rests upon the possibility to mirror independently valid relations between concepts. This form of justification echoes Russell's description of the foundational studies as described in the 1907 lecture *The Regressive Method of Discovering the Premises of Mathematics*.

> But when we push analysis farther, and get to more ultimate premises, the obviousness becomes less, and the analogy with the procedure of other sciences becomes more visible. The various sciences are distinguished in their subject matter, but as regards method, they seem to differ only in the proportions between the three parts of which every science consists, namely (1) the registration of 'facts', which are what I have called empirical premises; (2) the inductive discovery of hypotheses, or logical premises, to fit the facts; (3) the deduction of new propositions from facts and hypotheses[12].

This form of empiricism in the context of mathematics is strongly linked with the conceptual realism we described before. An analogy is indeed assumed to hold between mathematical reality and physical reality. Axioms are then intended to describe mathematical phenomena, as laws do in the case of nature, and in both cases mathematical, respectively, physical concepts are intended to act as a bridge between "reality" and its formalization.

Now, where to place the naturalness with respect to the division between concepts and reality suggested by the intrinsic-extrinsic dichotomy? Our suggestive answer is that the naturalness of an axiom is sustained both by intrinsic and extrinsic reasons; and this is a hint both of the weakness of this dichotomy and of the peculiarity of the notion of naturalness. Indeed the explicit reference to nature in the attribution of a natural character to a piece of mathematics may here refer to the nature of the concept of set, in the case of intrinsic reasons, or, in the case of extrinsic reasons, to a sufficiently stable set-theoretical reality that, like nature, we may be able to describe in an inductive way with a kind of reasoning that mimics an inference to the best explanation.

The attribution of naturalness to an axiom candidate seems to stress the acceptance of a realist perspective and the presence of a correspondence with a semantical level, whose essential aspects are described axiomatically. However, our discussion of naturalness will intend to convey that this descriptive aspect is only apparent. Indeed, we will argue that the attribution of naturalness is meant to be a prescriptive move of the mathematical work. As a matter of fact, an attentive analysis of the use of the

---

[12][Russell(1907)], p. 282.

term 'natural' in mathematical practice suggests that, even in keeping with a realist perspective, the justification of an axiom needs a dynamic framework, that the static conceptual realism cannot accommodate.

Before offering our view on naturalness, it is instructive to discuss some difficulties of the standard strategy of justification. Now 'standard' may be understood more theoretically, and precisely, as a justification that relays on conceptual realism and that makes use of the intrinsic-extrinsic dichotomy to classify the arguments in favor of new axioms candidates.

The limits of the standard strategy we discuss will help us showing the need of a new, less theoretically loaded, framework able to take into account more practical considerations, coming from the historical development of set theory. Indeed, in agreement with the modern perspective sketched in Section 1, we consider the process of justification of new axioms as an integral part of mathematical research.

However, it is important to clarify since now that we do not propose naturalness as a new criterion for the justification of new axioms, but as a linguistic indicator that points towards the revision of the standard strategy. As a matter of fact, the possibility to consider the naturalness of an axiom as an intrinsic and as an extrinsic justification is not meant to show that naturalness is a new form of justification, but that the intrinsic-extrinsic dichotomy fails in giving a ready-to-hand classification of the justification criteria. The reasons being, as we will see in the next section, that, on the one hand, the aforementioned dichotomy fails to be a dichotomy and, on the other hand, that it presupposes the outcome of the justification process it means to classify.

In showing the limits of the standard justification we will also discuss critically the basic assumption on which conceptual realism is based: the presence of a well defined concept of set, able to inform directly the criteria of justification. This latter criticism will not only affect a justification strategy based on intrinsic reasons, but the whole framework of conceptual realism. This is because the best understanding of extrinsic criteria we can offer relay on an analogy with natural science that sees in the regularity of mathematical phenomena the clue of the homogeneity of the underlying mathematical reality. Therefore the more or less emphasis of the conceptual component of set theory is only related to the more intensional or extensional perspective on the nature of set theory. In other words, although intrinsic reasons rely directly on the concept of set, we believe that also extrinsic reason deals, now indirectly, with the concept of set, by means of its extension. Indeed the stable properties observed in the everyday mathematical work are considered indicating the presence of a coherent and well-determined organization of mathematical reality, that therefore can be described abstractly by means of an objective concept of set, exactly as a physical model is supposed to describe reality.

We can now proceed in presenting two main difficulties of the standard strategy that we will name dogmas, since they can be seen to found the standard strategy, while being, in our opinion, unfounded.

3 First dogma

As anticipated we now proceed in showing the main difficulties of the standard strategy of justification. We start discussing the limits of the intrinsic-extrinsic dichotomy, epitomizing them in the following principle.

Fact 3.1. First dogma: *not only it is possible to give a clear distinction between intrinsic and extrinsic reasons and to apply them meaningfully in every situation, but moreover this dichotomy adds philosophical clarification to the process of justification of new axioms.*

We believe that a justification strategy solely relaying on intrinsic or extrinsic reasons, although widely used, is problematic in many respects, both from a theoretical and a practical point of view.

3.1 Theoretical difficulties

Let us start by trying to elucidate the following question: when can we say that an axiom is intrinsically justified? Since the source of such arguments is to be found in the nature of the concept of set, or at least in a uniform behavior of mathematical phenomena, we need to argue in favor of a link between a syntactic entity (an axiom) and a conceptual entity (the concept of set). Such a relation is in principle very difficult to ascertain, due to the general problem of finding a safe and faithful bridge between the formal and the informal sides of mathematics. Moreover, in the particular case of set theory, this task is even more difficult. Indeed the very nature of the concept of set is open to a never-ending sequence of specifications that can be hardly captured even at the level of second order logic. These problems apart, how it is possible to match axioms and concepts?

Vague as it may be, we may assume that axioms have conceptual contents, or in other words that they express propositions able to faithful describe some relevant aspects of the concept of set. But on which basis can we say that an axiom captures aspects of the notion of set? Without appealing to the opaque notion of Gödel's intuition – and for which Frege's old warning is apt: "We are all too ready to invoke inner intuition, whenever we cannot produce any other ground of knowledge"[13] – an interesting possible answer can be found in Boolos' famous article on the iterative concept of set ([Boolos(1971)]).

> It seems probable, nevertheless, that whatever justification for accepting the axiom of extensionality there may be, it is more likely to resemble the

---
[13] From §19 of the 1884 edition of [Frege(1950)].

3.1 Theoretical difficulties

justification for accepting most of the classical examples of *analytic* sentences, such as "all bachelors are unmarried" or "siblings have siblings" than is the justification for accepting the other axioms of set theory.[14]

The suggestion seems enlightening: when dealing with the intrinsic-extrinsic distinction are not we just proposing again the distinction between analytic and synthetic judgments?

We believe so, since we use similar conceptual tools and argumentative strategies for identifying a reason as intrinsic and a judgment as analytic – respectively, a reason as extrinsic and a judgment as synthetic. Interestingly, this idea has deep roots that can be traced back to Gödel's interpretation of analytic and synthetic, as described in the Gibbs Lecture.

> I wish to repeat that "analytic" here does not mean "true owing to our definitions", but rather "true owing to the nature of the concepts occurring [therein]", in contradistinction to "true owing to the properties and the behavior of things".[15]

We see here clearly outlined a notion of analytic judgment that refers to the nature of a concept, as intrinsic reasons do, opposed to a notion of synthetic judgment that refers to properties of things, that perfectly corresponds to the idea of justifying an axiom by means of the properties of the concept of set that the latter allows to prove.

At closer inspection the similitude is even more striking. For example, if we are interested in giving reasons for identifying a justification as intrinsic we may appeal to a form of direct link (not necessarily intuition) between an objective mathematical reality and our ability of formalizing it. As a consequence the distinctiveness of the essential features of a mathematical concept (or the reference to independently existent mathematical objects) makes analytic - in the concept of set – the axioms able to capture some fundamental aspects of the concept. Moreover, intrinsic justifications are supported by immediate reasons that recall the criteria for the truth of an analytic statement.

On the contrary, if we maintain that our relation with the concept of set is always mediated by formalization and thus that only formalized set theory is able to shape the concept of set, then any attempt to give intrinsic reasons for believing in an axiom runs into the problems of a circular argument typical of the justification of an analytical statement. Indeed, if axioms are essential for our understanding of the concept of set, then their justification rests ultimately on the axioms themselves.

The analogy between extrinsic reasons and synthetic judgments is even more compelling, since in both cases their justification relays on an external reality – conceptual

---
[14] [Boolos(1971)], p. 229.
[15] [Gödel(1990)], p. 321. The insertions in square brackets are by the editors of Gödel's Collected Works.

or concrete – able to express how the meaning of a formal expression (an axiom) relates to an informal domain (the concept of set).

The parallel we propose is meant to show that the use of intrinsic or extrinsic justifications presupposes the knowledge of the meaning of an axiom, or the way this meaning relates to mathematical reality. But this is problematic, since either before the use of these forms of justification we propose a full description of the concept of set and, respectively, an explanation of how meaning is capable of connecting axioms and concept, or any attempt to use intrinsic or extrinsic reasons loses much of its appeal[16]. However, even assuming to have attained such knowledge, once we have a full description of the notion of set, or of how the meaning of set-theoretical expressions relates to it, we doubt that the process of justification is still non-trivial and of some utility.

The connection between intrinsic and extrinsic justifications and analytic and synthetic judgments is not meant to totally disqualify such dichotomies, but only to stress that these notions should be handled with care. In trying to understand the argumentative strategies that operate in the justification of new axioms in set theory one rapidly gets to some of the most important and difficult problem of the philosophy of mathematics. Indeed, not only Gödel's intuition is a tool in need of philosophical clarification, but also the use of the intrinsic-extrinsic distinction should take into account the sharp criticism that, among others, Quine moved towards the analytic-synthetic dichotomy. As a consequence a careless use of the notions of intrinsic and extrinsic justification in a mathematical context poses more philosophical problems than the ones it helps to solve.

3.2 Practical difficulties

What we argued in the last section remains at a theoretic level. However we believe that also in practice it is not always clear how to apply intrinsic or extrinsic judgments. Let us start with the axioms of ZFC. These axioms are normally given for granted, once the main focus are *new* axioms. This is of course reasonable and it matches perfectly with Hilbert's description of the development of mathematics: the edifice of this science is built without firstly securing its foundations, but one gets back to them only when problems occur. However, as it is possible to see from next quote, the axioms of ZFC are not always considered as intrinsically justified; quite the opposite.

> I will start with the well-known axioms of Zermelo-Fraenkel set theory, not so much because I [...] have anything particularly new to say about them, but more because I want to counteract the impression that these

---

[16]We agree we Gödel that this is one of the most difficult task of mathematical logic: "The difficulty is only that we don't perceive the concepts of "concept" and "class" with sufficient distinctiveness, as is shown by the paradoxes." ([Gödel(1964)], p. 140).

3.2 Practical difficulties

axioms enjoy a preferred epistemological status not shared by new axiom candidates.[17]

Of course Maddy's naturalism is orthogonal to intrinsic reasons, but next quote, by Boolos, is taken from the same paper where the axiom of extensionality is considered analytic in the concept of set.

> Although they are non derived from the iterative concept, the reason for adopting the axioms of replacement is quite simple: they have many desirable consequences and (apparently) no undesirable ones.[18]

Therefore, if it is not a trivial matter to identify the intrinsic axioms even in the case of ZFC, then the property of being intrinsic seems to be a limit notion, more then a concrete property that we can attribute to new set theoretical axioms. As a matter of fact, the main existence of extrinsic reasons is by itself a sign of the difficulty we encounter in assigning intrinsic character to an axiom.

Moreover, the lack of consensus in the application of intrinsic reasons is not balanced by a clear strategy in the application of extrinsic ones. The following example is meant to show how uncertain can be to discern between two apparently equally extrinsically justified set theoretical principles. We will discuss the case of the the Axiom of Choice, that is often considered as the most extrinsically justified axiom among ZFC, and the Axiom of Determinacy, whose fruitful applications represented the success of the first step of Woodin's program.

Definition 3.2. (The Axiom of Determinacy (AD)) *Let $A \subseteq \omega^\omega$ (i.e. the set of sequences of natural numbers of length ω) and let $\mathcal{G}_A$ the game where players I and II choose, in turn, natural numbers*

$$\begin{array}{ccccc} I & x(0) & & x(2) & & x(4) & \ldots \\ II & & x(1) & & x(3) & & x(5) & \ldots \end{array}$$

and that ends after ω-many steps with the following winning conditions: player I wins when $x = \langle x(i) : i \in \omega \rangle \in A$ otherwise player II wins. Then AD is the following statement: for every $A \subseteq \omega^\omega$, the game $\mathcal{G}_A$ is determined; i.e., there is always a winning strategy[19] either for player I, or for player II.

This axiom that *prima facie* looks very distant from set-theoretic practice has tremendous consequences on many fundamental problems of modern set theory. After

---

[17] [Maddy(1988)], p. 482.
[18] [Boolos(1971)], p. 229.
[19] A winning strategy is a *function* $\sigma : \mathbb{N}^{<\omega} \to \mathbb{N}$ that tells a player what to play at his or her n-th move, considering what has been played before, and following which that player necessarily wins.

the bulk of interesting results that Woodin and others showed to hold in connection with this axiom, AD became a paradigmatic example of an axiom that rests on extrinsic reasons for its acceptance. Nevertheless AD is not compatible with all ZFC axioms. In particular AD implies the negation of the Axiom of Choice (AC), since it implies that all subsets of real numbers are Lebesgue measurable, while by means of AC it is possible to build a non-measurable subset of $\mathbb{R}$. Then how to decide between two apparently extrinsically justified and incompatible axioms?

If we look at the subsequent autonomous development of set-theoretical practice we can see that the possibility to have choice functions has been considered unavoidable, and so AD has been considered in need of a reformulation. As a matter of fact set theorists shifted their focus to restricted versions of this axiom; in particular to $AD^{L(\mathbb{R})}$ that is the Axiom of Determinacy restricted to the inner model constructible from all ordinals and reals. In the presence of large cardinals this structure is a model of ZF axioms, together with the Axiom of Dependent Choice: a principle of choice weaker then AC. Although AD has a lower consistency strength[20] then $AD^{L(\mathbb{R})}$[21] the latter has been preferred. This retreat from AD to $AD^{L(\mathbb{R})}$ has not been motivated by clear intrinsic or extrinsic reasons, but by a mixture of considerations of different forms motivated by the goal of accommodating a theory where the most of determinacy could hold together with the most of choice. In other terms, although AC is normally considered as extrinsic in the context of ZFC, when confronted to fruitful axioms that extend ZFC its intrinsic combinatorial value is put forward as a reason for not abandoning the freedom given by choice.

As for the theoretical difficulties we discussed before, here too we face, in practice, the problem of presupposing the outcome of the justification in the application of its criteria; in other words we assume the relevance of some elements from the outset, prior and independently of the criteria of justification we intend to use. Moreover, the case of AD and AC shows clearly that the historical development of set theory may influence the application of the criteria of justification, and consequently that these criteria are not fixed once for all but may vary in accordance to specific cases.

These considerations open the problem of the role of non-mathematical elements – may them coming from history or philosophy – in the process of selection of new axioms. Furthermore we think that a moral we can draw from the interplay between AC and AD is that the intrinsic-extrinsic dichotomy does not pertains to the level of forms of justifications, since these reasons are sensible to the context of their application and so their validity needs to be ascertained case by case. In other words, the discussion of AC and AD shows that the notion of extrinsic reason does not solve,

---

[20] AD is equiconsistent with the existence of infinitely many Woodin cardinals.

[21] Infinitely many Woodin cardinal and a measurable cardinal on top of them are needed for the proof of AD in $L(\mathbb{R})$. In contrast to the case of AD we have here an implication and not a relative consistency dependence. However, to our knowledge, this epistemological difference has never been offered as an argument for the primacy of one of the two principles.

alone, the problem of justification, since we then need other reasons to accept such a justification. But what kind of meta-justifications we are thus looking for?

The difficulties we found in the use of intrinsic and extrinsic reasons are by no means to be understood neither as the belief that such justifications are useless, nor as the denial that there are cases where this dichotomy can be meaningfully applied. What we discussed here is the wide range of nuances where a mixture of different arguments are needed in order to tip the balance towards one of the two sides. However, we believe that when the justification of new axioms of set theory is subject of a philosophical debate, then these categories loose much of their appeal and it may seem that supporters or opponents of new set-theoretical principles are not gaining much by their use.

4 Second dogma

Before presenting our view on naturalness, it is useful to discuss another aspect of the framework of justification, now in connection with the scope of new axioms. As we hinted in Section 1, besides conceptual realism the standard strategy of justification holds that the general concept of set is the reference of the meaning of new axioms. An example of this attitude can be found in the discussion of the justification of PD (i.e. the Axiom of Determinacy restricted to projective sets).

> Because of their richness and coherence of its consequences, one would like to derive PD itself from more fundamental principles concerning sets in general, principle whose justification is more direct.
>
> We know of one proper extension of ZFC which is as well justified as ZFC itself, namely ZFC + "ZFC is consistent". Extrapolating wildly, we are led to *strong reflection principles*, also known as *large cardinal hypotheses* [...] Reflection principles have some motivation analogous to that of the axioms of ZFC themselves, and indeed the axioms of infinity and replacement of ZFC are equivalent to a reflection schema[22].

This idea then pushes towards the search of global axioms and sustains the implicit idea that new advances in set theory can be achieved only by a further clarification of the general concept. Since we consider this attitude problematic, we will call it, again, a dogma.

Fact 4.1. *Second dogma: there is a direct link between the general concept of set and the solution of specific set-theoretical problems; hence it is only with respect to the general concept of set that we may justify a new axiom in set theory.*

---

[22] In [Martin and Steel(1989)], p. 72.

This dogma has the effect of making problematic the justification of local axioms, because we believe it is not the general concept but a specific instance of it that needs to be the ground of their justification.

Let us argue this point more in details. First of all it is useful to make explicit that we concentrate our analysis on the iterative conception of set, that is normally considered *the* conception of set that is able to determine and justify the axioms of set theory. The reason being that, typically, an argument in favor of an axiom that extends ZFC gives for granted, besides consistency, the existence of characteristic properties of the intended interpretation of ZFC, normally referred to as the cumulative hierarchy V. For example its so-called indescribability, that is the impossibility to give explicit properties, besides the ZFC axioms, that hold for the collection of all sets and not just for an initial segment of V. Indeed, the iterative conception is considered the conceptual underpinning of a cumulative hierarchy structure like V and it is been argued that this conception is able to deduce (in a Kantian sense) most of the ZFC axioms (see [Boolos(1971)]).

These connections between the conceptual and the formal levels can be made precise considering that on the one hand the Levy-Montague Reflection Principle (that is normally considered as expressing the indescribability of V) can be proved in ZFC to be equivalent to the conjunction of the axiom of infinity and the schema of replacement, and that on the other hand Zermelo's quasi-categoricity theorem for ZFC tells us both that the cumulative hierarchy is the right model of these axioms and that even second order logic cannot distinguish between inaccessible levels of the cumulative hierarchy.

It is also useful to specify what we mean by local and global. By global axiom we mean an axiom that deals with all sets in V or at least with an unbounded class of them, while with local axiom one that deals only with sets laying in a proper initial segment of the cumulative hierarchy. In other words global axioms deal with the height of the cumulative hierarchy, while local ones with its width. Therefore, our concerns about the use of the general concept of set for the justification of both local and global axioms can be restated asking: on which ground are we justified in using the same theoretical framework that gave rise to the cumulative hierarchy (i.e. the iterative conception) to argue in favor of axioms that do not influence the hight of the universe of sets, but only its width?

More precisely, if we accept the existence of a stable concept of set and we accept that ZFC axioms correctly (but partially, as shown by Zermelo's theorem) describe it, how can we justify, by means of the same notion, axioms that have consequences only on the lowest levels of the cumulative hierarchy? Although we assume for the sake of the argument that the iterative conception – able to justify ZFC[23] and to tell us

---

[23] As Boolos showed in [Boolos(1971)] this justification is far from being straightforward. Moreover we believe that the attempt to justify the axioms of ZFC in terms of the iterative conception does not sufficiently take into account the meaning of Zermelo's theorem. Indeed the idea of a cumulative hierarchy

what sets are – is so useful in justifying principles expressing the hight of the universe of sets like large cardinal axioms, how can we use arguments linked to the iterative conception for the justification of new local principles able to give a more detailed description of sets laying in an initial segment of the cumulative hierarchy, below the first inaccessible cardinal?

Pushing this point at a more conceptual level we find a related problem. As a matter of fact, even accepting that *the* concept of set can justify the ZFC axioms and the general properties of its intended model(s), on what theoretical ground can we argue that this same notion can determine the notion of "set existing in an initial segment of the cumulative hierarchy" in such a way that the arguments appealing to the former notion can be decisive in the argumentative process of the justification of axioms that deal with the latter notion?

Not only we believe that an argument in favor of a local axiom based on the general concept of set is, without further justification, a deceptive inference, but we also think that such an argument would face the following practical problem. If we call a local concept of set one that is intended to describe sets laying in an initial segment of the cumulative hierarchy, then a local concept should have different specifications than the global one. Indeed, exactly because of the indescribability of V, the main possibility to characterize an initial segment tells us that its axiomatization should be different from that of the universal class. As a consequence, if our aim is to give a (sufficiently) complete description of V by specifying step by step its initial segments, we will eventually face the possibility of outlining properties that are specific to particular sets and not to sets in general – at least the property of laying in an initial segment of V.

An interesting consequence of the rejection of the second dogma is the possibility of giving a substantive answer to what we may call the criticism of vagueness to set theory. By this we mean the following line of argument: the discovery of the widespread presence of independence phenomena in set theory (e.g. the independence of CH) tells us that the concept of set is a vague notion; hence there are good reasons for believing that questions like the Continuum Problem cannot be settled and, in particular, from the vagueness of the concept of set we may infer that CH has not a well-defined truth value. This argument is perfectly exemplified in [Feferman(1999)] and it has been criticized in [Martin(2001)] with structural considerations, similar to Zermelo's, addressed to the second half of the argument: from vagueness of the concept of set infer the lack of truth-values. On the contrary, if we realize the absence of cogency of the second dogma the argument *à la* Ferferman is blocked at its very

---

was proposed by Zermelo in his attempt give a consistency proof for ZF. Then we believe that any attempt to justify ZFC axioms in terms of the iterative conception seems distorted if one considers the origin of the latter from the idea of a cumulative hierarchy. However, it is not here the place where to discuss the appropriateness of the justification of the ZFC axioms in terms of the iterative conception. While agreeing on the deep connection between the ZFC axioms, the cumulative hierarchy and the iterative conception, we only want to point out the vacuity of the justification of axioms that can only fix one kind of structure in terms of a conceptual description of that kind of structure.

beginning. From the independence of CH it is not possible to infer the vagueness of the general notion of set. Although a parallel argument may be then proposed for the notion of countable set, we believe that the appeal to vagueness is, in this context, less persuasive, because on the one hand the ZFC axioms are not meant to formalize the notion of countable set, while on the other hand we have a better understanding of countable sets than we have of sets in general. Indeed, during the last fifty years it has been developed an intense and detailed study of the forcing method (a tool that applies to countable structures), that gave rise to the so-called Forcing Axioms: local axioms able to give a clear picture of the hereditarily countable sets and to decide, among many other things, the cardinality of the Continuum.

These are the reasons why we believe that it is not the general concept of set that can be the reference of the meaning of local axioms, and moreover that can be able to determine the width of the universe of sets and so to be the appropriate theoretical framework for the justification of principles that are meant to pursue Gödel's or Woodin's programs. It is then not surprising that are not global axioms like large cardinals, but local axioms like Forcing Axioms, that are able to give an answer to the problem of the cardinality of the Continuum.

5 Naturalness revisited

So much for the intrinsic-extrinsic dichotomy and the justifications based on the general concept of set. The criticism we moved to these kinds of justification showed their limits in a philosophical discussion on the acceptance of new axioms. However, we believe that these deficiencies are exemplar on the one hand to understand the use of the notion of naturalness in mathematics and, on the other hand, to suggest a different justification strategy.

In order to elucidate this point we start by widening our analysis and by making explicit our general view of the role and the weight that the notion of naturalness has assumed in contemporary mathematics. We start with the following table that displays the frequency of the use of the terms 'natural' and 'naturalness', between 1940 and 2009, in the texts of the *American Mathematical Society* database (MathSciNet).

Decade	*Total articles* (T)	*Occurrences* (N)	*Rate* $\left(\frac{N}{T}\right)$
1940 − 1949	40538	602	0.014
1950 − 1959	89158	1935	0.021
1960 − 1969	168567	4802	0.028
1970 − 1979	327427	11500	0.035
1980 − 1989	483143	21026	0.043
1990 − 1999	617522	34032	0.055
2000 − 2009	841470	47056	0.056

Further statistical evidence, in San [Mauro and Venturi(2015)], confirms that in the last sixty years the reference to natural components in mathematical works increased significantly, without marking an increment of the technical uses of the terms 'natural'– like in expressions as 'natural number', or 'natural transformation'. On the contrary the widespread presence of terms like 'naturally', or of expressions like 'it is natural to see that', clearly manifests a tendency of this notion towards informal components of mathematics.

It is also true that the expansion of the use of naturalness followed a development of mathematics toward abstraction and specialization. It is our understanding that, as in an attempt to bring to a more concrete and common ground the results of a field, the attribution of naturalness intends to stabilize aspects of mathematical work that fall short of an intuitive treatment. For this reason we believe that, contrary to the reference to nature that this notion explicitly brings with it, the attribution of naturalness manifests a prescriptive component that on the one hand aims at inverting the process of abstraction in favor of a more direct link with mathematical reality, while on the other hand relays on an habit of working with specific mathematical tools that can be acquired and intended only by people working in a specific field. Indeed, a fundamental character of naturalness in mathematics, as argued in San [Mauro and Venturi(2015)], is its dynamical character, that is, its variance in time: a clear indication of a substantial departure from the static meaning of natural as referent to nature.

In other words, although the reference to nature seems to relay on the acceptance of a realist framework, the use of naturalness judgments does not consist in the recognition of a descriptive character of a piece of mathematics, but instead in the prescriptive attribution of relevance with respect to a given theoretical context and against other pieces of mathematics, that may be thought to be equally relevant.

The role of context then plays a fundamental role in the attribution of naturalness. As a matter of fact calling something natural has the effect of specifying a point of view with respect to a subject matter, whenever the latter has degrees of freedom that allow for different clarifications. The prescriptive character of this act has the effect of making explicit which are the relevant components of a subject matter, and the act is performed precisely when a clear statement in this direction is needed; and this often happens when the abstractness of a field makes difficult the use of intuitive considerations. Moreover, the role of context acts also in determining the scope of naturalness. It is then not surprising that the increase of the reference of natural components in mathematics goes hand in hand with a specialization of the disciplines. Indeed, the smaller and more disconnected are the particular scientific communities of working mathematicians, the more natural a piece of mathematics will seem to a small group of researchers with a common background and working on similar problems.

Without entering here problems of sociology of mathematics, we only want to stress the role that the context plays in the recognition of naturalness. Hence, admitting a general framework in which recognition of natural components is meaningful (i.e. the ideas with respect to which a naturalness judgment may be attributed), the use of the notion of naturalness in mathematics brings with it prescriptive and historical (or contextual) components that, more than describing natural kinds, are meant to specify general ideas with respect to intentions and aims of mathematicians.

Now, coming back to the problem of naturalness of new axioms in set theory, the defects of the intrinsic-extrinsic dichotomy showed that an effective strategy of justification should not presuppose a completely specified concept of set or, respectively, a completely determined reality pointing to the same conceptual counterpart – if we want to save the value of the justification process – and, hence, that the reasons for accepting new axioms cannot be their descriptive character. On the contrary we have seen different pragmatic and historical reasons for justifying new fruitful principles, like $AD^{L(\mathbb{R})}$, that are meant to specify which are the relevant aspects of set-theoretical reality. The importance of contextual reasons is even more relevant in set theory, where the formal component is accompanied by the presence of an intended interpretation. In other words, in the axiomatization of set theory there are non-mathematical components that play a role in the attribution of meaning to formal sentences.

To summarize our view, we believe that a strategy of justification of new axioms should take into account three different aspects of formalization. At the lowest level we find the formal theory, where purely formal methods can be used to ascertain properties like consistency. At a different level we find the conceptual level, where the concept of set lays (i.e. the semantic counterpart of the formal theory). Contrary to conceptual realism, we believe that this level is not an independent realm where concepts are completely specified kinds that we may describe with axioms. Indeed we maintain a position that accepts the presence, also in mathematics, of open-concepts: entities with degrees of freedom open to further specifications[24]. At a third level we find general ideas that inform mathematical practice. Contrary to the conceptual level, it is here that we find the human component of mathematical work, able to connect syntax and semantics and to specify open concepts with respect to practice. Hence the naturalness of an axiom is to be found when there is accordance between the formal level and the ideal level. The ability of an axiom to capture general ideas that we find in our practice is then able to constitute the meaning of the axiom and thus to influence its semantical counterpart, specifying an open concept[25]. In other words, axioms are called natural when they are able to formalize our scientific practices and,

---

[24] This notion has been introduced by Waismann in a series of papers (see [Weismann(1951a)] and [Weismann(1951b)]) in the context of natural language. It has then been argued, in [Shapiro(2006)] that this notion is present and relevant also in a mathematics; the main example being the concept of function.

[25] Although we do not exclude the possibility of a concept revision, the history tells us that this happens only when contradictions have a logical character (i.e., contradiction as in the case of Russell's Paradox) and not mathematical. Consider the example of AC and AD and the fact that their incompatibility did not have the consequence of considering AC or AD false.

in turn, to modify the basic concept(s) of a theory. However the connections between the conceptual and the ideal level is active also in another direction. Indeed, the realist objectivization of a concept that acts in mathematical practice is able to modify the practice itself. A clear example is given by the effects that the iterative conception had on set-theoretical practice. The habit of thinking about sets as "sets laying in a cumulative hierarchy" has not only become a tacit thesis of set theory, but has suggested new axioms in terms of reflections principle able to describe the indescribability of the class of all sets. This dialectical movement between general ideas coming from mathematical practice, consistent axioms and open concepts lays at the heart of the development of set theory.

What we propose here is, thus, the analysis of the use of naturalness as an indicator of the need for a new justification strategy and the notion of open concepts as constituting a frame to overcome the static character of conceptual realism and the connected intrinsic-extrinsic dichotomy. The notion of naturalness indeed points at the presence of a moving target (as it is the case with the historical concept of set), without proposing stable criteria whose fixity may soon become obsolete, but suggesting a general accordance between theoretical aims and mathematical practice. Moreover, we believe that the notion of open concepts is also able to overcome the limits given by a justification strategy too often linked to a general concept of sets. Indeed the connection that the naturalness of an axiom manifests between formalization and mathematical practice does not necessarily rest on the recognition of fundamental properties of an alleged general concept of set, but may depends on a particular practice and on knowledge of a local notion of set (i.e. the conceptual counterpart of a collection of sets lying in an initial segment of the cumulative hierarchy).

Now, following our understanding of the use of the notion of naturalness in mathematics, the main question we should ask for the acceptance of set-theoretical axiom is the following: "with respect to which ideas, relevant for the historical development of set theory, for the aim of its formalization and its current practice, we may propose argument in favor of the naturalness of an axiom?"[26].

---

[26]We can find an antecedent of this perspective in [Hauser(2004)], although in connection with a radically different proposal: the use of a phenomenological standpoint for inquiring the human component of the reasons we adopt for choosing new axioms in set theory. We agree with Hauser that "we must abandon the one-sided view that the objective is something entirely alien to the subjective and that it ought to be studied with complete disregard of the mental life of the mathematician" ([Hauser(2004)], p. 112), but we think that the task of contemporary philosophy of mathematics is not to depict a new context where to argue in favor of the evidence of new axioms, but to make explicit the theoretical intentions of the proponents of new principles and discuss their accordance with the main theoretical ideas that motivate the formalization of set theory. Nonetheless, it is also interesting to note that the phenomenological proposal of [Hauser(2004)] is intended, similarly to ours, to elucidate the natural component of new axioms: "Rather one must examine how the mathematician is intentionally related to those facts because it is on these grounds that he accepts certain abstractions and idealizations as 'natural' or 'reasonable'." ([Hauser(2004)], p. 113). In some sense closely to our historical approach, in [Hauser(2013)] the author attempts a justification of strong axioms of infinity with respect to the metaphysical and theological ideas behind Cantor's theory of the Absolute.

Calling to the fore history, we need to make clear what we refer to. Since we do not intend here to sketch an historical picture of the development of set theory – and since [Ferreirós(1999)] is a very good reference on this subject – we defer to the following quote the task of giving a rough idea both of the presence of two main periods in the history of axiomatic set theory and of the individuation of the reasons that motivated the development of set theory as a foundational theory.

> Such a depiction [of set theory as a theory with no substantial antecedences] seems suitable for the matatheoretic period that set theory as a field lived from about 1950, [...] but *not* for the more properly theoretical and axiomatic period that antedated 1940. It was in this period, 1904 to 1940, that the core of understanding was gained of set theory, its axiomatic underpinning, the universe V [...] basically a matter of *understanding and clarifying the concepts of number and function.*[27]

The history of set theory after Cohen's results consists mainly in the development of an independent mathematical field, with its internal motivations. When viewing set theory by this perspective, the extension of ZFC is meant to pursuit the foundation of set theory considered, mostly, as the study of the different possible models of ZFC. Hence independent set-theoretical principles are either studied in order to understand their behavior in different models of ZFC, as happens for example in Hamkins's axiomatic treatment of the multiverse (see [Hamkins(2012)]), or with the aim of selecting the right models among the many possible universes of sets, as Arrigoni Friedman's Hyperuniverse's Program (see [Arrigoni and Friedman(2013)]) - among others proposals - aims to do.

Nevertheless, when we think of set theory as a foundation of mathematics, we should consider the theory that was studied before 1963, through the lens of the techniques and results developed after 1963. As argued in [Ferreirós(2011)], the role of set theory in clarifying the foundations of mathematics consisted in an attempt to develop a mathematical treatment of the most general notions of number and functions, in terms of the more primitive notion of *arbitrary set*[28]. Arbitrary sets are sets whose existence is independent from our possibility to define them. Therefore, an arbitrary set does not admit, by definition, a precise characterization or an explicit description, but its existence follows from some existential theorems like Cantor's on the uncountability of the real numbers. As a matter of fact, since our language is countable there will always be real numbers whose definition transcend the expressive power of our

---

[27][Ferreirós(2011)], p. 362.

[28]For reasons of space, we decide here not to discuss the difference between the idea of arbitrary set and the iterative conception of set. Although related, we believe the idea of arbitrary set to be wider then the iterative conception and more apt to play a regulative role with respect to the open concept of set. Nonetheless we acknowledge the importance of this theoretical clarification and we plan to elucidate this point in a future work.

language. The conception of mathematical objects that is connected with this notion is *quasi combinatorialism*, as described in [Bernays(1983)].

> But analysis is not content with this modest variety of platonism [to take the collection of all numbers as given]; it reflects it to a stronger degree with respect to the following notions: set of numbers, senquece of numbers, and function. It abstracts from the possibility of giving definitions of sets, sequences, and functions. These notion are used in a 'quasi-combinatorial' sense, by which I mean: in the sense of an analogy of the infinite to the finite.
>
> Consider, for example, the different functions which assign to each member of the finite series $1, 2, \ldots, n$ a number of the same series. There are n^n functions of this sort, and each of them is obtained by n independent determinations. Passing to the infinite case, we imagine functions engendered by an infinity of independent determinations which assign to each integer an integer, and we reason about the totality of these functions.
>
> In the same way, one views a set of integers as the result of infinitely many independent acts deciding for each number whether it should be included or excluded. We add to this the idea of the totality of these sets. Sequences of real numbers and sets of real numbers are envisaged in an analogous manner. From this point of view, constructive definitions of specific functions, sequences, and sets are only ways to pick out an object which exists independently of, and prior to, the construction.[29]

In [Ferreirós(2011)] there is an attempt to analyze which axioms of ZFC are able to formalize - and to what extent - the notion of arbitrary set, finding that most of them, with the exception of the Axiom of Choice, are very poor in capturing this notion. For the same reason we can easily discard $V = L$ as unnatural, since the restriction given by considering only constructible sets is exactly orthogonal to the notion of arbitrary sets. The fact that in the constructible universe AC holds, nonetheless, is not a hint of its ability to capture this specific notion, but exactly the opposite: the absence of arbitrariness that, as in the finite case, makes choice trivial.

We would like to conclude with a plan for future work. A class of axioms that may be analyzed in this setting is that of Forcing Axioms and, in fact, we plan to inquire their naturalness in a future work. The reason for choosing this type of axioms and not, for example, large cardinal axioms is firstly due to the absence in the literature of a sufficiently philosophical justification of these principles. Moreover, its specific character of local axioms represents a challenge for their justification that would not be perceivable in the case of large cardinals.

---

[29][Bernays(1983)], p. 264.

Concretely, we propose to reformulate the question about the naturalness of Forcing Axioms in the following way.

1. **Question 1:** Is the notion of arbitrary sets necessary for an intuitive motivation of Forcing Axioms?

2. **Question 2:** To what extent Forcing Axioms capture and sharpen this idea?

We believe that only giving a positive answers to the above questions we can argue in favor of the naturalness of these axioms. Indeed, following our idea that naturalness judgments hide a prescriptive component, we believe that first we should individuate the relevant aspects of set theory that we intend to formalize, and only subsequently we may argue in favor of the naturalness of an axiom, in terms of its pertinence with the goals of axiomatization – in this particular case the formalization of the notion of arbitrary set.

Acknowledgments

This research was financially supported by FAPESP grant number 2013/25095-4 and by the BEPE grant number 2014/25342-4. I would like to thank the organizers and the participants of the conference "7th French PhilMath", where parts of the content of this paper were presented, for the useful discussions.

References

[Arrigoni and Friedman(2013)] T. Arrigoni and S. Friedman. The hyper universe program. *Bulletin of Symbolic Logic*, 19(1):77–96, 2013.

[Bagaria(2004)] J. Bagaria. Natural axioms on set theory and the continuum problem. *CRM Preprint*, 591:19, 2004.

[Bernays(1983)] P. Bernays. On mathematical platonism. In P. Benacerraf and H. Putnam, editors, *Philosophy of mathematics: selected readings*, pages 258–271. Cambridge University Press, 1983.

[Boolos(1971)] G. Boolos. The iterative conception of set. *Journal of Philosophy*, 68 (8):215–231, 1971.

[Feferman(1999)] S. Feferman. Does mathematics need new axioms? *American Mathematical Monthly*, 106:106–111, 1999.

[Ferreirós(1999)] J. Ferreirós. *Labyrinth of thought. A history of set theory and its role in modern mathematics*. Birckhäuser, 1999.

REFERENCES

[Ferreirós(2011)] J. Ferreirós. On arbitrary sets and ZFC. *Bullettin of Symbolic logic*, 17(3):361–393, 2011.

[Frege(1950)] G. Frege. *The Foundations of Arithmetic: A Logico-Mathematical Enquiry into the Concept of Numbers*. Blackwell, 1950.

[Frege(1980)] G. Frege. *Philosophical and mathematical correspondence*. Basil Blackwell, 1980.

[Friedman(2006)] S. Friedman. Stable axioms of set theory. In J. Bagaria and S. Todorcevic, editors, *Set Theory: Centre de Recerca MatemÃ tica, Barcelona, 2003-2004*, pages 275–283. Birkhäuser Verlag, 2006.

[Gödel(1938)] K. Gödel. The consistency of the axiom of choice and of the generalized continuum hypothesis. *Proceedings of the National Academy of Sciences of the United States of America*, 24:556–557, 1938.

[Gödel(1964)] K. Gödel. Russell's mathematical logic. In S. Feferman, J. Dawson, S. Kleene, G. Moore, and J. Van Heijenoort, editors, *Collected works. Volume II 1938-1974*, pages 119–141. Cambridge University press, 1964.

[Gödel(1983)] K. Gödel. What is Cantor's continuum problem? In P. Benacerraf and H. Putnam, editors, *Philosophy of mathematics selected readings*, pages 470–485. Cambridge University press, 1983.

[Gödel(1990)] K. Gödel. *Collected works. Volume II 1938-1974*. Oxford University Press, 1990.

[Hamkins(2012)] J. Hamkins. The set-theoretic multiverse. *Review of Symbolic Logic*, 5:416–449, 2012.

[Hauser(2004)] K. Hauser. Was sind und was sollen (neue) axiome? In G. Link, editor, *One hundred years of Russell's paradox*, pages 93–118. Waler de Gruyter, 2004.

[Hauser(2013)] K. Hauser. Cantor's absolute in methaphysics and mathematics. *International Philosophical Quarterly*, 53(2):161–188, 2013.

[Hellman(1989)] G. Hellman. *Mathematics without numbers*. Oxford University Press, 1989.

[Linnebo(2008)] Ø Linnebo. Structuralism and the notion of dependence. *Philosophical Quarterly*, 58:59–79, 2008.

[Maddy(1988)] P. Maddy. Believing the axioms I. *Journal of Symbolic Logic*, 53(2): 481–511, 1988.

[Maddy(1990)] P. Maddy. *Realism in Mathematics*. Oxford University Press, 1990.

[Maddy(1997)] P. Maddy. *Naturalism in mathematics*. Clarendon Press, 1997.

[Maddy(2011)] P. Maddy. *Defending the axioms: on the philosophical foundations of set theory*. Oxford University Press, 2011.

[Martin(2001)] D. Martin. Multiple universes of sets and indeterminate truth values. *Topoi*, 20:5–16, 2001.

[Martin(2005)] D. Martin. Gödel's conceptual realism. *Bulletin of Symbolic Logic*, 11(2):207–224, 2005.

[Martin and Steel(1989)] D. Martin and J. Steel. A proof of Projective Determinacy. *Journal of the American Mathematical Society*, 2(1):71–125, 1989.

[Mauro and Venturi(2015)] L. San Mauro and G. Venturi. Naturalness in mathematics. In Lolli, Panza, and Venturi, editors, *Philosophy of mathematics; from logic to practice. Boston studies in the philosophy of science*. Springer, 2015.

[Russell(1907)] B. Russell. The regressive method of discovering the premises of mathematics. In D. Lackey, editor, *Essays in Analysis*, pages 272–283. George Allen & Unwin, 1907.

[Shapiro(2006)] S. Shapiro. Computability, proof, and open-texture. In A. Olszewski, J. Wolenski, and R. Janusz, editors, *Church's thesis after 70 years*, pages 420–455. Ontos Verlag, 2006.

[Weismann(1951a)] F. Weismann. Analytic-Syntethic III. *Analysis*, 11(3):49–61, 1951a.

[Weismann(1951b)] F. Weismann. Analytic-Syntethic IV. *Analysis*, 11(6):115–124, 1951b.

[Woodin(2001)] W. H. Woodin. The Continuum Hypothesis. Part I. *Notices of the AMS*, 48(6):567–576, 2001.

Logic and Epistemology

Cómo pensar sobre otra cosa

Axel Arturo Barceló Aspeitia
abarcelo@filosoficas.unam.mx

Índice

1. **Resolviendo la paradoja de las tres monedas** 151
2. **Extendiendo la propuesta: Las paradojas del prefacio y de la lotería** 157
3. **La Paradoja Sorites** 163
4. **Conclusiones** 166

Permítaseme empezar con una paradoja que, hasta lo que yo sé, no ha sido estudiada en la filosofía y que llamaré la paradoja de las tres monedas:

> Cuenta la historia que era muy difícil hacer que Proclo pagara sus deudas, pese a ser un sofista muy rico. En una ocasión, Glaucón le había prestado un dracma y no había logrado cobrar. Una tarde, Glaucón se enteró de que Proclo acababa de cobrar tres dracmas por uno de sus trabajos, así que lo abordó en la plaza para cobrar su dinero. Pero, una vez más, Proclo no le haría fácil su trabajo a Glaucón.
>
> – Mira, Glaucón – le respondió sacando de su bolsillo tres monedas – éstos son los tres dracmas que tengo. Me dices que te debo uno ¿podrías decirme cual?
>
> – No entiendo tu pregunta.
>
> – Quiero que me digas cual de estas monedas es la que te debo. ¿Cómo es posible que digas que te debo una moneda, pero no puedas decirme cuál?
>
> – Ninguna moneda en particular, Proclo, . . . cualquiera.
>
> – Entonces – respondió Proclo señalando dos de las monedas en su mano – puedo darte una de estas dos y – ahora señalando a la tercera – no tengo que darte ésta otra ¿verdad?
>
> – Tienes razón Proclo.

– Entonces no tengo que darte esta moneda.

– Así es.

– Y lo mismo sucede con esta otra moneda – continua Proclo señalando a una de las otras dos monedas –. Tampoco tengo que dártela ¿verdad? En tanto que puedo darte una de las otras dos.

– Creo que tengo que afirmar que tienes razón Proclo – respondió Glaucón.

– Y esta última moneda – arguyó Proclo, señalando la tercera moneda en su mano – no es diferente a las otras dos, ¿verdad? Tampoco tengo que dártela, porque puedo darte cualquiera de las otras.

– Así parece.

– Entonces, no tengo que darte esta moneda, pero – continuó, señalando una de las otras monedas en su mano – tampoco tengo que darte esta, ni – señalando la tercera y última moneda en su mano – esta otra tampoco. Por lo tanto, reconoces que no debo darte ninguna de estas monedas que traigo ¿verdad?

Glaucón se quedo callado, sin saber que responder, por lo que Proclo continuó hablando – Bueno, Glaucón, ya que no debo darte ninguna de estas monedas que traigo conmigo, podemos dar por concluido este asunto y yo seguiré por mi camino. Hasta luego.

El objetivo de este texto es ofrecer una manera de resolver ciertas paradojas clásicas – la del prefacio, la lotería y el sorites –, motivando una revisión de nuestra manera de entender la relación entre aceptación e inferencia. Para lograr esto, he introducido una paradoja novedosa pero de muy fácil solución. Lo que pretendo hacer después es extender la solución de esta paradoja a otras paradojas más interesantes, como la de la lotería. Esta solución está basada en la existencia de lo que llamaré las "condiciones de normalidad" de una aceptación. La idea fundamental detrás de este concepto es que cuando uno acepta una afirmación, lo hace bajo la suposición de que ciertas condiciones implícitas se cumplen, de tal manera que dicha aceptación está condicionada a que la situación descrita en la afirmación sea, por decirlo de alguna manera, normal. Por ejemplo, si acepto la afirmación de que "los ratones no hablan" lo hago bajo el supuesto (entre otros) de que *hablar* se está entendiendo de manera normal, es decir, como un sistema de comunicación complejo, con vocabulario, sintaxis y semántica, no en el sentido mas general de sistema de comunicación en base de sonidos. Diferentes afirmaciones pueden requerir diferentes condiciones de normalidad, y no será raro que dos afirmaciones sean aceptables bajo diferentes e incompatibles condiciones de normalidad, de tal manera que aunque cada una de ellas sea aceptable por separado, no puedan serlo de manera conjunta. Mi tesis central en este texto es que esto es precisamente lo que sucede en las paradojas del prefacio, la lotería y el sorites: todas ellas involucran una inferencia en la cual las condiciones de aceptación de las premisas son incompatibles con las condiciones de aceptación de la conclusión y, por lo tanto, pese a que la conclusión parece seguirse de las premisas (pues la proposición expresada en la conclusión *independientemente de sus condiciones de normalidad* es consecuencia lógica de las proposiciones expresadas en las premisas *independientemente de sus condiciones de normalidad*).

En todas las paradojas que reviso en este texto, una conclusión contradictoria parece seguirse lógicamente de una serie de premisas aceptables. En consecuencia, existen tres maneras de resolver paradojas de este tipo: mostrando o bien (i) que la conclusión no es realmente contradictoria, o (ii) que no se sigue de las premisas, o que (iii) alguna de dichas premisas no es realmente aceptable. En cada caso, además, es necesario explicar o bien (i) porqué la conclusión, pese a no ser contradictoria, nos parece serlo, o (ii) porqué, aunque no se sigue, *parece* seguirse de las premisas, o bien (iii) porqué, aunque alguna de las premisas no es aceptable, todas *nos parecen* aceptables. Soluciones de cada uno de estos tipos existen en la literatura, y mi solución es del segundo tipo: trataré de mostrar que, en todos los casos, la conclusión no se sigue de las premisas.

1. Resolviendo la paradoja de las tres monedas

Empecemos entonces con la paradoja de las tres monedas. Cualquiera que tenga por lo menos el mínimo de conocimiento de alguna lógica intensional habrá reconocido de inmediato el paso falaz en la argumentación de Proclo: pasar de P, Q y R a $(P \vee Q \vee R)$. Como es bien sabido, el operador ?◇ es débil y por lo tanto esta inferencia es inválida: en general, de que esté permitido P y esté permitido Q, no se sigue que esté permitido P y Q. Por ejemplo, yo me puedo casar con Ana y me puedo casar con María, pero de ello no se sigue que me puedo casar con Ana y María. Un paso falaz similar se da al principio de la paradoja, cuando Proclo trata de pasar de $\forall \diamond F$ a $\diamond \forall F$, es decir de que no hay ninguna moneda que deba entregarle Glaucón a que no debe entregarle ninguna moneda a Glaucón.

Si bien creo que esta manera de diagnosticar y resolver la paradoja es correcta, prefiero no usar explícitamente las herramientas de la lógica intensional en este texto por dos razones: en primer lugar, porque creo que la solución puede presentarse también en términos mas intuitivos y no quiero dirigir este trabajo sólo a aquellos con conocimientos o inclinación por la lógica formal, y segundo – aunque esto puede que no quede claro mas que a los que tengan estos antecedentes lógico-formales – porque no quiero dar la impresión de que la solución depende esencialmente de la modalidad involucrada, en este caso, la de permisibilidad. [1] Por el contrario, y como trataré

---

[1] Una reciente propuesta para resolver estas paradojas apelando a la modalidad de permisibilidad se encuentra en (Kroedel 2013). Una crítica muy clara a los problemas que este tipo de propuestas involucran

de mostrar en el resto de mi capítulo, creo que el diagnóstico y solución que daré generalizan a otros casos en los que no hay una modalidad explícita.

Para facilitar la exposición de la solución démosle nombre a las tres monedas en la mano de Proclo. Llamemos a las monedas "Hugo", "Paco" y "Luis". En un primer momento, Proclo, señalando dos de las monedas en su mano – supongamos que señala a Paco y a Luis - le dice "puedo darte una de estas dos y – ahora señalando a la tercera, es decir, a Hugo, continua - no tengo que darte ésta otra ¿verdad?" Al igual que Glaucón, no podemos sino asentir a la pregunta de Proclo. Ahora bien, reflexionemos sobre lo que sucede cuando aceptamos que, efectivamente, Proclo no debe entregarle Hugo a Glaucón; es fácil darse cuenta de que aceptamos precisamente porque reconocemos que hay una manera en la cual Proclo puede pagar su deuda y así cumplir su compromiso con Glaucón, sin entregar a Hugo, a saber, entregándole a Paco o a Luis. En otras palabras, aceptamos que Proclo no tiene que entregarle Hugo a Glaucón, pero siempre y cuando le entregue a Paco o a Luis. Es decir, reconocemos que Proclo puede no entregarle una de las monedas, siempre y cuando le de una de las otras. Ahora bien, nótese que aunque no hagamos explícita esta condición en nuestra respuesta afirmativa a la pregunta de Proclo, ésta es esencial para nuestra aceptación de la permisibilidad de que Proclo no le entregue Hugo a Glaucón. Esto es así porque, dadas las circunstancias, si Proclo no le entrega Hugo a Glaucón, lo normal sería que le entregara a Paco o a Luis. A esta condición es lo que de ahora en adelante voy a llamar la **condición de normalidad** detrás de nuestra aceptación: aceptamos que Proclo no tiene que entregarle a Hugo a Glaucón, bajo la condición (de normalidad) de que le entregue a Paco o a Luis[2].

Nótese que el mismo proceso tiene lugar con cada una de las otras dos monedas. Aceptamos que Proclo no tiene que entregarle a Paco a Glaucón, bajo la condición de normalidad de que le entregue a Hugo o a Luis, y aceptamos que Proclo no tiene que entregarle a Luis a Glaucón, bajo la condición de normalidad de que le entregue a Hugo o a Paco. En otras palabras, cada vez que juzgamos si es obligatorio o no que Proclo haga o no alguna cosa, lo que en realidad evaluamos es todo un curso de acción. Cuando aceptamos que está permitido que Proclo no entregue a Hugo, lo que juzgamos como permitido es un curso de acción complejo que no se reduce a Proclo no entregando a Hugo, sino que incluye también y de manera esencial el que le entregue a Paco. Es decir, aceptamos que está permitido que no entregue a Hugo, sólo cómo parte de un curso de acción permitido donde, aunque no entrega a Hugo, sí entrega a Paco o a Luis. En otras palabras, en nuestra aceptación, hay algo explícito – aceptamos que Proclo no tiene que entregarle a Hugo a Glaucón – y algo implícito –

---

se encuentra en (Littlejohn 2013). Mi teoría de la relación entre aceptación, e inferencia pretende llenar el hueco en una propuesta como la de Kroedel.

[2]En este ejemplo, la condición de normalidad es fácilmente accesible a la conciencia. Sin embargo, no quiero que quede la impresión de que así debe ser. Por el contrario, creo que una de las razones por las cuales las condiciones de normalidad son implícitas es porque muchas de ellas son de difícil acceso a la conciencia.

1 Resolviendo la paradoja de las tres monedas

el que le entregue a Paco o a Luis.

La idea básica detrás de la noción de condición de normalidad es que, cuando evaluamos un enunciado, no consideramos *todas* las maneras en que la proposición expresada podría ser verdadera, sino sólo las que nos parecen mas naturales. Por ejemplo, si nos preguntamos si debemos golpear al desconocido con el que nos cruzamos en la calle, fácilmente descartamos como poco relevante la posibilidad de que el desconocido en cuestión se dirija a cometer un crimen horrible. En otras palabras, cuando pensamos en un desconocido, racionalmente pensamos que no es un un criminal peligroso que debe ser detenido por los medios que sean necesarios. Cuando consideramos el enunciado "En la florería de mi tía venden lirios", por poner otro ejemplo, no sólo pensamos que tenemos una tía que tiene una florería, sino también que en dicha florería es una florería *normal*, es decir, una en la que no se venden sólo lirios. En sentido estricto, una florería que vende lirios podría vender lirios solamente, pero lo racional es considerar la posibilidad mas natural, lo que los lógicos formales llaman los mundos posibles mas cercanos donde la proposición es verdadera.[3] Así pues, cuando Glaucón considera la posibilidad de que Proclo no le de una moneda, estima como lo más natural el que no le de dicha moneda porque le da otra, no porque no le da ninguna moneda. Proclo fuerza dicha estimación sobre Glaucón al decirle "puedo darte una de estas dos y – ahora señalando a la tercera – no tengo que darte ésta otra ¿verdad?". Es por ello que Glaucón piensa, al considerar si Proclo debe entregarle a Hugo, que lo normal sería que le entregara a Paco o a Luis en su lugar.

Ahora bien, si aceptamos que Proclo no tiene que entregarle a Hugo a Glaucón, que no tiene que entregarle a Paco, y que no tiene que entregarle a Luis ¿porqué no hemos de aceptar que no tiene que entregarle a Hugo ni a Paco ni a Luis, es decir, que no tiene que entregarle ninguna de las monedas en su mano? Creo que la respuesta debe ser ya obvia. Si bien podemos reconocer fácilmente un posible curso de acción en el que Proclo puede pagar su deuda sin entregar a Hugo (porque, en su lugar, le entrega a Paco o a Luis), podemos reconocer un posible curso de acción en el que Proclo puede pagar su deuda sin entregar a Paco (porque, en su lugar, le entrega a Hugo o a Luis) y podemos reconocer un posible curso de acción en el que Proclo paga su deuda sin entregar a Luis (porque, en su lugar, le entrega a Hugo o a Paco), no podemos reconocer ningún posible curso de acción en el que Proclo paga su deuda sin

---

[3] Algunos teóricos han propuesto una formalización de esta idea apelando a mundos posibles, una relación de cercanía entre dichos mundos posibles y una función que ordena los mundos posibles por qué tan ideales son de acuerdo a la normatividad relevante. Por ejemplo, si estamos modelando la obligatoriedad que emana de un código legal dado, un mundo W_1 sería más ideal que otro W_2 si W_1 obedece mas provisiones de dicho código que W_2. Así, una proposición p es obligatoria si y sólo si los mundos posibles donde p es verdadera son mas cercanos al mundo real son mas ideales que los mundos mas cercanos donde p es falsa. La idea es ir mas allá del modelo tradicional de obligatoriedad, donde una proposición p es obligatoria si y sólo si es verdadera en todo mundo donde se cumplen con la normatividad relevante. Este nuevo modelo encaja con mi propuesta en que, cuando evaluamos si una proposición dada es obligatoria, no consideramos todas las maneras en que la proposición podría ser verdadera, sino sólo las que nos parecen mas naturales, es decir, aquellas donde se cumplen sus condiciones de normalidad.

entregar una de las monedas, ya sea a Hugo o a Paco o a Luis.

Para entender mejor como funcionan las condiciones de normalidad en la inferencia, analicemos otro caso posible. Supongamos que después de que Glaucón reconoce que Proclo no tiene que entregarle a Hugo, ni tiene que entregarle a Paco, Proclo le hubiera preguntado si de ello se sigue que no tiene que entregarle ni a Hugo ni a Paco. La respuesta de Glaucón seguramente sería positiva. Efectivamente, Proclo puede no entregarle a Hugo ni a Paco a Glaucón y aun así pagar su deuda, siempre y cuando le entregue a Luis. En otras palabras, podemos bien aceptar que Proclo no tiene que entregarle ni a Hugo ni a Paco, pero sólo bajo la condición de normalidad de que le entregue a Luis. En este caso, la inferencia de que Proclo no tiene que entregarle a Hugo, ni tiene que entregarle a Paco, a que no tiene que entregarle ni a Hugo ni a Paco, no es falaz. La razón se encuentra en las condiciones de normalidad: Recordemos que la condición de normalidad de que Proclo no entregue a Hugo es que entregue a Paco o a Luis, y la condición de normalidad de que Proclo no entregue a Paco es que entregue a Hugo o a Luis. Ahora bien, para que aceptamos que Proclo no entregue ni a Hugo ni a Paco, deben cumplirse las condiciones de normalidad detrás de cada una de las premisas del argumento, es decir, tiene que entregar a Paco o a Luis y a Hugo o Luis. La única manera en que, tanto la conclusión como ambas condiciones de normalidad se cumplan es que Proclo entregue a Luis. En otras palabras, en el contexto de este argumento, la conclusión tiene como condición de normalidad que Proclo le entregue a Luis a Glaucón. En otras palabras, cuando ejecutamos la inferencia de que Proclo no tiene que entregarle a Hugo, ni tiene que entregarle a Paco, a que no tiene que entregarle ni a Hugo ni a Paco, lo que conjuntamos no es sólo la parte explícita de las premisas, sino también sus condiciones de normalidad. Esquemáticamente:

1. No tengo que entregar a Hugo [bajo la condición de normalidad de que entrego a Paco o a Luis]

2. No tengo que entregar a Paco [bajo la condición de normalidad de que entrego a Hugo o a Luis]

    Por lo tanto, no tengo que entregar ni a Hugo ni a Paco [bajo la condición de normalidad de que entrego a (Paco o a Luis) y entrego a (Hugo o a Luis) sin entregar ni a Hugo ni a Paco, es decir, bajo la condición de normalidad de que entrego a Luis][4]

Nótese como la conclusión hereda la parte que es consistente con la conclusión de la conjunción de las condiciones de normalidad de las premisas. En otras palabras, la condición de normalidad de no entregar ni a Hugo ni a Paco es entregar a Luis porque esa es la única manera de que se cumplan la conclusión y las condiciones de

---

[4]Para seguir este tipo de inferencias al lector le puede servir pensar en la relación entre una proposición y su condición de normalidad como una implicación material cuyo antecedente es la conjunción de las condiciones de normalidad y cuyo consecuente es la proposición en cuestión.

normalidad de las premisas. Sin embargo, no podemos hacer lo mismo para obtener la conclusión que quiere Proclo, precisamente porque las condiciones de normalidad de las premisas son inconsistentes con la conclusión, es decir, no hay manera que se cumplan las condiciones de normalidad de todas las premisas y la conclusión. No es posible entregar a Hugo o a Paco, a Paco o a Luis y a Hugo o a Luis, y al mismo tiempo no entregar ni a Hugo ni a Paco ni a Luis. Por eso, no podemos inferir de que Proclo puede no entregar a Hugo, puede no entregar a Paco y puede no entregar a Luis, a que puede no entregar a ninguno de los tres.[5]

1. No tengo que entregar a Hugo [bajo la condición de normalidad de que entrego a Paco o a Luis]

2. No tengo que entregar a Paco [bajo la condición de normalidad de que entrego a Hugo o a Luis]

3. No tengo que entregar a Luis [bajo la condición de normalidad de que entrego a Hugo o a Paco] Por lo tanto, no tengo que entregar ni a Hugo ni a Paco ni a Luis [bajo la condición de normalidad de que entrego a Paco o a Luis, a Hugo o a Luis, a Hugo o Paco y ni a Hugo ni a Paco ni a Luis]

En general, uno puede asentir a una serie de enunciados $P_1, P_2, P_3, \ldots P_n$ sin comprometerse con una de sus consecuencias lógicas C, si (la conjunción de) las condiciones de normalidad de $P_1, P_2, P_3, \ldots P_n$ son inconsistentes con C.[6]

Nótese como apelar a las condiciones de normalidad es esencial para explicar así la falacia detrás de la paradoja de las tres monedas. Nótese además que el número de monedas en la mano de Proclo es completamente irrelevante para la paradoja y su solución. Pudimos haberla formulado con dos, cinco, veinticinco o doce mil monedas. Es más, pudimos haberla formulado incluyendo todos los dracmas de la Grecia antigua. (¿Podríamos haberla formulado con un número infinito de monedas? Creo que sí, pero este punto no es relevante ahora, así que no nos detendremos en él) Es más, me parece que el fenómeno va mas allá de asuntos de dinero[7] y deudas o de permisos y

---

[5] Hay otra manera, completamente equivalente de diagnosticar y resolver la paradoja. Podemos decir que ésta comete una falacia de equivocación, pues hay dos lecturas del enunciado "Proclo no tiene que darle ninguna de las monedas en su mano a Glaucón", una (en la que significa que ninguna de las monedas que Proclo tiene en su mano es la que tiene que darle a Glaucón) que efectivamente se sigue de lo que Glaucón le aceptó a Proclo, pero no tiene la consecuencia indeseable de llevar a Glaucón a inconsistencia y otra (en la que significa que es falso que Proclo tiene que darle a Glaucón alguna de las monedas en su mano) que sí tiene esta consecuencia, pero no se sigue de lo que Glaucón aceptó. Por cuestiones de espacio, no desarrollaré esta manera de resolver la paradoja.

[6] Nótese que esto es compatible con que un sujeto acepte C tanto como $P_1, P_2, P_3, \ldots P_n$ Esto es así porque el sujeto puede asentir a C bajo otras condiciones de normalidad que no tengan nada que ver con las condiciones de normalidad de $P_1, P_2, P_3, \ldots P_n$. Sin embargo, en ese caso, no habrá inferido C de dichas creencias.

[7] Después de todo, la paradoja de las tres monedas se podría haber resulto también haciendo una distinción entre dinero - que es lo que Proclo le debe realmente a Galucón - y monedas - que es lo que usamos para representar el dinero.

obligaciones. En realidad, tiene más que ver con qué significa aceptar un enunciado, como trataré de mostrar en lo que queda del capítulo.

En el caso de la paradoja de las tres monedas, creer que Proclo le debe entregar a Galucón una moneda para cumplir con su deuda no significa que haya tal cosa como la moneda que uno crea es la que se le deba entregar para cumplir con dicha deuda. En general, uno puede creer que hay un X que es F sin que haya tal cosa como el X que uno crea que es F. Uno puede bien creer que un vaso de agua calmará su sed sin que haya tal cosa como el vaso del cual uno crea que es el que calmará su sed. Uno puede saber que uno de los sospechosos es el asesino sin que haya tal cosa como el sospechoso que uno sabe es el asesino; etc.

Algo análogo sucede en el caso de las aceptaciones. El que uno acepte que Proclo le debe entregar a Galucón una moneda para cumplir con su deuda no significa que haya alguna moneda que uno deba aceptar sea la que le deba entregar para cumplir con dicha deuda. En general, uno puede aceptar que hay un X que es F sin que haya tal cosa como el X tal que uno acepte que dicho X sea F. Esto se debe a que las condiciones de normalidad de los enunciados existenciales son muy distintos de las de los enunciados particulares. Cuando uno hace una afirmación existencial, como "Me debes un dracma" o "nací en un pueblo muy pequeño", normalmente no existe un objeto en particular del que uno esté hablando o su identidad es considerada poco relevante. En contraste, cuando alguien dice de un objeto o grupo de objetos X que tienen cierta propiedad F, normalmente lo hace para contrastarlo con otros objetos que no son F (o lo son de una manera más débil, obvia o menos importante). Igualmente, cuando uno considera un enunciado de la forma "X es F" (o "Los X son F") no piensa en todas las posibilidades que lo harían verdadero, sino en las mas naturales, y ellas usualmente excluyen casos en los que es verdadero de manera vacua, por ejemplo, porque no hay ningún F.

Ilustremos esta tesis con un ejemplo. Supongamos que nuestro amigo Víctor es un activista comprometido con la lucha por los derechos de las minorías sexuales, y que ha dedicado a dicha lucha mucho de su vida. Además, aunque en menor grado, le preocupan los problemas que genera la explotación demográfica y por ello ha llegado a la conclusión de que ninguna familia debería tener más de un hijo. Si a Víctor se le preguntara si cree que las parejas gay debieran no tener más de un sólo hijo (en un contexto en el cual su compromiso con los derechos de las minorías sexuales no fueran conocido y asumido), tenemos la fuerte intuición de que su respuesta no podría ser "sí", pese a que de que ninguna familia debería tener más de un hijo se sigue que las parejas gay – en tanto familias – tampoco deberían tener más de un hijo. En general, si creemos que todos los X son Y (y que tal información es relevante para la conversación) no solemos afirmar enunciados más débiles como que un X en particular es Y o que un grupo particular de X es Y. Es por ello que uno puede creer que todos los X son Y, y rechazar que un X en particular o un grupo particular de X sean Y sin caer en inconsistencia.

Nótese además que si, en vez de preguntarle sobre parejas gays, le hubiéramos preguntado sobre cualquier otro tipo *particular* de familia si creía que no debía tener más de un hijo, pasaría exactamente lo mismo. Dado que Víctor no cree que ningún tipo *particular* de familia debería tener a lo más un hijo, sino que *ninguna* familia *en general* debería tener más de un hijo, lo correcto sería que rechazara la afirmación de que, por ejemplo, las familias monopaternales no deberían tener más de un hijo, o las familias con padres de escasos recursos, etcétera. Esta tensión entre la aceptación de enunciados particulares y una creencia general que parecía ser inconsistente con ellas es lo que está detrás de la paradoja de las tres monedas — y de otras paradojas conocidas, como trataré de mostrar en este texto: Así como Víctor puede tener la creencia general de que las familias no deberían tener más de un hijo, sin aceptar de ningún tipo particular de familia que no debería tener más de un hijo, en la paradoja de las tres monedas, Glaucón puede creer racionalmente que se le debe entregar una moneda pero no aceptar de ninguna moneda que es la que se le debe entregar (pues no hay tal cosa como la moneda que crea que sea la que Proclo le debe entregar).

Reconozco que pocos tomarían la paradoja de las tres monedas como una paradoja genuina o especialmente difícil de resolver. Sin embargo, como trataré de mostrar en el resto de la platica, me parece que una fenómeno del mismo tipo está detrás de algunas de las paradojas más famosas de la filosofía analítica reciente, en particular, la paradoja del prefacio, la de la lotería y la del *sorites*. Esta simple paradoja me ha servido, mas que nada, para introducir la noción de condición de normalidad en un contexto poco problemático. Es por ello que me he detenido tanto tiempo en ella, antes de pasar a paradojas mas interesantes y genuinas.

2. Extendiendo la propuesta: Las paradojas del prefacio y de la lotería

Admito que la paradoja de las tres monedas por si misma es poco interesante. Sin embargo, creo que fijarnos qué hay detrás de ella sirve de muy buen propedéutico para tratar luego otras paradojas más interesantes. Empecemos por una muy simple: la famosa paradoja del prefacio, la cual es muy sencilla de formular. Supongamos que un autor –llamémosle "Miguel"– escribe un libro de no-ficción. Miguel es un investigador concienzudo y ha hecho una labor formidable investigando su tema, pero también es sensato y sabe que pese a que lo que ha escrito es el resultado de una investigación seria y profunda, lo más probable sea que haya cometido por lo menos algún error y, en consecuencia, en su prefacio reconoce que por lo menos algo de lo que sostiene en el libro muy probablemente sea falso. No hay nada extraño, ni insensato en esto. Sin embargo, la actitud de Miguel frente a su propio libro parece ser paradójica. Por un lado, tiene la creencia existencial de que por lo menos una de las oraciones que sostiene en su libro es falsa, pero si le preguntáramos, de cualquier oración en su libro si es verdadera, sin duda el autor afirmaría su verdad sin chistar. En otras palabras, sean $P_2, P_3, \ldots P_n$ las oraciones contenidas en su libro, el autor aceptaría que "P_1 es verdadera", "P_2 es verdadera"... o "P_n verdadera". Esto parece significar que piensa que todas y cada una de las afirmaciones de su libro son verdaderas, lo cual es inconsistente con su propia afirmación en el prefacio de su obra de que por lo menos alguna de dichas afirmaciones es falsa. Esquemáticamente, el argumento tiene la siguiente forma:

1. La primera afirmación del libro es verdadera. [Premisa]

2. La segunda afirmación del libro es verdadera. [Premisa]

3. La tercera afirmación del libro es verdadera. [Premisa]

   ...

N. La n afirmación del libro es verdadera. [Premisa]

$N + 1$. Hay n afirmaciones en el libro. [Premisa]

$N + 2$. Por lo menos una de las afirmaciones del libro es falsa. [Premisa]

$N + 3$. De 1 a $N + 1$ se sigue que todas las afirmaciones del libro son verdaderas.

$N + 4$. Pero $N + 2$ y $N + 3$ se contradicen entre sí.

La paradoja del prefacio nos enfrenta con un dilema vergonzoso ambos de cuyos cuernos son inaceptables. O bien el autor de ningún libro puede realmente reconocer que debe haber por lo menos un error en su libro (y decirlo sólo es falsa molestia) o bien todo autor introduce en sus libros algo que no cree que es verdadero. Sin embargo, nos parece muy claro que es posible que uno puede racionalmente llenar un libro de afirmaciones que considera verdaderas y, sin embargo, reconocer que por lo menos una de ellas puede ser falsa, sin caer en una contradicción. ¿Qué sucede en esta situación entonces?

Mi propuesta es ver la paradoja del prefacio como completamente análoga a la paradoja de las tres monedas. En ambas, me parece, hay una aparente tensión entre un juicio existencial positivo y una serie de juicios particulares negativos que parecen excluir la instanciación del existencial. En el caso de la paradoja de las tres monedas, la tensión se daba entre el juicio existencial "Proclo le debe una moneda a Galucón" y la serie de juicios particulares de la forma "Proclo no le debe dar esta moneda a Galucón", "Proclo no le debe dar esta otra moneda a Galucón", etc., que parecen excluir la posibilidad de que el existencial se instancíe (ya que si se instanciara, debería de ser en esta moneda, o en esta otra, etc.) Como burlonamente preguntaba Proclo de manera retórica, ¿cómo es posible que digas que te debo una moneda, pero no puedas decirme cuál?

2 Extendiendo la propuesta: Las paradojas del prefacio y de la lotería   159

La paradoja del prefacio tiene exactamente la misma estructura. Tenemos un juicio existencial – hay por lo menos una afirmación falsa en este libro – en aparente tensión con una serie de juicios particulares – esta afirmación es verdadera, esta también, y esta también... – que parecen excluir la instanciación del existencial. Es como si Julián le preguntará a Miguel: ¿Cómo es posible que digas que hay por lo menos una falsedad en el libro, pero no puedas decirme cuál? Y así como tiene una estructura similar, su diagnóstico y solución también es completamente análogo. Basta reconocer que cada vez que el autor del libro acepta una de sus propias afirmaciones como verdadera, lo hace bajo la condición de normalidad de que no es infalible, es decir, que puede, ha y volverá a cometer errores y afirmar cosas que son falsas. En otras palabras, acepta cada cosa que ha dicho bajo la condición de normalidad de que en otras ocasiones habrá dicho o dirá cosas falsas. En particular, en el caso del libro, el autor acepta de cada afirmación que hace que es verdadera, bajo la condición de normalidad de que por lo menos *alguna otra* de las afirmaciones que hace en él es falsa.

Nótese como este caso es completamente análogo al caso de las tres monedas. Así como Glaucón aceptaba de cada una de las monedas en la mano de Proclo que no tenía que dársela, bajo la condición de normalidad de que le diera *alguna otra* moneda, así también el autor del libro acepta de cada una de las afirmaciones que hace en el libro que es correcta, bajo la condición de normalidad de que por lo menos *alguna otra* no lo sea. Es por ello que, de que el autor acepte de cualquiera de las afirmaciones en el libro que cree que es verdadera no se sigue que el autor crea (o esté comprometido a aceptar) que todas ellas son verdaderas. Esto se debe a que, si juntamos las condiciones de normalidad de cada una de las premisas en el argumento, veremos que son inconsistentes con la conclusión de éste. Imaginemos que el autor hace 300 afirmaciones en su libro. Es fácil ver que el autor acepta que su primera afirmación es verdadera sólo bajo el supuesto de que por lo menos una de las otras 299 afirmaciones no lo sea. Análogamente, acepta que su segunda afirmación es verdadera, sólo bajo el supuesto de que alguna de las otras no lo sea. Así sucesivamente con la tercera, la cuarta, la enésima afirmación. Si acepta una es porque piensa que debe haber *otra* que sea falsa. En otras palabras, detrás de la aceptación de que la afirmación n es verdadera, está la condición de normalidad de que hay una afirmación distinta a n en el libro que no lo es. Sin embargo, no hay ninguna situación aceptable para el autor en la cual se cumplen el conjunto de las condiciones de normalidad de las afirmaciones particulares – es decir, que alguna de las otras afirmaciones es falsa – y la conclusión del argumento – que todas las afirmaciones en el libro son verdaderas. Por ello es que, aunque la conclusión se sigue deductivamente de las premisas, quien acepte las premisas no está racionalmente comprometido a aceptar la conclusión.

Aplicando lo visto en la sección anterior sobre obligatoriedad al caso de la aceptación o creencia, mi propuesta es que cuando evaluamos una proposición para aceptarla o rechazarla, lo que hacemos es evaluar una situación más compleja, que consideramos es la manera más natural en la que se cumple la proposición.[8] Continuando con uno de los ejemplos ya mencionados. Si se nos pide evaluar el enunciado "La florería de mi tía vende lirios", lo que evaluamos no es solamente la verdad de la proposición "La florería de mi tía vende lirios", sino mas bien una situación mas compleja donde tenemos una tía que tiene una florería con lirios y otras flores, los precios que da son normales, no es el frente de una operación internacional de lavado de dinero o contrabando de divisas escondidas en macetas, etc. Cuando asentimos al enunciado de que la florería de nuestra tía vende lirios, lo que está detrás de nuestra afirmación es la consideración de que el mundo es muy probablemente tal y como lo describe dicha situación. Si dicha situación nos parece real o por lo menos muy probable, aceptamos el enunciado. En otras palabras, la aceptación de un enunciado es derivada, no de la aceptación de una proposición ligada semánticamente al enunciado, sino de la aceptación de una situación más compleja; dicha situación es más compleja que la proposición en tanto contiene mayores compromisos. Estos compromisos extras que la situación añade a la proposición es lo que he llamado condiciones de normalidad. Es por ello que para poder combinar dos proposiciones en una inferencia, necesitamos considerarlas en una sola situación (que comúnmente es la suma de las dos situaciones en las que las aceptábamos por separado), y al reunirlas en dicha situación, sumamos sus condiciones de normalidad.[9]

En consecuencia, el que exista una situación aceptable en la que P es verdadera y otra en la que Q es verdadera, no garantiza que exista una situación aceptable en la que ambas P y Q sean verdaderas. Sólo si ambas son verdaderas en una misma situación (es decir, si sus condiciones de normalidad son compatibles) podemos combinarlas en una inferencia. Es por ello que la regla de conjunción no se cumple siempre, sino solamente cuando se satisfacen ciertas condiciones (entre ellas, el que las condiciones de normalidad de los conjuntos sean consistentes entre sí y con la conjunción). Es por ello que, en el argumento de la paradoja, que el autor acepte las premisas de 1 a N, no lo compromete a que acepte también la conjunción total $N+3$ (todas las afirmaciones del libro son verdaderas.). Así se evita comprometerse a la contradicción de aceptar tanto $N+3$ como $N+2$ (por lo menos una de las afirmaciones del libro es falsa), y se resuelve la aparente paradoja.[10]

---

[8]A decir verdad, no es suficiente que la proposición en cuestión sea verdadera en la situación, sino que se requiere algo más fuerte, algo como que la proposición describa bien la situación. Evito esta complejidad a lo largo de mi texto porque la diferencia no es muy importante para los casos que aquí discuto. Una idea muy similar es desarrollada por Mitchell S. Green en (1999).

[9]Nótese que la tesis no es una doble implicación, es decir, que no dice que baste que las condiciones de normalidad de las premisas sean consistentes para que sea racional realizar la inferencia. La consistencia es una restricción mínima a la racionalidad de la inferencia que se desprende de la existencia de condiciones de normalidad en la aceptación y evaluación de proposiciones, pero no creo que sea la única; aunque sí es la única que trataré en este texto, pues ella es suficiente para tratar con las paradojas que me ocupan aquí.

[10]Es importante notar también que dado que las condiciones de normalidad nos dan condiciones necesarias mas no suficientes para la inferencia, de mi propuesta no se sigue que sería correcto para el autor llegar al enunciado 299 y decir "Dado que sé que por lo menos un enunciado de los que contiene mi libro es falso,

2 Extendiendo la propuesta: Las paradojas del prefacio y de la lotería　　　161

Una paradoja de estructura muy similar es la paradoja de la lotería, según la cual hay una tensión entre creer que un boleto de la lotería ganará, y rechazar cualquier afirmación de que algún boleto particular será el ganador. En palabras de Mauricio Zululaga (2005):

> Supongamos que hemos comprado un billete de lotería, alguno de los cuales deberá ganar. Supongamos además que la hay 100 billetes de lotería. La probabilidad de que uno de los billetes gane es muy baja: 0,01. Supongamos que usted ha comprado un billete de lotería. De acuerdo con la baja probabilidad de que su billete gane, usted está justificado para afirmar que cree que perderá –suponiendo que uno está justificado para creer que algo será el caso, si la probabilidad de que ocurra es mayor a 0,5–. Pero bajo estas condiciones usted también está justificado para creer que cada uno de los billetes perderá. Así es racional creer que ninguno de los billetes ganará, porque para cada uno de ellos usted ha podido establecer que la probabilidad de que gane es menor a 0,5. Pero, *ex hypothesi*, ha de haber un billete de lotería que gane, así que usted cree algo contradictorio. Usted está justificado para creer que ningún billete ganará y, al mismo tiempo, cree que uno lo hará.

Una vez más, cada vez que aceptamos de un boleto que no será el ganador lo hacemos bajo la condición de normalidad de que *otro* boleto ganará, es decir, lo que aceptamos es una situación en la que el boleto en cuestión no gana, pero otro sí. Por ello, la conjunción de todas las condiciones de normalidad de todas las afirmaciones sobre cada boleto son incompatibles con el que ningún boleto gane (a decir verdad, cada una de las proposiciones de cada una de las creencias particulares es ya incompatible con el que ningún boleto gane). Las tres paradojas que hemos mencionado hasta ahora tienen la misma estructura: pretenden que haya una inconsistencia entre aceptar una proposición existencial (1) y una serie exhaustiva de aceptaciones singulares sobre el dominio del existencial de dicha proposición (2):

1. S acepta que hay (por lo menos) un X que es F.

2. Para todo x en X, S acepta que x no es el/uno de los F.

En el caso de las tres monedas, éstas son[11]:

1. Glaucón acepta que Proclo le debe entregar una de las monedas en su mano.

2. Para toda moneda x en la mano de Proclo, Glaucón acepta que esa moneda x no es la que Glaucón le debe entregar.

---

he pasado por todos menos este último y de todos estoy seguro es correcto, debo concluir que este último debe ser falso."

[11]Las X son las monedas en la mano de Proclo, y F es la propiedad de deber ser entregada a Glaucón.

En el caso del prefacio, éstas son: [12]

1. El autor acepta que por lo menos una de las afirmaciones en su libro es falsa.

2. Para toda afirmación x en su libro, el autor acepta que esa afirmación x no es una de las las que son falsas.

Finalmente, en el caso de la lotería, éstas son: [13]

1. Aceptamos que por lo menos uno de los boletos de la lotería ganará.

2. Para todo boleto x de la lotería, aceptamos que ese boleto x no es el que ganará.

Gracias a que tienen la misma estructura, las tres paradojas se resuelven de la misma manera, mostrando que de (1) y (2) no se sigue que el agente está siendo irracional al aceptar o comprometerse a aceptar una contradicción. Para ello, basta darse cuenta de que lo que el agente acepta en (1) no es inconsistente con lo que acepta en (2). En el caso de la paradoja de las tres monedas, habíamos mostrado que aunque Glaucón acepta de todas las monedas en la mano de Proclo que éste no debe entregársela, esto no lo compromete a que deba aceptar que Proclo no debe entregarle ninguna moneda. En otras palabras, de (2) no se sigue que

3. Glaucón está comprometido a aceptar que Proclo no le debe entregar ninguna de las monedas en su mano.

De (1) y (3) sí se sigue que Glaucón está siendo irracional en su aceptación de una contradicción explícita (pues en (1) acepta que Proclo le debe entregar una de las monedas en su mano y en (3) debe aceptar que no le debe entregar ninguna de las monedas en su mano). Afortunadamente, (3) no es una descripción adecuada de lo que Proclo ha aceptado; (2) lo es, y de (2) no se sigue (3).

De manera similar, en el caso del prefacio, hay una tesis (3) similar:

3. El autor está comprometido a aceptar que ninguna de las afirmaciones en su libro es falsa.

Una vez más, de (1) y (3) sí se sigue que el autor está siendo irracional en su aceptación de una contradicción explícita, y también en este caso (3) no es una descripción adecuada de lo que el autor ha aceptado; (2) lo es, y de (2) no se sigue (3).

Igualmente, en el caso de la lotería, tenemos la tesis

3. Aceptamos que ninguno de los boletos de la lotería ganará.

Tal que, aunque de (1) y (3) sí se seguiría que estamos siendo irracionales en nuestra aceptación de una contradicción explícita, (3) no es una descripción adecuada de lo que el autor ha aceptado; (2) sí lo es, pero de (2) no se sigue (3).

---

[12] Las X son las afirmaciones en el libro, y F es la propiedad de ser falsa.
[13] Los X son los boletos de lotería, y F es la propiedad de ganar la lotería.

En otras palabras, en general, de (2) no se sigue que

3. S está comprometido a aceptar que ningún X es F.

Que sería lo necesario para derivar la irracionalidad del agente involucrado en cada paradoja. En otras palabras, en todos los casos la paradoja se resuelve rechazando la inferencia de, para todos los X, S cree que X es F, a S cree que todos los X son F.

3. La Paradoja Sorites

Me parece que apelar así a las condiciones de normalidad tiene muchas ventajas, entre ellas la de poder dar un diagnóstico y solución similar para un gran número de paradojas y acertijos filosóficos, desde la paradoja de Moore a las presuntas fallas de clausura epistémica, etc. Por supuesto, no puedo mostrar esto aquí, aunque vale la pena mencionar que Nancy Nuñez está explorando esta manera de explicar las aparentes fallas de clausura epistémica. Sin embargo, no quiero dejar de pasar la oportunidad de mostrar como apelar a condiciones de normalidad podría usarse para resolver otra famosa paradoja, aparentemente muy distinta las antes presentadas: la paradoja de sorites[14].

En una de sus versiones más conocidas, se dice que la paradoja se produce porque mientras el sentido común sugiere que los montones de arena tienen las siguientes propiedades, éstas son inconsistentes entre sí:

P. Dos o tres granos de arena no son un montón.

Q. Un millón de granos de arena juntos sí son un montón.

R. Si n granos de arena no forman un montón, tampoco lo serán $(n+1)$ granos.

En otras palabras, si bien el sentido común sugiere aceptar como verdaderas estas tres proposiciones, de su conjunción se sigue una contradicción, a saber, (S) que un millón de granos de arena juntos son y no son un montón. Esto es un problema porque, idealmente, quisiéramos una teoría de los montones (y términos similares) que respetara nuestras intuiciones de sentido común sin comprometernos a aceptar una contradicción.

A primera vista, la paradoja del sorites no parece asemejarse a las paradojas de la lotería y del prefacio. Sin embargo, quiero defender que, contra toda apariencia, sí tiene la misma forma y por ello, también podemos resolverla apelando a las condiciones de normalidad de las premisas para mostrar como quien acepta las premisas no está obligado epistémicamente a aceptar la conclusión contradictoria. Para mostrar esto,

---

[14]Mi tratamiento de la paradoja sorites está inspirado en el trabajo de Delia Graff Fara (2000).

permítaseme re-plantearla de una manera que realce la semejanza formal con las otras paradojas aquí tratadas.

Por principio de cuentas, creo que es un supuesto implícito importante del *sorites* que si n granos de arena forman un montón, cualquier grupo de mas de n granos de arena también será un montón. Bajo este supuesto, la conjunción de las premisas P y Q son equivalentes a lo siguiente:

1. Hay un límite, entre tres y un millón de granos de arena, que distingue lo que es un montón de lo que no es.

Es decir, debe haber un punto, entre tres y un millón de granos de arena, en el que el número de granos empieza a ser un montón.[15] Ahora bien, ¿dónde se encuentra dicho punto límite? La premisa R nos dice que, para todo número n, el límite no se encuentra exactamente entre n y $n+1$. En otras palabras, no importa dónde busquemos, una vez que nos enfocamos en un punto dentro de la serie sorítica de tres a un millón, no lo encontraremos ahí. Así pues, podemos re-formular la premisa R como un millón de premisas de la siguiente manera:

2. Si 1 grano de arena no forma un montón, 2 granos tampoco son un montón, es decir, la diferencia entre lo que es un montón y lo que no lo es no es la diferencia entre 1 grano y 2 granos.

3. Si 2 granos de arena no forman un montón, 3 granos tampoco son un montón, es decir, la diferencia entre lo que es un montón y lo que no lo es, no es la diferencia entre 2 granos y 3 granos.

4. Si 3 granos de arena no forman un montón, 4 granos tampoco son un montón, es decir, la diferencia entre lo que es un montón y lo que no lo es, no es la diferencia entre 3 granos y 4 granos.

. . .

999,999. Si 999,999 granos de arena forman un montón, 999,998 granos también son un montón[16], es decir, la diferencia entre lo que es un montón y lo que no lo es, no es la diferencia entre 999,998 granos y 999,999 granos.

1,000,000. Si un millón de granos de arena forman un montón, 999,999 granos también son un montón, es decir, la diferencia entre lo que es un montón y lo que no lo es, no es la diferencia entre 999,999 granos y un millón de granos.

---

[15]Dicho límite no tiene que ser de un sólo grano, es decir, puede haber una zona de penumbra entre lo que es un montón y lo que no lo es; sin embargo, es importante notar que la existencia de dicha penumbra no resuelve la paradoja, sino que solamente la mueve hacia un nivel superior. De ahí que, aunque en este texto formulo la paradoja en términos de un límite preciso, la misma paradoja (y mi solución) se reproduzcan a cualquier nivel superior.

[16]He formulado estas últimas premisas de forma contrapositiva a las primeras porque suena más normal cuando consideramos grandes cantidades, pero la formulación es obviamente equivalente a la original.

Dado que la premisa 1 no hace sino sintetizar la información contenida en las premisas P y Q, y las premisas de 2 a 1,000,000 no hacen más que hacer explícitos los casos particulares contenidos en la premisa R original, esta nueva formulación de la paradoja (con el millón de premisas de 1 a 1,000,000) es equivalente a la formulación tradicional, pero tiene la ventaja de poner de manifiesto sus similitudes formales con las paradojas del prefacio y la lotería (y las tres monedas de Proclo). Al igual que en ellas, tenemos una premisa existencial (la premisa 1) y un número muy grande de premisas particulares (de 2 a un millón) que parecen contradecirla. Mientras que la primera premisa nos dice que el número mínimo de granos de arena que forman un montón se encuentra entre tres y un millón, las premisas que siguen nos dicen que la diferencia no es entre tres y cuatro, o cuatro y cinco, o cinco y seis, etc. Esto significa que el sorites tiene la misma estructura que habíamos identificado antes en las paradojas anteriores: todas ellas pretenden que haya una inconsistencia entre aceptar una proposición existencial (1) y una serie exhaustiva de aceptaciones singulares sobre el dominio del existencial de dicha proposición (2):

1. S acepta que hay (por lo menos) un X que es F.

2. Para todo x en X, S acepta que x no es el/uno de los F.

En el caso del sorites, X es un número entre tres y un millón, mientras que F es la propiedad de ser el número máximo de granos de arena que aún no forman un montón.

Aplicando el mismo esquema que aplicamos a las paradojas anteriores, tenemos que la paradoja sorítica se basa en el error de creer que es inconsistente aceptar el existencial (1) y las premisas particulares de (2) a (1,000,000); y al igual que ellas, se resuelve recociendo que de la aceptación de (1) y (2) no se sigue que el agente está siendo irracional al aceptar o comprometerse a aceptar una contradicción. Para ello, basta darse cuenta de que lo que el agente acepta en (1) no es inconsistente con lo que acepta en (2). Al igual que en las paradojas anteriores, hay una tesis intermedia (3) de la forma 'Hay un X tal que S acepta de X que es el F' tal que aunque aceptar (2) y (3) implica cierta inconsistencia por parte del sujeto, aceptar (1) y (2) no. Esto se debe a que de (2) no se sigue (3), porque las condiciones de normalidad de las premisas en (2) son inconsistentes con (3). En este caso, la falsa tesis intermedia es:

3. Hay un número n de granos de arena que aceptamos es el número límite de granos de arena que forman un montón.

Cada vez que aceptamos de un número de granos que éste no corresponde al límite entre lo que es un montón y lo que no lo es, lo hacemos bajo la condición de normalidad de que dicho límite existe, sólo que se encuentra en *otro* lado. Cuando aceptamos que el límite no es cuatro, por ejemplo, lo hacemos bajo la condición de normalidad de que *hay* un límite mas adelante, de tal manera que la conjunción de todas las condiciones de normalidad de todas las afirmaciones sobre cada número en la serie sorítica son incompatibles con el que ningún número de granos sea el número máximo de granos que aún no forman un montón (a decir verdad, cada una de las proposiciones de cada una de las creencias particulares es ya incompatible con el que ningún número de granos sea el número máximo de granos que aún no forman un montón).

En otras palabras, el error detrás de la paradoja sorítica es pensar que quién acepta que hay una línea divisoria entre lo que cae dentro de la extensión de un término y lo que cae fuera de él debe aceptar también de algún lugar que ahí se encuentra dicha línea divisoria. Espero haberlos convencido de que no es así, y que por lo tanto, nuestros juicios de sentido común no nos llevan a ninguna contradicción.

4. Conclusiones

En este artículo he propuesto una nueva manera de concebir la aceptación de proposiciones y he mostrado como adoptarla nos permite resolver de manera relativamente sencilla algunas paradojas aparentemente tan disímbolas cómo la del prefacio, la lotería y el *sorites*. La idea central de dicha propuesta es que, cuando aceptamos una afirmación lo hacemos bajo lo que he llamado *condiciones de normalidad*. Estas condiciones contienen información implícita que restringe, de manera derrotable, la manera en que interpretamos las afirmaciones que aceptamos. Es por ello que para que podamos combinar información contenida en diferentes afirmaciones, que hemos aceptado por separado, como premisas en una inferencia epistémicamente vinculante, sea necesario que la conclusión pueda aceptarse de manera conjunta con todas ellas. En otras palabras, no basta que la conclusión se siga de manera lógica de las premisas, sino que también es necesario que la conclusión sea aceptable bajo el supuesto de que las condiciones de normalidad de las premisas se satisfacen. Esto es precisamente lo que pasa en paradojas como la de la lotería, el prefacio y el *sorites*. Tenemos una consecuencia contradictoria que se sigue lógicamente de varias premisas completamente aceptables, pero cuyas condiciones de normalidad no son consistentes con dicha conclusión. Si tengo razón, esto significa que dicha inferencia, pese a ser lógicamente válida no es epistémicamente vinculante, es decir, podemos aceptar las premisas y rechazar la conclusión contradictoria sin caer en inconsistencia.

Referencias

[Barwise(1983)] Jon Barwise & John Perry. *Situations and Attitudes*. MIT Press, 1983.

[Fara(2000)] Delia Graff Fara. Shifting Sands: An Interest-Relative Theory of Vagueness. *Philosophical Topics*, 28(1):45–81, 2000.

REFERENCIAS

[Gómez-Torrente(2010)] Mario Gómez-Torrente. The Sorites, Linguistic Preconceptions, and the Dual Picture of Vagueness. *R. Dietz y S. Moruzzi (eds.), Cuts and Clouds. Essays in the Nature and Logic of Vagueness, Oxford University Press*, 228–253, 2010.

[Green(1999)] Mitchell S. Green. Attitude Ascription's Affinity to Measurement. *International Journal of Philosophical Studies*, 7(3):323–348, 1999.

[Kroedel(2013)] Tomas Kroedel. The Lottery Paradox, Epistemic Justification, and Permissibility. *Logos & Episteme*, 4(1):103–111, 2013.

[Littlejohn(2013)] Clayton Littlejohn. Don't know, don't believe: reply to Kroedel. *Logos & Episteme*, 4(2):231–238, 2013.

[Zuluaga(2005)] Mauricio Zuluaga. El Problema de Agripa. *Ideas y Valores*, 54(28): 61–88, 2005.

Algunos problemas de epistemología social: Probabilidades grupales dinámicas

Eleonora Cresto
CONICET / (Consejo Nacional de Investigaciones Científicas y Técnicas, Argentina)

Índice

1. Introducción   169
2. Conocimiento común y conocimiento distribuido   171
3. El conocimiento grupal como concepto dinámico   174
4. Probabilidades grupales   176
5. Algunos refinamientos   180
6. Conclusiones   184

1. Introducción

Este trabajo tiene como objetivo final proponer y modelar un concepto satisfactorio de probabilidad grupal. La pregunta por un concepto adecuado de probabilidad grupal puede verse como un aspecto particular del problema más general de cómo producir una agregación (combinación) adecuada de actitudes intencionales individuales en el marco de un colectivo, para que éste pueda ser considerado *agente* grupal con pleno derecho. Es notorio que no hay acuerdo sobre cómo debe proceder dicha agregación, ni para el caso de juicios proposicionales ni, sobre todo, para el caso de juicios de probabilidad personal, o grados de creencia[1]. Si Juan, María y Rosa tienen diferentes

---
[1] Para una presentación general del problema de la agregación de probabilidades en el marco más general de la agregación de juicios, véase el capítulo 7 de Grossi y Pigozzi (2014). En Dietrich (2010) se encuentra una propuesta positiva interesante, que retomaremos en la Sección 5. Para una bibliografía comentada sobre cómo combinar distribuciones de probabilidad véase Genest y Zidek (1986).

grados de confianza en la pericia de Raúl para completar a tiempo un trabajo asignado, y entre los tres tratan de consensuar sus opiniones, ¿cuál es la confianza que debería otorgarle el colectivo, en tanto entidad grupal, a la afirmación de que Raúl termina el trabajo a tiempo? La manera de encarar la agregación de probabilidades puede diferir, además, según se trate de un panel de expertos con mayor o menor pericia que emiten su dictamen sobre un tema técnico (donde lo que nos importa es que el colectivo se acerque a la afirmación correcta), o pares con iguales derechos que expresan sus opiniones subjetivas (donde lo que nos importa sobre todo es reflejar con *equidad* las creencias de los individuos), entre otras posibilidades.

La posición que voy a adoptar aquí deja abierta la respuesta final por el mecanismo de agregación (aunque discutiré algunas propuestas), pero en cambio trata de resolver un aspecto previo del problema. A la hora de pensar en probabilidades grupales, la pregunta por el mecanismo de agregación recién aparece cuando se ha solucionado un asunto más fundamental, a saber, cuando se ha tomado una decisión respecto de cuáles son las *probabilidades individuales que cuentan*, cuáles son las funciones que de hecho debemos agregar. En este sentido, no es lo mismo agregar probabilidades previas que probabilidades posteriores; si decidiéramos agregar probabilidades posteriores, a su vez, hay que tener claro sobre qué pedimos que los agentes individuales condicionalicen. Argumentaré aquí que lo que debe agregarse son funciones individuales condicionalizadas *sobre el conocimiento grupa*l. Ahora bien, el punto de partida de este trabajo es la constatación de que por 'conocimiento grupal' se suelen entender cosas muy diferentes; en particular, hay al menos dos maneras bien diferenciadas en que podemos atribuir conocimiento proposicional a un grupo: el conocimiento común, y el conocimiento distribuido. Con lo cual habría al menos dos maneras de condicionalizar. Según entiendo, la distancia que encontramos en los casos típicos entre el conocimiento común y el conocimiento distribuido nos sugiere un concepto de conocimiento grupal esencialmente dinámico; una concepción adecuada de las probabilidades grupales heredará dicho carácter dinámico.

En la Sección 2 presentaré algunas herramientas estándar de lógica epistémica que permiten caracterizar tanto al conocimiento común como al conocimiento distribuido. En otro lado he desarrollado un argumento que justifica, como ideal normativo, la exigencia de que los agentes se 'apropien', por así decir, del conocimiento distribuido del grupo, de modo que conocimiento común y distribuido coincidan. Aquí no reproduciré dicho argumento ni profundizaré en consideraciones filosóficas relacionadas, ya que nos llevarían demasiado lejos[2] En cualquier caso, es interesante explorar qué herramientas técnicas tenemos para modelar el pasaje de uno a otro. En la Sección 3 explicaré cómo la maquinaria de las 'lógicas de anuncios públicos' (*public announcement logics*) pueden cerrar la brecha entre ambos. En la Sección 4 presentaré mi perspectiva sobre probabilidades grupales dinámicas, en un marco kripkeano enriquecido con probabilidades; veremos cómo usar también las 'lógicas de anuncios públicos' en

---

[2]Véase Cresto (2015).

el ámbito probabilístico. En la Sección 5, finalmente, exploraré algunas propuestas de agregación de probabilidades, y estableceré conexiones con varios resultados técnicos de la bibliografía reciente. En la Sección 6 ofreceré algunas conclusiones.

2. Conocimiento común y conocimiento distribuido

Consideremos un lenguaje proposicional L_E enriquecido con operadores epistémicos para n agentes diferentes, en el modo usual, de modo que podamos expresar:

(i) El agente i sabe que ϕ : $K_i\phi$

(ii) Los agentes $1\ldots n$ saben que ϕ : $E_{\{1\ldots n\}}\phi$

(iii) $E^{k+1}\phi =_{\text{def}} EE^k\phi$

(iv) El grupo compuesto por los agentes $1\ldots n$ tiene conocimiento común de que ϕ : $C_{\{1\ldots n\}}\phi = \wedge_{k=0}^{\infty} E^k\phi$

donde i es cualquier agente, y ϕ es una oración cualquiera del lenguaje (véase Fagin *et al* (1995)).

En otras palabras, sea '$E_{\{1\ldots n\}}\phi$', para cualquier oración ϕ, la afirmación de que todos en el grupo $\{1\ldots n\}$ saben que ϕ. Cuando se sobreentienda de qué agentes estamos hablando abreviaré la expresión anterior simplemente como '$E\phi$'. A falta de un rótulo mejor, llamémoslo *conocimiento compartido* de que ϕ. Desde luego, los agentes pueden tener *conocimiento compartido* del hecho de que tienen *conocimiento compartido*, como en (iii). '$C\phi$', por su parte, representa el *conocimiento común* (del grupo dado) de que ϕ: esto significa que no solamente los agentes del grupo saben que ϕ, sino que saben que saben que ϕ; también saben que saben que saben que ϕ, y así infinitamente; el conocimiento compartido aquí ha sido infinitamente iterado.[3] El concepto de conocimiento común evidencia una cohesión particular del colectivo, que no tenemos en el caso del conocimiento compartido; puede haber conocimiento compartido entre individuos que no conforman un *grupo* en ningún sentido interesante. En esta línea, se ha argumentado algunas veces que las creencias de un genuino agente grupal requieren siempre de algún tipo de intencionalidad común.[4]

La semántica estándar de L_E en términos de estructuras de Kripke es bien conocida. Consideremos una estructura $M = <W, R_1, \ldots, R_n \ldots, v>$, en la que W es un conjunto de mundos posibles, $R_1, \ldots, R_n$ son relaciones de accesibilidad entre

---

[3] Para una explicación más detallada del concepto de conocimiento común véase por ejemplo Geanakoplos (1994), y Vanderschraaf y Sillari (2007).

[4] Véase por ejemplo Tuomela (1992), o Gilbert (1989). El objetivo de estos autores es en verdad el concepto de creencia grupal, más que de conocimiento. Sin embargo, en la medida en que el conocimiento grupal pueda identificarse con un tipo de creencia grupal particular, sus consideraciones pueden importarse también al presente ámbito de discusión.

mundos (una para cada agente), y v es una función de valuación. Como es usual, supondré que las proposiciones son conjuntos de mundos posibles. Como en cualquier modelo kripkeano, las relaciones de accesibilidad regulan el comportamiento de los operadores K_i de conocimiento, para cada agente i[5]. Presupondremos que R_i es por lo menos reflexiva, para todo i, lo que garantiza la factividad del conocimiento (es decir, garantiza que los agentes sólo conozcan oraciones verdaderas). Además, para cada agente i y mundo w, sea $R_i(w)$ la proposición más fuerte conocida por i en w (con lo cual dicho conjunto de mundos estará incluido en cualquier otra proposición sabida por i); esto es, para todo agente i y todo mundo w:

$$R_i(w) = \{w' : wR_iw'\}$$

Así, si una oración ψ expresa exactamente la proposición '$R_i(w)$', entonces desde luego el agente sabe que ψ en w, i.e.: si $R_i(w) = \{w' : M, w' \models \psi\}$, entonces $M, w \models K_i\psi$.

Por su parte, la relación de accesibilidad que caracteriza al conocimiento común es la clausura transitiva de la unión de todas las relaciones de accesibilidad de los agentes individuales, es decir, la clausura transitiva de la unión de todas las R_i que teníamos antes. A la relación de accesibilidad que gobierna al operador C la llamaremos RC; correspondientemente, $RC_{1,...n}(w)$, para todo w, será el conjunto más fuerte que representa todo lo que el grupo $\{1, ... n\}$ conoce en w. Nuevamente, omitiremos la referencia a agentes $1 ... n$ cuando sea claro contextualmente de qué grupo se trata. Así,

$$RC(w) = \{w' : wRCw'\}$$

El modelo estándar, como es claro, nos permite hacer afirmaciones sobre lo que diferentes agentes saben en diferentes mundos. Con un poco mayor de precisión, una vez que fijamos la terminología básica, diremos que $Know_i(w)$ representa el *conjunto de proposiciones* que el agente i conoce en w, para algún w en M (en un tiempo dado); como es evidente, se tratará del conjunto de proposiciones que contiene a $R_i(w)$:

$$Know_i(w) = \{A \in 2^W : R_i(w) \subseteq A\}$$

Por otro lado, digamos que $Know_G(w)$ representa lo que el grupo sabe en w, para algún w en M (en un tiempo dado). Si el conocimiento grupal proposicional es interpretado como conocimiento común (lo cual aún está por verse que sea una buena idea), entonces, en w, el grupo conoce todas las proposiciones que contengan $RC(w)$:

$$(GCK) \quad Know_G(w) = \{A \in 2^W : RC(w) \subseteq A\} \quad \text{para un } w \text{ y } W \text{ dados.}$$

---

[5] En Cresto (2012) sugiero considerar una secuencia de relaciones de accesibilidad para cada agente, ya que allí presupongo que existe una jerarquía de operadores de conocimiento distintos para órdenes superiores. El modelo se complementa luego con secuencias de funciones de probabilidad evidencial. En esta oportunidad voy a prescindir de dichas herramientas, por razones de simplicidad.

2 Conocimiento común y conocimiento distribuido

Es importante mencionar que el conocimiento común satisface transparencia, independientemente de cómo sean las relaciones de accesibilidad R_i de cada individuo. Es decir que, para toda oración ϕ, todo mundo w y todo modelo, obtenemos:

$$\models C\phi \longrightarrow CC\phi$$

Este es un resultado inmediato, dada la semántica ofrecida para el operador C. Hay un sentido, pues, en el que podríamos decir que el conocimiento común funciona como una suerte de autoconciencia grupal 'cuasi-hegeliana', para decirlo un tanto metafóricamente, según la cual cada agente individual se ve involucrado en un proceso de reconocimiento infinito sobre las actitudes epistémicas de los otros individuos.

Obsérvese además que las siguientes fórmulas son válidas, en todo modelo:

$$\models CE\phi \longrightarrow C\phi$$

$$\models CK_i\phi \longrightarrow C\phi \text{ para todo } i \text{ y todo } \phi$$

La moraleja es que, toda vez que pidamos conocimiento común sobre otras actitudes proposicionales, tendremos inmediatamente conocimiento común sobre los objetos de las actitudes de primer orden. Otra manera de decirlo podría ser: la autoconciencia grupal no puede limitarse a otras actitudes de primer orden. Si nos comprometemos con la idea de que todo auténtico conocimiento grupal proposicional requiere de autoconciencia grupal en el sentido fuerte mencionado más arriba, entonces todo conocimiento grupal pasa a ser conocimiento común. No hay intermedios.

Por otro lado, siguiendo nuevamente a Fagin *et al.* (1995), diremos que en un mundo dado w un grupo de individuos tiene *conocimiento distribuido* de que ϕ (lo que abreviaremos '$D\phi$') si la *intersección* de todas las relaciones de accesibilidad de los agentes nos deja llegar sólo a mundos-ϕ desde w. A la relación de accesibilidad que gobierna el operador D la llamaremos RD. Así,

$$RD(w) = \{w' : wRDw'\}$$

No es difícil conceder que al menos en algunas circunstancias, nuestra comprensión pre-teórica de qué sea el conocimiento grupal proposicional se acerca notablemente al concepto de conocimiento distribuido. Esto ocurre, por ejemplo, toda vez que atribuimos a un agente grupal actitudes intencionales que no coinciden con la de ninguno de sus miembros. Supongamos que Pedro y Arianna forman parte de un panel de expertos; Pedro sabe que, dado el estado de la red pluvial, si caen más de 15 mm de agua en pocos minutos el centro de La Plata se inunda; Arianna por su parte ha hecho la predicción de que habrá lluvias muy intensas (digamos, caerá un mínimo de 30 mm de lluvia en muy poco tiempo) sobre La Plata durante el fin de semana. Ni Pedro ni Arianna están por sí mismos en condiciones de emitir un alerta meteorológico, pero el equipo formado por ambos sí lo está (y de hecho el alerta puede efectivamente materializarse si hubiera una tercera persona - digamos, un vocero que no pertenece al grupo, o una computadora - encargada de procesar los informes de los miembros). Es bueno recordar aquí que, según algunos autores, la *posibilidad* de que un grupo genere actitudes intencionales que no coincidan con las de sus miembros resulta crucial para poder hablar de *agencia* grupal propiamente dicha, como diferente de una mera agregación de individuos (de manera paradigmática, cf. Pettit (2001, 2003); List y Pettit (2011)). Esto parece motivar la idea de que el conocimiento grupal en w algunas veces debería capturarse como el conjunto de proposiciones que incluyen $RD(w)$:

$$(GDK) \quad Know'_G(w) = \{A \in 2^W : RD(w) \subseteq A\} \quad \text{para un } w \text{ y } W \text{ dados.}$$

3. El conocimiento grupal como concepto dinámico

La existencia de conocimiento distribuido no trivial (esto es, el que no coincide con el conocimiento de los individuos) nos permite identificar un desiderátum para los agentes ideales: a saber, nos pide que ajusten sus estados epistémicos de modo de poder incorporar en ellos el conocimiento distribuido del grupo. He discutido en otras oportunidades la justificación filosófica de este desiderátum[6]. Pero aún sin formular un argumento a este efecto, obsérvese que resulta en principio razonable suponer que es deseable para los agentes individuales, *ceteris paribus,* apropiarse del conocimiento grupal. Dicha apropiación puede representarse con la ayuda de un conjunto de reglas dinámicas que capturen la manera en la que un modelo inicial evoluciona a través del tiempo, generando así una *secuencia* de modelos.

Para implementar esta idea deberemos enriquecer el lenguaje con operadores de la Lógica de Anuncios Públicos (*Public Announcement Logic*), con su semántica usual. Me referiré a dicho lenguaje como L_{EP}. Consideremos pues una secuencia $S = <M_0, \ldots M_k>$, con dominio común W, tal que:

(a) Nuestra secuencia comienza con una estructura M_0, o modelo base.

(b) Encontramos luego sucesivos requerimientos de cambios sobre M_0. Tales cambios pueden originarse a partir de 'anuncios' o comunicaciones hechas de buena fe. El efecto típico de tales anuncios es la modificación de las relaciones de accesibilidad de algunos miembros del grupo. Permitir que la información 'circule', digamos, puede provocar cambios en el estado epistémico de uno o más agentes, generando una secuencia de modelos $M_1 \ldots M_k$. Así, necesitamos encontrar, para cada oración ϕ sobre la cual el grupo tiene conocimiento distribuido en el modelo base (en un mundo w), una sucesión de oraciones $\psi_1 \ldots \psi_n$ de L_{EP} tales que, luego del anuncio de la oración $[\psi_i]$, ψ_i se vuelva verdadera, hasta llegar a ϕ. Más precisamente: para cada oración ϕ y mundo w: si $M_0, w \models D_{1\ldots n}\phi$, existen oraciones $\psi_1 \ldots \psi_k$ tales que:

---
[6]Cresto (2015).

3 El conocimiento grupal como concepto dinámico

$$M_1, \quad w \models [\psi_1]\phi_1$$
$$M_2, \quad w \models [\psi_2]\phi_2$$
$$\vdots$$
$$M_k, \quad w \models [\psi_k]\phi_k, \text{ donde } \phi_k = C\phi$$

Así pues, para cualquier oración ϕ_i y mundo w:

Si $M_i, w \models D_{\{1...n\}}\phi_i$, entonces habrá algún $j \geq i$ tal que $M_j, w \models C_{\{1...n\}}\phi i$ En el límite, todo el conocimiento distribuido de M_0 deviene conocimiento común en algún modelo de la progresión.

Esta tarea dota a las lógicas de anuncios públicos de una nueva justificación teórica, y a la vez de nuevos objetivos formales. Por un lado, de ahora en más podemos interpretarlas como herramientas para monitorear, dentro de un sistema de modelos, el grado de progresión hacia la satisfacción plena de los compromisos epistémicos de los individuos involucrados. Por otro lado, y ahora desde un punto de vista técnico, las lógicas de anuncios públicos ahora enfrentan el desafío de encontrar maneras eficientes de construir una ruta desde '$D\phi$' a '$C\phi$', para cada oración ϕ y cada mundo, dado que tendremos diferentes maneras de definir la ruta mencionada, o, en otras palabras, diferentes secuencias S.

(c) Finalmente, nuevas interacciones de los agentes individuales con el medio pueden resultar en cambios ulteriores en sus relaciones de accesibilidad, esta vez para reflejar la adquisición de nuevo conocimiento *fáctico* (esto es, conocimiento sobre proposiciones fácticas), lo cual a su vez generará nuevo conocimiento distribuido. Así, eventualmente obtendremos un nuevo modelo $M'_0 \in S'$, con lo cual todo el proceso recomienza.

Así, la responsabilidad epistémica ideal del grupo y la de los individuos que lo componen mantienen entre sí una tensión dialéctica: encontramos por un lado una presión inflacionaria, que apunta a permitir (y fomentar) la generación de creencias y conocimientos grupales que vayan más allá de las actitudes intencionales individuales, pero una vez que las tenemos aparece inmediatamente otra presión en sentido opuesto (deflacionaria), que apunta a la apropiación de las mismas por parte de los individuos.

Esta manera de concebir la naturaleza de la agencia grupal tiene un corolario inmediato para el *conocimiento* grupal. Sugiero aquí que el conocimiento grupal constituye un fenómeno complejo que no puede ser reducido enteramente a conocimiento común o distribuido - ni tampoco, en realidad, a ninguna otra actitud proposicional, en sentido estricto. El conocimiento grupal en un momento particular, y en un mundo w, no es entonces un conjunto de proposiciones, sino más bien una *expectativa de apropiación* de un conjunto particular de proposiciones. Para decirlo de modo diferente: el conocimiento grupal, en un mundo w, es una posible trayectoria, o tarea a realizar, más que un objeto fijo. Esta es desde luego una manera informal y algo metafórica de expresarnos, pero podemos dar una formulación más precisa:

Para un modelo base M_0, y un mundo w:

$$(GK) \quad Know''_G(w) = \langle\langle RD_0(w), RC_0(w)\rangle, \mathcal{F}\rangle$$

donde '$RD_0(w)$' es la proposición más fuerte conocida distribuidamente en w y M_0; '$RC_0(w)$' es la proposición más fuerte de la que se tiene conocimiento común en w y M_0; y '$\mathcal{F}$' refiere a una familia de posibles secuencias $S_i = \langle M_0, \ldots, M_j \rangle$, definidas del modo indicado anteriormente, que nos llevan de '$RD_0(w)$' a '$RC_0(w)$', en el siguiente sentido:

Para cada S_i, existe un $M_j \in S_j$ tal que $RC_j(w)$ es la proposición más fuerte de la que se tiene conocimiento común en w en el modelo M_j; y $RC_j(w) = RD_0(w)$.

Así, el primer miembro del par '$\langle (RD_0(w), RC_0(w) \rangle$' exhibe la distancia que los agentes ideales deben recorrer para llegar a su meta. Cada trayectoria en $\mathcal{F}$ puede verse como una secuencia de pasos (más o menos eficientes) para efectivamente conseguir esta meta, a saber, para conseguir que el conocimiento distribuido coincida con el conocimiento común.

Es importante advertir que esta secuencia de pasos no puede servir como un conjunto de instrucciones a ser conscientemente implementadas por los agentes reales: es el *teórico* el que ve la posibilidad de construir dicha sucesión de modelos. De hecho ningún agente puede satisfacer por sí mismo los requisitos, ya que su cumplimiento depende del comportamiento agregado de todos los miembros del grupo. Y, desde luego, simplemente decirles a los agentes que se esfuercen por comunicar todo lo que saben resulta impracticable. Una manera de sobreponerse a esta dificultad es pedirles que revelen sus probabilidades personales acerca de los hechos que el grupo quiera establecer en un momento dado. Si modelamos las probabilidades individuales en un marco kripkeano, podemos tenerlas también como objeto de los anuncios. Un marco más rico para hablar del conocimiento de un grupo, entonces, incluirá no solamente operadores de conocimiento, sino *asignaciones de probabilidad*. Esa es la tarea de nuestra próxima sección.

4. Probabilidades grupales

Sea $M_P = \langle W, R_1, \ldots, R_n, P_1, \ldots, P_n, v \rangle$ un modelo kripkeano, donde $W, R_1, \ldots, R_n$ y v son como fue explicado anteriormente, mientras que $P_1, \ldots, P_n$ son medidas de probabilidad previa con aditividad finita para agentes 1 a n, respectivamente, sobre subconjuntos de W. Voy a suponer que W es contable, y que las funciones de probabilidad previa satisfacen *regularidad*, esto es, para toda proposición A y todo agente $i : P_i(A) = 0$ sii $A = \emptyset$[7].

---

[7] El requisito de regularidad resulta un supuesto natural en el presente contexto, dado que los elementos de W son en sentido intuitivo posibilidades genuinas (o, por construcción, el teórico los omitiría en primer

4 Probabilidades grupales

Definimos luego la *probabilidad evidencial* (o posterior) en un mundo w, para un agente i, como la probabilidad previa de i condicionada sobre la proposición más fuerte conocida por i en w:

$$P_i, w(A) = P_i(A|R_i(w)) \qquad \text{para cualquier } A \text{ y } w$$

Adviértase que dicha probabilidad condicional está siempre bien definida, dado que hemos supuesto regularidad, y dado que R_i es al menos una relación reflexiva, para todo i. Por otra parte, siguiendo a Williamson (2014), abreviaremos el conjunto de mundos en el que la probabilidad evidencial es r (para algún r en [0,1]) como:

$$[P_i(A) = r] = \{w : P_{i,w}(A) = r\}$$

Dicha abreviatura nos proporciona una herramienta para transformar funciones metalingüísticas en proposiciones entendidas como conjuntos de mundos. Debemos también extender nuestro lenguaje de manera que ahora seamos capaces de expresar oraciones acerca de atribuciones de probabilidad. Sea L_{EPP} dicho lenguaje. En estas páginas no voy a hacer explícita la manera de construir dicha extensión. Simplemente subrayaré en cada caso la proposición relevante (tal como es expresada en el metalenguaje), como abreviatura de alguna oración de L_{EPP} que exprese exactamente dicha proposición[8].

En este marco, es fácil identificar una manera inmediata de lidiar con las probabilidades *del grupo* en un mundo dado. Por analogía con el caso individual, podemos capturarlas por medio de un conjunto de medidas individuales condicionadas sobre la proposición más fuerte conocida por el grupo en ese mundo:

$$(PG) PG_{\{1...n\},w}(A) = \{P_1(A),\ldots,P_n(A)\}|G_{\{1...n\}}(w)$$
$$PG_{\{1...n\},w}(A) = \{P_1(A|G_{\{1...n\}}(w)),\ldots,P_n(A|G_{\{1...n\}}(w))\}$$

para cualquier proposición A y cualquier mundo w, donde '$G_{\{1...n\}}(w)$' es la proposición más fuerte conocida por el grupo en w.

Nuevamente, omitiremos la referencia a los agentes cuando no haya riesgo de confusión.

Ahora bien, según vimos en secciones anteriores, hay al menos dos sentidos en los que podemos hablar de 'la proposición más fuerte conocida por el grupo en un

---

lugar). En cuanto a la exigencia de que W sea contable, en verdad nada esencial se sigue de ello. Si relajamos este supuesto, simplemente deberemos especificar un cuerpo sigma apropiado sobre W como dominio de la función de probabilidad (mientras que si W es contable podemos suponer que éste coincide con el conjunto potencia de W). En cuanto al requisito de aditividad finita, obsérvese que, debido al requisito de regularidad, si W es infinito (numerable) $P_i(\{w\})$ será positivo, para cualquier mundo w y cualquier agente i, de modo que no podemos tener aditividad contable.

[8] Al igual que ha ocurrido con los operadores de conocimiento, aquí presento sólo una versión simplificada del modelo que considero más apropiado. Así pues, consideraré solo un operador de probabilidad para cada agente i, y no una jerarquía tipificada de P_s.

mundo'. De manera correlativa, tendremos dos maneras diferentes de condicionalizar:

$$(PG1) PD_w(A) = \{P_1(A), \ldots, P_n(A)\} | RD_{\{1\ldots n\}}(w)$$

$$= \{P_1(A|RD(w)), \ldots, P_n(A|RD(w))\}, \text{ para todo } A \text{ y } w$$

$$(PG2)\ PC_w(A) = \{P_1(A), \ldots, P_n(A)\} | RC_{\{1\ldots n\}}(w)$$

$$= \{P_1(A|RC(w)), \ldots, P_n(A|RC(w))\}, \text{ para todo } A \text{ y } w$$

Aquí la ecuación (PG1) es el resultado de condicionalizar sobre conocimiento distribuido, mientras que (PG2) condicionaliza sobre conocimiento común. (PG1) y (PG2) se adecuan de manera natural a la concepción de los grupos como entidades inestables, o dinámicas, que sugerimos en la sección anterior. En el límite, la imposición de requerimientos normativos sobre los individuos epistémicamente responsables (como dijimos antes, el requisito de que los individuos "se apropien" del conocimiento del grupo) hará que (PG1) y (PG2) coincidan.

Recuérdese que el conocimiento grupal puede interpretarse como una ruta a ser recorrida, una ruta que conecta el conocimiento proposicional distribuido con el común. De la misma manera, las probabilidades grupales también pueden ser interpretadas como una familia de trayectorias (esto es, una familia de secuencias de modelos) que conectan diferentes maneras de condicionalizar medidas individuales. Así, para un modelo de base M_0 y un mundo w, tendremos:

$$(PG \text{ Dinámica}) \qquad PGDin_w = \langle (PD_w, PC_w), \mathcal{F}_p \rangle$$

PG visto desde una concepción dinámica será pues, una vez más, un par ordenado, donde

- PD_w es la probabilidad distribuida del grupo en w;

- PC_w es la probabilidad común del grupo en w, y

- '$\mathcal{F}$' refiere a una familiar de posibles trayectorias que nos llevan de la una a la otra:

    como antes, las trayectorias nos llevan de un modelo M_0 a un modelo M_k en una posible secuencia de modelos, tal que la probabilidad común grupal en M_k coincide con la probabilidad grupal distribuida en M_0.

Como antes, cada trayectoria $\mathcal{F}$ puede identificarse con una secuencia de modelos generada por sucesivos anuncios públicos, pero la secuencia de oraciones relevantes $\psi_1 \ldots \psi_k$ ahora consistirá en oraciones que expresan las probabilidades *posteriores* de agentes particulares en el grupo. En otras palabras, las oraciones relevantes $\psi_1 \ldots \psi_k$ para anuncios públicos (aquellas que determinan una trayectoria particular) serán probabilísticas.

Incidentalmente, obsérvese que esta trayectoria probabilística constituye lo que podríamos llamar un *generador de puntos ciegos* (a *blindspot generator*), en el sentido de que, los agentes no pueden tener conocimiento común de sus discrepancias probabilísticas. Así, en varias instancias, los agentes que revelen mutuamente sus probabilidades posteriores estarán forzados a *modificar* o corregir tales asignaciones,[9] a la luz de la nueva información adquirida. De hecho la dinámica que encontramos aquí es una generalización obvia del teorema bien conocido de Aumann (Aumann 1976), según el cual los agentes no pueden tener conocimiento común del hecho de que tienen probabilidades posteriores diferentes, si parten de las mismas probabilidades previas[10]. Si la relación básica de accesibilidad no es una relación de equivalencia los cambios serán típicamente más lentos que si contáramos con particiones; pero, en cualquier caso, el conjunto de mundos que representen la proposición más fuerte conocida por cada agente se reducirá progresivamente hasta que coincida con el conocimiento distribuido del grupo, inclusive si los agentes comienzan con probabilidades previas diferentes. Desde luego, si las probabilidades previas no coinciden, las probabilidades posteriores tampoco necesitan coincidir.

Este es un punto interesante, porque nos advierte sobre el hecho de que, cuando tratamos con probabilidades, la comunicación no solamente genera la posibilidad de aprender lo que ignorábamos, sino que también genera la posibilidad de involucrarnos en auto-corrección, ¡a pesar de que las relaciones de accesibilidad que estamos considerando son solamente epistémicas, y no doxásticas! (con lo cual uno esperaría que nadie pueda estar equivocado acerca de lo que sabe).

Conceptualizar la trayectoria mencionada en términos de generación de puntos ciegos es interesante, además, porque nos permite establecer conexiones entre ciertas afirmaciones probabilísticas y otras proposiciones estructuralmente incognoscibles aparentemente muy dispares, tales como las proposiciones mooreanas (afirmaciones del tipo "Ahí hay un conejo, pero no lo creo"), para poder eventualmente adoptar una perspectiva unificada sobre todas ellas

---

[9]Existe un fenómeno de punto ciego epistémico, o doxástico, cuando un agente no puede pronunciar una aserción (normalmente una aserción en primera persona, aunque hay casos más generales) sin producir un enunciado anómalo; en presencia de algunos supuestos sobre la racionalidad de los hablantes, dicho enunciado anómalo puede convertirse en una contradicción formal. En Van Ditmarsch et al (2008) se discuten algunas conexiones interesantes entre la generación de puntos ciegos y la lógica epistémica dinámica.

[10]Véase por ejemplo Geanakoplos y Polemarchakis (1982), para una aproximación dinámica del teorema de Aumann.

5. Algunos refinamientos

La propuesta anterior puede ser refinada de varias maneras. En primer lugar, podríamos pedir que los conjuntos en las ecuaciones $(PG1)$ y $(PG2)$ sean conjuntos convexos de medidas de probabilidad, en cuyo caso las probabilidades grupales podrían ser representadas por sub-intervalos en $[0,1]$ (recuérdese que un conjunto de probabilidades es convexo si para cualesquiera $P1$ y $P2$ que pertenezcan al conjunto, para toda proposición A y todo número real α en $[0,1]$: $\alpha P1(A) + (1-\alpha)P2(A) = P3(A)$, tal que $P3$ también pertenece al conjunto). Se podría objetar, sin embargo, que los conjuntos (sean o no convexos) no nos proveen de verdaderos métodos de *agregación*. En otras palabras, podría objetarse que el problema de la agregación de probabilidades debe entenderse como el problema de encontrar una *función única* que pueda asignarse al grupo como un todo. Esto es por supuesto discutible, y en estas páginas no voy a argumentar a favor de dicha idea, pero en lo que sigue la presupondré por mor del argumento. Para responder a esta preocupación consideremos una función de agregación F sobre las medidas condicionales correspondientes:

$$(PG') \qquad PG'_w(A) = F[\{P_1(A|G(w)), \ldots, P_n(A|G(w))\}]$$

Lo cual resultará en:

$$(PG1') \quad PD'_w(A) = \quad F[\{P_1(A|RD(w)), \ldots, P_n(A|RD(w))\}]$$
$$(PG2') \quad PC'_w(A) = \quad F[\{P_1(A|RC(w)), \ldots, P_n(A|RC(w))\}]$$

según condicionalicemos sobre conocimiento común o conocimiento distribuido. Si adoptamos la llamada 'doctrina Harsanyi' sobre las probabilidades previas (esto es, la idea de que los agentes racionales no pueden desacordar sobre sus probabilidades previas, de modo que $P_i = P_j$, para todo agente i,j), entonces F se vuelve redundante, por supuesto.

Existen muy diferentes propuestas en la bibliografía sobre cómo definir F[11]. Para mencionar sólo algunas soluciones bien conocidas, podríamos pedir que F sea una:

(a) Agregación lineal (ej., la media aritmética):

$$F(P_1 \ldots P_n) = w_1 P_1 + \cdots + w_n P_n$$

(b) Agregación geométrica:

$$F(P_1 \ldots P_n) \propto P_1^{w1} \ldots P_n^{wn}$$

(c) Agregacion supra-bayesiana

$$F(P_1 \ldots P_n)(A) = F(A|P_1 \ldots P_n)$$

---
[11] Véase Genest y Zidek (1986).

(d) Agregación multiplicativa [donde 'π_i' refiere a la probabilidad posterior de i]

$$F(P_1\ldots P_n)(A) \propto \alpha \frac{\pi_1\ldots\pi_n}{P_1\ldots P_n} F(P_1\ldots P_n)$$

Es claro que según cómo sea F podremos instanciar diferentes propiedades. En lo que sigue exploraré algunas consecuencias de pedir que F sea la media geométrica de las funciones del conjunto, con pesos iguales. Esto garantizará que tanto $(PG1')$ como $(PG2')$ satisfagan una condición conocida como 'Bayesianismo externo' ('*external Bayesianity*'), según la cual la agregación es conmutativa con la actualización bayesiana. Esta propiedad es una consecuencia directa del hecho de que la media geométrica de un conjunto de cocientes es igual al cociente de un conjunto de medias geométricas.

Otras consecuencias son menos evidentes. En lo que sigue mostraré, en primer lugar, que $(PG1')$ resulta equivalente a la propuesta de Dietrich (2010), que se presenta como un tipo de agregación multiplicativa (**Proposición I**); dada esta equivalencia, las propiedades estudiadas por Dietrich para su propuesta se aplican también a $(PG1')$, como es obvio, aunque Dietrich use una terminología diferente. Claro que, en el presente contexto, $(PG1')$ no agota todo lo que hay que decir sobre probabilidad grupal: es un punto de partida, más que un punto de llegada. $(PG2')$ nos espera al final del camino. Una vez puestas las cosas de este modo, puede probarse que la estrategia *general* de condicionalizar sobre lo que el grupo sabe y luego agregar las funciones resultantes no necesita decirnos nada sobre las probabilidades posteriores individuales de los agentes (i.e., las probabilidades individuales condicionalizadas sobre lo que *cada agente* sabe). Esto es particularmente claro para $(PG2')$ (**Proposición II**). Argumentaré que, en la medida en que las probabilidades posteriores individuales desempeñan algún papel en $(PG1')$, esto se debe exclusivamente al hecho de que dichas probabilidades posteriores *están implícitas* en el conocimiento distribuido $RD(w)$. A su vez, la explicación de este fenómeno es que un grupo si puede tener conocimiento distribuido del hecho de que diferentes individuos cuenten con diferentes probabilidades posteriores, aun cuando *sus probabilidades previas coincidan* (**Proposición III**). Este resultado puede verse como una suerte de contraparte, para el conocimiento distribuido, del teorema clásico de Aumann (1976) para conocimiento común.

Proposición I: *Sea* $M = \langle W, R_1, \ldots, R_n, P_1, \ldots, P_n, v \rangle$ *como se definió anteriormente, y sea F una función de agregación de opinión geométrica, con pesos iguales. Para cualquier* $w \in W, (PG1') = (PG1Dietrich) = (PG1Post)$, *donde las instrucciones provistas por* $(PG1')$, $(PG1Dietrich)$ *y* $(PG1Post)$ *son como sigue:*

(PG1′) Condicionalice cada función de probabilidad previa $P_1, \ldots, P_n$ sobre $RD(w)$, y luego combine las medidas resultantes;

(PG1Dietrich) combine las probabilidades previas $P1, \ldots, P_n$ y luego condicionalice sobre $RD(w)$.

(PG1Post) combine las probabilidades *posteriores* de cada agente (en w) y luego condicionalice sobre $RD(w)$.

Más precisamente, si F es una función de agregación de opinión geométrica:

(PG1′) $F[\{P_1(A|RD(w)), \ldots, P_n(A|RD(w))\}] =$

(PG1Dietrich) $F[\{P_1(A), \ldots P_n(A)\}]|RD(w) =$[12]

(PG1Post) $F[P_{1,w}(A), \ldots, P_{n,w}(A)\}]|RD(w)$

La prueba es inmediata, de modo que la dejo en manos del lector.

Proposición II: *Sea* $M = \langle W, R_1, \ldots, R_n, P_1, \ldots, P_n, v \rangle$ *como antes, y sea F una función de agregación de opinión geométrica, con pesos iguales. Para cualquier mundo* $w \in W$ *en el que $RD(w)$ y $RC(w)$ no coincidan:* $(PG2') = (PG2*) \neq (PG2Post)$, *donde las instrucciones provistas por* $(PG2'), (PG2*)$ *y* $(PG2Post)$ *son como sigue:*

(PG2′) Condicionalice cada función de probabilidad previa $P_1, \ldots, P_n$ sobre $RC(w)$, y luego combine las medidas resultantes;

(PG2∗) Combine las probabilidades previas $P_1, \ldots, P_n$ y luego condicionalice sobre $RC(w)$;

(PG2Post) Combine la probabilidad posterior de cada agente (en w) y luego condicionalice sobre $RC(w)$.

Más precisamente: si F es una función de agregación de opinión geométrica (con pesos iguales):

(PG2′) $F[\{P_1(A|RC(w)), \ldots, P_n(A|RC(w))\}] =$

(PG2∗) $F[\{P1(A), \ldots P_n(A)\}]|RC(w) \neq$

(PG2Post) $F[\{P_1(A|R_1(w) \cap RC(w)), \ldots, P_n(A|R_n(w) \cap RC(w))\}]$

excepto en el caso en que $RD(w) = RC(w)$.

Demostración. Una vez más, la equivalencia entre $(PG2')$ y $(PG2*)$ es inmediata. En cuanto a $(GP2Post)$, obsérvese que:

(1) $F[\{P_1(A|R_1(w)), \ldots, P_n(A|R_n(w))\}]|RC(w) =$

(2) $\dfrac{F[\{P_{1,w}(A \cap RC(w)), \ldots, P_{n,w}(A \cap RC(w))\}]}{F[\{P_{1,w}(RC(w)), \ldots, P_{n,w}(RC(w))\}]} =$

(3) $F[\{P_1(A \cap RC(w)|R_1(w)), \ldots, P_{n,w}(A \cap RC(w)|R_n(w))\}] =$

(4) $\dfrac{F[\{P_1(A \cap RC(w) \cap R_1(w))}{P_1(R_1(w))}, \ldots, \dfrac{P_n,(A \cap RC(w) \cap R_n(w))\}]}{P_n(R_n(w))} =$

(5) $F[\{P_1(A|R_1(w)), \ldots, P_n(A|R_n(w))\}]$.

(6) Supóngase ahora que $(PGPost) = (PG2')$

(7) Entonces para cualquier $i : R_i(w) = RC(w)$ [de (5)]

(8) Por lo tanto $RD(w) = RC(w)$. □

De (1) a (2) simplemente explicitamos la demanda de $(PG2Post)$ de condicionalizar la agregación sobre $RC(w)$. Pero tomar la media geométrica de la probabilidad *posterior* de cada agente (en w) y condicionalizar sobre $RC(w)$) equivale en verdad a tomar la media geométrica de cada función, condicionalizada *tanto* sobre $RC(w)$ como sobre la proposición más fuerte conocida por cada agente i, o $R_i(w)$) (paso (3)). Esto a su vez, por definición de probabilidad condicional, equivale a agregar cocientes del modo en que se muestra en (4). Pero la proposición más fuerte conocida por cada agente está incluida en la proposición más fuerte de la cual el grupo tiene conocimiento común, con lo cual podemos deshacernos de $RC(w)$ (paso (5)). Por lo tanto, de hecho lo que obtenemos es simplemente una agregación de probabilidades posteriores; el conocimiento común se ha evaporado de la escena. Con lo cual (PG2 Post) claramente no necesita coincidir con (PG2'), esto es, con el resultado de condicionalizar sobre el conocimiento común.

Desde luego, las seis formulaciones [$(PG1')$, $(PG1Dietrich)$, $(PG1Post)$, $(PG2')$, $(PG2*)$, y $(PG2Post)$] coincidirán si el conocimiento común y el conocimiento distribuido también coinciden, esto es, si todos los agentes conocen la misma proposición más fuerte en un mundo dado. Esta es evidentemente una situación idealizada que nos retrotrae a los requerimientos normativos de secciones anteriores. En el límite, la presión deflacionaria discutida anteriormente *disuelve la diferencia* entre la agregación de probabilidades previas y posteriores.

Finalmente, la **Proposición III** nos dice que las probabilidades posteriores individuales están *implícitas* en $RD(w)$, en el sentido preciso de que un grupo puede ciertamente tener conocimiento distribuido del hecho de que los agentes cuenten con diferentes probabilidades posteriores, *aún si las probabilidades previas coincidieran:*

Proposición III: *Un grupo puede tener conocimiento distribuido del hecho de que sus miembros cuentan con diferentes probabilidades posteriores, aún si tuvieran las mismas probabilidades previas.*

Demostración. Por razones de simplicidad, supóngase que $M = \langle W, R_1, \ldots, R_n, P_1, \ldots, P_n, v \rangle$, tal que $R_1 \ldots R_n$ son diferentes relaciones de equivalencia sobre $W \times W$, y $P_i = P_j$, para todo i, j. Sea $B = R_i(w)$, para algún individuo i y algún mundo w. Esto garantiza que $P_{i,w}(B) = 1 \neq P_{j,w}(B)$, para algún $j \neq i$ (ya que por hipótesis no todas las relaciones de accesibilidad coinciden). Por lo tanto $M, w \models K_i(P_i(B) = 1)$ & $K_j(P_j(B) = r)$, para $r < 1$. Por ende $M, w \models D[(P_i(B) = 1)$ & $(P_j(B) = r)]$. □

De modo que el grupo, *qua* grupo, puede 'acordar en desacordar', aunque no lo pueden hacer sus miembros. Aquí el conocimiento distribuido funciona como una actitud proposicional de segundo orden sobre las actitudes *probabilísticas* de primer orden de los individuos. La presión deflacionaria hará posible que el desacuerdo se evapore, si hay suficiente acuerdo en el punto de partida (esto es, en las probabilidades previas).

En síntesis, en general, condicionalizar sobre el conocimiento del grupo no es equivalente a agregar probabilidades posteriores individuales. Las dos ideas coinciden para la versión $(PG1')$ de la propuesta, pero esto se debe simplemente a que, para cualquier mundo w, el grupo puede tener conocimiento distribuido de dichas probabilidades posteriores; *qua* grupo, el colectivo puede 'acordar en desacordar' sobre sus probabilidades.

En cualquier caso, obsérvese que informar públicamente las probabilidades posteriores de cada quien eventualmente permitirá, en una situación idealizada, cerrar el hiato entre $(PG1')$ y $(PG2')$. Con lo cual advertimos que, a efectos del concepto de probabilidad grupal, el verdadero papel de las probabilidades posteriores individuales consiste en posibilitar la progresión hacia nuestro desiderátum.

6. Conclusiones

En estas páginas defendí que, a la hora de agregar probabilidades, las funciones que cuentan son las funciones de probabilidad de los individuos condicionalizadas sobre el conocimiento del grupo. Esta propuesta no es sino la manera natural de generalizar el concepto de probabilidad evidencial: los agentes, ya sean individuales o colectivos, condicionalizan sobre lo que conocen.

Con este objetivo en mente, comencé por identificar dos sentidos en los que puede decirse que un grupo conoce un conjunto de proposiciones, y a partir de allí propuse un sentido dinámico, no proposicional, de conocimiento grupal. El conocimiento grupal en un momento dado no es un conjunto fijo de proposiciones, sino una *expectativa de adquisición* de proposiciones; de esta manera, el conocimiento grupal en un momento dado estará determinado no sólo por lo que cada agente conoce, sino por lo que cada agente *ignora* acerca de lo que los demás conocen. Según argumenté, esta idea puede representarse en términos de una familia de posibles trayectorias entre el conocimiento distribuido y el conocimiento común. Las lógicas de anuncios públicos proveen aquí las herramientas indicadas para recorrer la distancia entre ambos tipos de conocimiento proposicional, lo cual permite dotarlas de nuevos objetivos conceptuales y de nuevos desafíos formales. De un modo análogo, las probabilidades grupales pueden pensarse en términos de una familia de posibles trayectorias entre dos tipos de condicionalización. En este punto es interesante advertir que toda trayectoria probabilística de este tipo es generadora de puntos ciegos; conceptualizarla de este modo favorece la obtención de una perspectiva unificada sobre fenómenos aparentemente disímiles.

Finalmente, examiné algunas consecuencias de combinar las probabilidades individuales condicionalizadas sobre el conocimiento grupal mediante una función de agregación geométrica. En particular, mostré que las probabilidades posteriores individuales no desempeñan ningún papel en la estrategia *general* de obtener probabilidades grupales por condicionalización sobre el conocimiento del grupo, aunque sí pueden desempeñar un papel importante a la hora de delinear el mencionado conjunto de trayectorias. Sin embargo, por otro lado puede probarse que la estrategia de utilizar una función de agregación geométrica sobre medidas condicionalizadas sobre el conocimiento distribuido del grupo es equivalente a un procedimiento de agregación multiplicativo, que toma explícitamente en cuenta las medidas posteriores individuales. La explicación es interesante: se puede probar que el grupo de hecho puede tener conocimiento distribuido de las probabilidades posteriores de los agentes individuales. Esta es una suerte de contrapartida del teorema de Aumann (1976); a diferencia del conocimiento común, el conocimiento distribuido permite 'acordar que desacordamos'; claro está, el acuerdo no lo suscribe ningún agente concreto, sino el grupo en tanto tal.

Referencias

[Aumann(1976)] Robert Aumann. Agreeing to Disagree. *The Annals of Statistics*, 4: 1236–1239, 1976.

[Cresto(2012)] Eleonora Cresto. A Defense of Temperate Epistemic Transparency. *Journal of Philosophical Logic*, 41 (6):923–955, 2012.

[Cresto(2015)] Eleonora Cresto. El conocimiento grupal de agentes epistémicamente responsables. *Veritas*, 60 (3):460–482, 2015.

[Dietrich(2010)] Franz Dietrich. Bayesian Group Belief. *Social Choice and Welfare*, 35:595–626, 2010.

[Fagin et al.(1995)Fagin, Halpern, Moses, y Vardi] Ronald Fagin, Joseph Y. Halpern, Yoram Moses, y Moshe Y. Vardi. *Reasoning About Knowledge*. Cambridge, Massachusetts: MIT Press, 1995.

[Geanakoplos(1994)] John Geanakoplos. Common Knowledge. *Handbook of Game Theory, Vol.2, ed. Robert Aumann and Sergiu Hart. Elsevier Science B.V.*, 2: 1438–1496, 1994.

[Geanakoplos y Polemarchakis(1982)] John Geanakoplos y H. Polemarchakis. We Can't Disagree Forever. *Journal of Economic Theory*, 28(1):192–200, 1982.

[Genest and Zidek(1986)] Christian Genest and James Zidek. Combining Probability Distributions: A Critique and an Annotated Bibliography. *Statistical Science*, 1: 114–148, 1986.

[Gilbert(1989)] Margaret Gilbert. *On Social Facts*. London and New York: Routledge, 1989.

[Grossi y Pigozzi(2014)] Davide Grossi y Gabriella Pigozzi. *Judgment Aggregation: A Primer Synthesis Lectures on Artificial Intelligence and Machine Learning*. Morgan & Claypool Publishers, 2014.

[List y Pettit(2011)] Christian List y Phillip Pettit. *Group Agency: The Possibility, Design, and Status of Corporate Agents*. Oxford: Oxford University Press, 2011.

[Pettit(2001)] Philip Pettit. *Deliberative Democracy and the Discursive Dilemma*. Philosophical Issues (Supp. Nous), Vol 11, 2001.

[Pettit(2003)] Philip Pettit. Groups with Minds of Their Own. *Frederick Schmitt (ed.), Socializing Metaphysics. New York: Rowman and Littlefield*, pages 167–193, 2003.

[Pivato(2008)] Marcus Pivato. *The Discursive Dilemma and Probabilistic Judgment Aggregation*. Munich Personal RePEc Archive, 2008.

[Tuomela(1992)] Raimo Tuomela. Group Beliefs. *Synthese*, 91:285–318, 1992.

[Van Ditmarsch et al.(2008)Van Ditmarsch, Van der Hoek, y Kooi] Hans Van Ditmarsch, Wiebe Van der Hoek, y Barteld Kooi. *Dynamic Epistemic. Logic*. Springer, 2008.

[Vanderschraaf y Sillari(2007)] Peter Vanderschraaf y Giacomo Sillari. *Common Knowledge*. Stanford Encyclopedia of Philosophy. URL: http://plato.stanford.edu/entries/common-knowledge/, 2007.

[Williamson(2014)] Timothy Williamson. Very improbable knowing. *Erkenntnis*, 79: 971–999, 2014.

Remarks on the use of formal methods

Décio Krause*   Jonas R. Becker Arenhart

Research Group in Logic and Foundations of Science
Department of Philosophy
Federal University of Santa Catarina

Contents

1	**Introduction**	**188**
2	**On what presupposes what**	**190**
3	**Adequacy**	**194**
4	**Explication**	**200**
5	**Summing up**	**206**

Abstract

In this paper we discuss the relation between formal theories and the informal concepts they aim to formalize. Three questions are addressed and discussed: i) what are the resources required in order to develop a formal system? Usually, there is a sense that much of what one aims at formalizing is already required beforehand, so that a kind of explanatory circularity arises. We propose that only the resources of a constructive metatheory is required. ii) Then, we discuss the idea that some formal system may adequately capture an informal concept. It could be supposed that once a formal theory is developed, it could be judged as to its accuracy in describing the target concepts it formalized. There is a sense in which this can be done, through the use of the so-called squeezing arguments. We present such strategy, but the cases where such an argument for adequacy are available are still very rare. iii) The third point we shall develop concerns the cases when the formal system is clearly at odds with the informal notion it attempts to formalize. This is rather the rule in formalization, and we argue that it should

---
*Partially supported by CNPq.

be seen in positive lights in many cases. The Carnapian notion of explication helps us clarify what is going on in these situations. It is in these cases that the use of formal methods may be employed to discovery and to the elucidation of informal concepts. We exemplify the points with some case studies and, due to the pluralism of logics, we argue that the very role of formal techniques involving theories and concepts still need to be studied and considered with due care.

1 Introduction

The use of formal methods is thought to bring clarity and rigor to many distinct fields, from mathematics to philosophy. Using formal languages to clarify philosophical problems is even thought of as an adequate method for philosophy of science(see Lutz [Lutz(2011)], for a recent defense). However, employing the tools of formal methods also generates questions that are difficult to answer, questions that exemplify typical philosophical conundrums that the sheer use of formal methods cannot answer. In this paper, we present some of these questions. We divide the main questions in three groups.

The first question, which typically intrigues beginners as well as veterans in logic courses is this: in order to develop and formally study logic, don't we need to use and understand logic beforehand? From another, but related, perspective: in order to develop formal versions of arithmetics, don't we need to have knowledge of that same arithmetics beforehand? Or, more specifically: in trying to define the number 'two' in formal arithmetics, for instance, don't we already use the very concept of 'two'? There is a feel that one is begging the question somehow, or running through explanatory circles, if one is to use a notion that must be understood in order to clarify that very same notion. Formal tools may only work to the purpose of presenting what one already knows. Finally, the same happens with a fundamental notion such as the notion of identity: some (as the authors of this paper) claim that we can develop logics where the notion of identity does not make sense for certain objects. But in making such a move, are we not assuming that identity holds for everything right from the start? The concepts we *use* to develop a formal theory can be said to be captured and enlightened by that same theory?

For the second question, we selected a related (albeit different) problem, which may be put thus: what is the precise nature of the relation between the understanding of an informal concept we have, mainly intuitively and pre-theoretically, with the concept that stems from the formalization we develop in order to systematize that very concept? There is not a single answer to this question, as we shall see, but we hope to give a first attempt at bringing to light the connections involved. A formal system may relate very differently with the informal notion it attempts to capture, ranging from an attempt to faithfully capture most of its distinct shades to attempts at a complete revision of informal uses through the regimentation provided by formalization. For

1 Introduction

instance, consider logical consequence, as captured in classical logic. Does it anyhow attempt to capture how we actually reason, or merely some aspects of our inferential practices which are deemed somehow relevant for certain purposes? In case of disagreement between typical inferential practices and a system of logic, which should have precedence, the formal or the informal? Those are difficult questions, eluding simple answers.

For another related problem, the third that we shall touch on, what if a formal system has unexpected consequences, contradicting our intuitive judgements about the formalized concepts? Should those consequences be dismissed as outright incorrect, showing thus the incorrectness of the formalization, or should they be seen as opening new perspectives for the investigation of the field? What, if anything, is a correct formalization of a concept? Should those consequences be taken to represent features of the target system or are they better understood as a kind of artifact of the formalism? Most of the times, formal theories have unexpected consequences that perhaps only a thorough study of the formalization could bring to light. We must learn to distinguish interesting consequences that may be theoretically fruitful from others that are mere artifacts of the formalization. Of course, this is easier to state than to do.

As the previous paragraphs make clear, then, this is a paper about the relation between an intuitive informal concept and its formal version(s) in an appropriate formal system. We shall explore three aspects of this relation, as encapsulated in these three questions. In this sense, it is being supposed here that the systems to be discussed are thought of as applied, in the sense that they all attempt to mimic aspects of the workings of a natural practice or phenomenon, instead of being developed for sheer mathematical curiosity. In this sense, for instance, the problem with logic, when applied to the study of inferences, is how to relate the inferences we already make in natural language with the resulting systematization of usual systems of logic. In the case of arithmetics, we would be interested in understanding the relation between the results established in formal systems with the informal arithmetics of usual mathematical practice. So, the issues we shall discuss are related with the so-called 'canonical application' of a system.

Of course, nothing prevents someone from developing a formal system (a logical calculus, for instance) just for the sake of pure mathematics, without having applications in mind. But in general this is not the way logicians and those working on the foundations of science proceed. In most of the paradigmatic cases, they have something in mind — a previous field of knowledge they intend to lay in clear foundations — and for doing that, the axiomatic method is in fact a precious tool. The interesting fact is that the formal theory they create may present unsuspected characteristics, never dreamt before. This may either bring interesting new fields for investigation or else insurmountable problems; the resulting system may be employed to enlighten the practice, or else be rejected in terms of its discrepancies with it. In order to approach the scheme, we shall consider that the scientist firstly has a 'theory' of a certain field of knowledge in mind, which we shall call *informal theory* (IT), say Galilean mechanics, Darwin's evolution theory, informal arithmetics, Cantorian set theory. The question to be considered is this: in axiomatizing one of these theories, do we 'preserve' the original one? In other words: does the axiomatized version correspond precisely to *the same* theory we had before? Can we attempt to grant a kind of identity between the formal and the informal version? Is it always desirable? We shall see that in most cases we cannot ensure that perfect adequacy, and perhaps the best we can do is to argue that the axiomatized version *captures* relevant aspects of the original theory, while leaving others aside.

This paper is supposed to be a preliminary attempt at clearing the grounds for a fuller discussion, putting the terms and the problem to be investigated in grounds that are as clear as possible. We believe that the examples and the argumentation we shall provide will suffice for starting the discussion.

2 On what presupposes what

Let us begin by addressing our first question: how to spell the relation between the concepts we use — informally and pre-theoretically — in order to develop a formal axiomatized theory, and the concepts that are being captured by such a formalization?[1] Is it the case that we are employing the very concepts we want to formalize in order to develop the formal theory, providing for a kind of explanatory circle?

We shall be concerned here with the formalization of logic and *scientific theories*, although this concept is rather vague.[2] It is widespread in the literature that scientific knowledge is, in great measure, conceptual knowledge; we approach science, at least when we are concerned with theories, by means of concepts represented in mathematical structures, which on their turn are described by the language of the theory (see [Krause and Arenhart(2016)]). In mathematics, say, in arithmetics, we use the notions of natural number, of zero, of successor of a natural number, of addition and multiplication of natural numbers, of prime numbers, and so on. In its informal development, we don't know beforehand which concepts we shall use as basic concepts; we introduce them as we go on, as they are being required. Calculus, Analytic Geometry[3] are other examples of informally developed mathematical theories. In the field of empirical sciences we find most of the paradigmatic cases of intuitively developed theories, with few concerns for axiomatic organization; very few empirical theories can be said to have been introduced axiomatically (perhaps Newton's mechanics is one of the few

---

[1] We shall suppose that the axiomatic theories we are talking about are formalized, in the sense that formal languages are being employed in the axiomatization.

[2] In fact, 'scientific theory' may also be the subject of the kind of discussion we are providing for other concepts in this paper. There are many approaches to provide for formal counterparts for the very notion of scientific theory. See also [Krause and Arenhart(2016)] for a discussion.

[3] In the sense of the study of geometry by means of systems of coordinates.

2 On what presupposes what

exceptions).

When we proceed axiomatically, we select some basic concepts to start with, and the postulates describe their inner relations. In the case of axiomatic arithmetics (call it PA for Peano Arithmetics), we can define 'one' as the sucessor of zero (a primitive term), 'two' as the sucessor of the sucessor of zero, and so on. But in defining, say, 'two', are we not presupposing the notion of two already? In a certain sense, yes. In formulating PA, we need to acknowledge, say, that we have *two* concepts, *two* distinct symbols in the basic language, and so on.

An interesting suggestion about what is needed in order to develop a formal theory and its relation with pre-theoretical concepts is advanced by Newton da Costa. In his book *Ensaio Sobre os Fundamentos da Lógica* [da Costa(1980), p.57] (see also [Krause and Arenhart(2016), chap.3]), da Costa suggests that the formalization of a field of knowledge is the most appropriate way to enlighten the workings of reason in that field. This is done, roughly, by the rigorous presentation of a underlying logic and of the internal relations holding between the basic concepts, articulated through the postulates. In order to axiomatize any field of knowledge, to ground the informal knowledge we already have, what is presupposed is a pragmatical Principle of Reason which he calls *The Principle of Constructivity*, which reads as follows:

> "[t]he whole exercise of reasoning presupposes that it has a certain intuitive ability of constructive idealization, whose regularities are systematized by intuitionistic arithmetics, with the inclusion of its underlying logic."

The idea is rather simple. In order to reason while we are still providing for formal systems of logic and others, one still must employ reason itself; there is no way to operate on a vacuum of rational principles. However, reason itself shows its operation through the workings of logic. How then? In order to avoid a circle, da Costa suggests there is a minimal nucleon of rational activity presupposed in every activity of reason itself. This very minimum is encapsulated by the *pragmatic* principles of reason.[4] In particular, according to da Costa, in order to employ language and symbolic activity one must presuppose the inferences and the mathematics available in intuitionistic arithmetic and the underlying intuitionistic logic. That is a minimal core required for a formal system to get off the ground. As an analogy, in the development of formal language and linguistic activity, intuitionist arithmetics plays a role similar to material (inhaltlich) arithmetic plays in Hilbert's program (see Hilbert's classic [Hilbert(1967)]), although intuitionistic arithmetic already seems to surpass in much the resources Hilbert was willing to grant his finitistic mathematics, at least as conceived by himself in his published material.

Of course, of most relevance here is also the distinction between object language and metalanguage. The metalanguage encapsulates the minimal resources mentioned,

---

[4] We shall have occasion to meet other of such principles in the next section.

while the object language, by itself, needs not obey such restrictions. By relying only on such apparently meagre resources one is entitled to develop formal systems of classical logic, classical mathematics, and even of many typical branches of empirical sciences, when those theories are presented as a syntactical object (more on this soon). So, we certainly need to use concepts such as 'two' in order to develop a formal system of arithmetic, but this is part of the resources of the metalanguage, which makes the development of formal systems available.

However, there is more even to logic than mere construction of formal systems. Typically, more than a syntactical approach to logical consequence through proofs and deductions, there is also a semantic approach, through model theory. Ideally, the very notion of valid argument that is the typical target notion in logic courses is couched intuitively in terms of truth preservation: a valid argument is one in which the truth of the premises somehow grants the truth of the conclusion. So, the semantic notions are important too (some people say they precede syntactical notions of deduction, but this is an issue we shall not enter in here). Similarly, in empirical sciences, this shift to the semantical approach is achieved through the model theoretic approach to scientific theories (see again the details in [Krause and Arenhart(2016)]). In this case, much richer resources than intuitionistic arithmetics are needed.

But how to proceed? Well, in the case of logic, and even in the case of the semantical approach to empirical sciences, the usual resources are the set theoretical structures. That is, structures developed inside a typical set theory such as Zermelo-Fraenkel. In such theory, we provide for the structures where the formal languages are interpreted, and a notion of truth is defined following Tarski's approach, which allows us to define a semantic notion of logical consequence. Of course, this is not truth *per se*, but rather a set theoretical definition of truth.[5] Other definitions are available, and the issue of whether such a definition really captures every aspect of truth and semantical features of natural languages is an open one (again, we shall not discuss these issues here). So, in the end, in order to define logical consequence, one needs to define truth in a model, which on its turn requires set theory reasonably developed. Well, but set theory operates on a notion of logical consequence also, so what is at issue?

Perhaps the solution to the problem may be provided by Kenneth Kunen's suggestion that "formal logic must be developed *twice*" [Kunen(2009), p.191]. The idea may be put as follows: in order to study model theory and characterize the semantics for formal languages (including the semantic concept of logical consequence in precise terms), we must develop the model theory inside a set theory, which on its turn requires logic previously developed. As he says,

> "[w]e start off by working in the metatheory (...). As usual, the metatheory is completely finitistic, and hence presumably beyond reproach. In

---

[5]Some would say, a syntactical substitute for truth, given that the defined truth predicate is operated on the syntax of set theory; see also da Costa [da Costa(1980)] chap.2, sec. IX.

the metatheory, we develop formal logic, including the notion of formal proof. We must also prove some (finitistic) theorems about our notion of formal proof to make sense of it. (...) Then we go on to develop [say] ZFC [The Zermelo-Fraenkel set theory with the Axiom of Choice], and *within* ZFC, we develop all of standard mathematics, including model theory, most of which is not finitistic. To develop model theory, we again must develop formal logic." (*ibid.*)

In other words, and generalizing the claim to other branches of knowledge, in developing formal arithmetics (PA), we of course make use of the notion of 'two', but in the finitistic metatheory, which is in general an informal mathematical setting.[6] The informal use of number two doesn't cause conflicts with the formal notion being developed, except in the sense we shall consider in a few moments that perhaps the defined notion does not fit the intuitive one as expected (if any such fitting is expected at all).

The real trouble appears when the notions we model through such a process seem to contradict the very concepts employed in order to model them. As a simple example, consider the case of identity. The notion of identity is one of the most debatable (see [Bueno(2014)], [Krause and Arenhart(2015)]). Some authors say that we cannot rule identity out of any context, for we are using it all the time, say in order to identify two distinct occurrences of a symbol. It is a requisite of the Constructive Principle of Reason it seems, so that identity is required in order to develop any kind of symbolic manipulation. So, although we may develop systems at the object level that may violate some principles of identity, it is debatable whether identity may be violated at every level, without precluding the working of reason itself. We tend to agree with this opinion if we restrict the remark to the meta-level. While we deal with symbols and formalizations, the Constructivity Principle requires that identity applies without limits, given that linguistic symbols are its objects. However, for other domains of objects, to which more sophisticated theories apply, it may be the case that identity may be restricted. So, at the object level we may develop theories that somehow contradict tenets of the meta-level. This is not irrational, given that the objects of the meta-level principles are finitary linguistic entities.

Something similar may be said about paraconsistent logic. Some of these logics, recall, violate a version of the Principle of Non-contradiction, which roughly states that a proposition and its negation cannot be both true. However, at the meta-level, paraconsistent logicians don't assume that the principle gets violated. For instance, they never assume that some expression is *and* is not a formula. The Principle of Constructivity would be useless to Reason unless one may prohibit such claims. As Kunen remarked, the finitary metatheory is "beyond reproach".

---

[6]Of course this metamathematics can also be formalized, which can be done in an informal (finitistic) meta-metamathematics, and so on.

In brief, there is a minimal apparatus that is required in order to develop a formalism. With that minimal apparatus, strong systems such as set theory may be developed, and inside set theory the relation between model theory and proof theory may be studied. In the case of empirical science, a similar remark may be made. A strong framework such as set theory is required in order to develop the models of empirical theories such as quantum mechanics, and it is inside that framework that one can study such models. Models do not exist in a theoretical vacuum.

Notice that the meagre resources provided by the constructive principle of reason are certainly surpassed by the results of the formalizations that one may develop inside such frameworks as set theory. Perhaps one of the best examples is quantum mechanics. As Susskind and Friedman have said,

> "[t]here is an obvious reason why classical mechanics is intuitive: humans, and animals before them, have been using it many times every day for survival. But no one ever used quantum mechanics before the twentieth century." [Susskind and Friedman(2014), p.xix]

We could say, following these ideas, that evolution did not equip us for quantum mechanics, and this may be the reason why we have so much difficulty to understand it (if this is even possible, of course, if we are to believe Feynman[7]). So, we proceed as we can. In principle reasoning with finitary means, but soon achieving strong systems where we can reconstruct at least part of what we had previously presupposed. Or so it seems. Concerning the number two, we have realized that there are a plenty of possible non equivalent reconstructions. Do they address the same entity? Should we prefer one of the formalizations? These are questions to be considered next.

3 Adequacy

What is the aim of formalization? This question hides so many difficulties that it is not possible to make justice to everything already said about it in a single volume. Here we shall present the main lines of how the trouble could be approached in a philosophically interesting way.

To begin with spelling the main difficulties, we mention Newton da Costa again. In his discussion of the pragmatic principles of reason, besides the Principle of Constructivity, with which we already had the opportunity to meet in the previous section, he advances also a *Principle of Adequacy* (see [da Costa(1980), chap.1,sec.VII]). Roughly speaking, in a rational context (which always means a scientific theory for da Costa), an underlying logic must be operating, and in every case, *the Principle of Adequacy* demands that the underlying logic to be chosen must be the one that better

---

[7]It is well known that Richard Feynman said that no one understands quantum mechanics due to its radically outside of scope results.

suits the features of the context. That is, it is supposed that there is a logic that is better equipped to deal with the features of the context, and it is precisely this one that should be adopted. This logic, furthermore, is unique, according to another rational principle, the *Principle of Uniqueness*: in every context, the underlying logic is unique (again, the discussion is to be found in [da Costa(1980), chap.1,sec.VII]).[8]

The main idea underlying such principles seems to be clear: without a logic there is simply no way we may draw inferences and reason properly in a given context. Furthermore, it is not the case that any logic will do: it must be the one that better suits the domain of discourse. This demand seems to originate in the fact that otherwise we would run the risk of being led to inadequate reasoning: it could happen that some conclusions of our inferences do not match with the features of the context. Uniqueness is also a reasonable demand: there is no hope for fruitful debate when distinct and incompatible rules of inference are being employed by the contenders. In this case, there is simply no court of appeal to grant the legitimacy of any piece of reasoning, and anarchy is to be expected. So, adequacy, understood as a demand of reason, seems really desirable.

However, how precisely to cash such notion of adequacy? Is it plausible to say that at least in some cases a formal system perfectly captures every feature of the informal (target) concept it attempts to formalize? Is this complete formalization even desirable in every case?

Notice that da Costa was not the first to hit at the requirement of adequacy. In his pioneer discussion about logical consequence, Tarski [Tarski(1956), p.409] wrote:

> The concept of *logical consequence* is one of those whose introduction into the field of strict formal investigation was not a matter of arbitrary decision on the part of this or that investigator; in defining this concept, efforts were made to adhere to the common usage of language of everyday life.

It seems that Tarski may also be taken to be in the business of trying to adequately capture informal concepts into formal ones. The target concepts must somehow constrain the formalized versions up to a point where one is able to recognize that the later is indeed a rigorous version of the former, and here lies the issue of adequacy. But this is still not an answer to the problem of how to characterize adequacy. In fact, both Tarski and da Costa are not to be misunderstood here, demanding that every formal system perfectly captures the target concepts. The continuation of Tarski's paper is very clear about the role and influence of the informal concept in the formalization. There are "difficulties" that present themselves in such formalizations. It is worth quoting Tarski [Tarski(1956), p.409] again:

---

[8]In his discussion of these principles, da Costa seems to imply that a context is individuated by its underlying logic, so that in a certain sense the logic is constitutive of the context. In fact, a change of logic implies a change of context. Although this remark brings interesting consequences for what we say here, we shall not explore this issue here.

> With respect to the clarity of its content the common concept of consequence is in no way superior to other concepts of everyday language. Its extension is not sharply bounded and its usage fluctuates. Any attempt to bring into harmony all possible vague, sometimes contradictory, tendencies which are connected with the use of this concept, is certainly doomed to failure. We must reconcile ourselves from the start to the fact that every precise definition of this concept will show arbitrary features to a greater or lesser degree.

That is, in general the informal concept that is our target of formal studies is not a clear cut concept awaiting to be captured and formalized. There is in fact a plurality of notions that are involved in the informal notion of consequence, for instance, many of them inconsistent among themselves. A formalization will have to arbitrarily set some borders and delimit the concept to be formalized. What holds for logical consequence also holds for the formalization of other concepts as well, including the approach to the logical connectives, identity, and many others. There is some idealization involved, and the relation between the idealized formal version and the target notion is what is relevant for us. It seems that there is no hope that a notion that is so vague and unbounded can be captured by a formalization, which must be precisely delimited. So, does that mean that the search for adequacy is a quest without hope?

Not really. There is a chance that among the many distinct (and conflicting) senses comprising the informal notion to be formalized, one may also be able to select one among them that, between certain limits, may be well specified in the scope of informal language. If that selection or precisification of the informal notion is possible, an attempt at proving adequacy may be sought, as we shall see. In fact, that selection of a more workable notion of the informal concept is already performed in every case where formalization is employed. It is not just any rough informal notion that is formalized; rather, it is an idealized version of the informal notion that one attempts to formalize.

For an illustration of what is meant by the idealization of the informal notion, consider the study of arguments and our attempts at elaborating logical theories to separate the good arguments (the ones we call valid) from the bad ones (the invalid). To begin with the elaboration of such a theory, we never let any collection of meaningful sentences whatever to count as an argument. In general, we select arguments consisting only of declarative sentences, but not of orders or questions, for instance. That selection already delimits the field, but it does so for a good reason: it is not clear that orders or questions may be said to be true or false, and in general, validity is cashed in terms of such semantic notions. Notice that, informally, the notion of an argument may well comprise orders, for instance, but one does not take it as a failure that an account of validity limits itself to declarative sentences. Other studies may focus on orders and imperatives and attempt a treatment of validity for such cases also.

3 Adequacy

Continuing with our rough description of the kind of precisification involved in the development of a theory of arguments and their validity (which is only an example of the typical procedure of how we select one among many notions in the informal context, remember), we go on to characterize validity. In general validity is supposed to be a matter of truth preservation from premises to conclusion: it is impossible for the premises to be true, while the conclusion is false. When spelled out in more details, this idea is typically understood as comprising a modal component, precisely in the claim of the impossibility. But what are we to understand when we say that it is *impossible* for the premises to be true, while the conclusion is false? Aren't premises and conclusion simply true or false? As it is well known, this modality is explained in terms of logical form and truth in an interpretation of the non-logical vocabulary. It is here that the use of artificial languages in the place of natural languages comes as a useful tool too. They work as surrogates for parts of the vocabulary in natural languages, with a more or less direct correspondence between the artificial language and the natural language.

Logical form is accounted for by a distinction, in the sentences comprising an argument, between the logical and the non-logical terms. So, in order to account for that, we somehow select logical operators that will make for the logical structure or form of the sentences, while non-logical components will be the varying component, the ones that may be variously interpreted in distinct interpretations.[9] By separating a logical part from a non-logical part and allowing that truth is to be understood as relative to an interpretation of the non-logical vocabulary, we spell how the modality of validity is to be taken: there is no interpretation in which the premises are true while the conclusion is false (the issue is developed in Shapiro [Shapiro(2005)]). Notice that the very idea that some pieces of language work as logical constants also require a lot of idealization. In classical logic, for instance, we assume that conjunction implies nothing whatsoever about the temporal order of the conjuncts.

So, by introducing those notions of logical form we leave out of our account inferences that could count as intuitively valid for many people, such as "Floor is taller than Tarja, so it follows that Tarja is smaller than Floor". Given that there is only non-logical vocabulary in this argument, it is possible to write its form as "Tfl, so Slf", where T stands for the binary relation "is taller than", S for the binary relation "is smaller than", and f and l are the names of Floor and Tarja, respectively. Also, depending on how the logical vocabulary is understood, distinct inferences will be valid.

Much more could be said about this idealization of the notion of truth preservation. For instance, notice that truth in an interpretation is already developed with the apparatus of an informal set theory. So, the features of interpretation that are taken to be relevant are purely extensional. In this sense, the most important features of distinct

---

[9] The choice of what counts as "logical" vocabulary is also guided by considerations concerning the purpose of the theory, and there is much idealization here too.

interpretations are the cardinalities of their domains. This is already taking us very close to the familiar Tarskian notion of interpretation in model theory, and very far from the manifold of notions that comprise the intuitive notion of logical consequence (see also Shapiro [Shapiro(2005)]). In fact, the intuitive notion, as so elaborated, is already a worked out concept, a concept that needs to be prepared in order for us to discuss whether it is captured successfully or not by a formalization.

Once much of the fluctuating ideas involved in an informal concept are determined, such as we have highlighted in the previous discussion for the case of logical consequence, can we attempt to prove that a formalization of that notion is successful? That is, once we have a concept of logical consequence that is still informal, but that was idealized to a point that it resembles a lot our typical model theoretical notion of validity, can we grant that those notions somehow match?

Following Smith [Smith(2011)], there is a general strategy in order to try to argue for the adequacy of the resulting formulation: the so-called *squeezing arguments*. The general idea is motivated by its attempt at generalizing Church's thesis, the claim (roughly speaking) that the extension of the intuitive notion of computable function is the same as the set of recursive functions. The previous is an intuitively given concept, while the later is a mathematically characterized class of functions. Notice that computable functions are those functions intuitively characterized as allowing us to calculate, on a step-by-step basis, the value of any given input. The steps are not to be random or arbitrary, but rather each step in the calculating procedure is to be determined by antecedent steps in the calculation. The idea is that there is a kind of algorithmic procedure for the calculation, which leads one step by step to the result, no ingenuity required. According to Church's thesis, this intuitively characterized class of functions is the same as the rigorously defined class of recursive functions.

Of course, Church's thesis is called a 'thesis' precisely because one cannot offer a proof of the identity claim involved. However, there is evidence that the two classes of functions are co-extensional. No one, for instance, managed to find a computable function that could not also be defined through the use of recursive functions. Squeezing arguments attempt to go one step ahead of the statement of a thesis and offer a proof that an informal concept is co-extensional with a formally defined one. In rather rough terms, the idea may be described as follows (we follow Smith [Smith(2011)]). Consider a concept I, and its extension $|I|$. Now, the first step consists in finding precisely defined concepts S and N such that

$$|S| \subseteq |I|$$

and

$$|I| \subseteq |N|.$$

Once that is done, comes the squeezing step; in order to do that, it must also be possible to prove that $|N| \subseteq |S|$ (which presumably may be a theorem in the rigorous theory in which concepts S and N are defined). Granted the squeezing step, we would have (assuming basic facts about concept extensions) that

$$|S| \subseteq |I| \subseteq |S|,$$

which leads us to conclude that $|S| = |I| = |N|$. In this case, our informal concept I is proved to be extensionally identical with the well defined concepts S and N.

Consider now the case of valid arguments as we have roughly elaborated above, that is, with validity couched in terms of truth preservation, with logical form and truth in an interpretation involved (see Smith [Smith(2011), pp.24-25]). Suppose that the language is restricted to a first-order language. This informal elucidation of validity we take to be our concept I, for the intuitive part in the squeezing argument. Now, consider S as the property of being derivable in any formulation of natural deduction for classical first-order logic. Notice that S is precisely defined. One can then argue that $|S| \subseteq |I|$. That is, if one can prove α from Γ in a system of natural deduction, then clearly α follows from Γ in the intuitive sense we have presented. This comes from the fact that the deduction rules are sound according to the intuitive notion of validity, and chaining inferences preserves truth. So, one part of the squeezing argument is done.

For the second part, define N as the property of having no countermodel in the set of natural numbers. This is to be defined in the traditional Tarskian model theory, so that this notion is also precisely defined in mathematical terms. The idea is that when we interpret the non-logical vocabulary in the domain of natural numbers, there is no way to perform this interpretation in such a way that the formulas in Γ are true in the standard model theoretical sense while α is false (idem). If an inference has a countermodel in the set of natural numbers (if N does not hold for it), then it is not the case that it is valid in the intuitive sense (I does not hold for it). Of course, once that a countermodel is found in the domain of natural numbers one may also find an intuitive interpretation (in the sense described above) according to which the premisses are true but the conclusion is false. So, by contraposition, if I holds for this inference, then N also holds, or, $|I| \subseteq |N|$.

So, we have $|S| \subseteq |I| \subseteq |N|$. The last step for a squeezing argument requires that we show that $|N| \subseteq |S|$. But this is a precisely established result, we know from first-order logic that one can prove that whenever there is no countermodel in the set of natural numbers for a derivation of α from Γ, then there is a proof of α from Γ in the selected system of natural deduction. This finishes the squeezing argument.

That is a kind of argument that answers our question: how can adequacy be determined? Well, when a squeezing argument is available, it seems, it is possible to grant that extensionally, at least, the informal argument coincides with well defined ones. That is an answer as precise as possible, however, it does not work when the intuitive concept to be formalized is just too rough. As we remarked, there must be some elaboration of the informal concept, it must be shaped in order to be accommodated in a formal system (see again the discussion in Shapiro [Shapiro(2005)] and Smith [Smith(2011)]). This kind of informal elaboration may not always be possible, and even when it is, it may be difficult to establish a squeezing argument. In these cases, perhaps, adequacy is to be understood as a goal that may be attained even more partially than it is with squeezing arguments. It is to these situations that we now turn.

4 Explication

Perhaps the best example of a monster we create when axiomatizing is provided by first-order arithmetics, PA. A second gigantic monster is of course set theory. We shall speak of both in the sequence. But in general, the thesis can be generalized to any informal theory being axiomatized.

Before the rise of first-order logic, no one had heard about non-standard models, countable models of analysis, and so on. In the same sense, even after set theory being axiomatized by Zermelo and that other set theories were developed, the nature of 'models' of set theory was not completely known, if ever thought about. Today we find results which shock the non specialist in logic, but whose conclusions cannot be put aside (Skolem's and Banach-Tarski's paradoxes are typical examples).

Until the middle of the XIXth century, "logic" was synonym of "Aristotelian logic". Except for few insights such as Leibniz's, who noted that the Aristotelian schema of syllogisms was not enough for capturing most of the mathematical reasoning, the discipline Logic was identified most with Aristotle's logic and with the improvements achieved in the Middle Ages by many important logicians. It is well known De Morgan's example of an argument that cannot be deduced in the theory of syllogisms, namely, the De Morgan Argument (DMA) problem, which claims that it is not possible do show syllogistically that "Every man is an animal" entails that "Every head of a man is the head of an animal" (*apud* [Valencia(2004), p.489]).

Things began to change with Boole, De Morgan, Peirce, Schröder and others, into the tradition of what we know presently as Algebraic Logic, which was sedimented in the XXth century as one of the most relevant parts of Logic. Almost in parallel, a *linguistic tradition* was also developed by Frege, Russell, Peano and others. This tradition imposed itself, and a logic became identified with a language. Frege developed the first logical system able to deal with (at least) arithmetics, which was his target. But Frege's logic is what we call today a higher-order system, close to set theory. The distinction between higher-order and first-order systems was made clear only in 1928 in the book by Hilbert and Ackermann [Hilbert and Ackerman(1950)], although the basic ideas seem to go back to Löwenheim in 1915. In the 1920s and 1930s, we have acquired a deeper knowledge of formal systems and their limitations and applications, including the distinction between syntactical and semantical aspects of axiomatized theories, the different levels of language we commonly use (objects language, metalanguage, metametalanguage, and so on).

The natural turn to empirical sciences was anticipated by Hilbert in his 6th problem of his celebrated list of 23 Problems of Mathematics, a kind of legacy left by the XIXth century to the XXth. There, he noted the necessity of analyzing the basic axioms of the theories of physics. This problem of course does not require a *solution*, but a continuous task of putting the theories into an axiomatic framework. There are of course advantages in doing this: the basic concepts over which we can erect the theory are explicitly mentioned, the postulates over which the theory bases itself are put in a clear way, and syntactical and semantical aspects of the resulting theory can be analyzed.

Some examples come to mind and may help us in explaining what we intend to say. First, consider Mary William's axiomatization of Darwinian theory of evolution. Her purpose was to give "the foundation for a precise, concise and testable statement of Darwin's theory of evolution (i.e., anagenisis)" [Williams(1970)]. It is not important to reproduce her axioms here, but just mention that the resulting theory concerned only with the part of Darwinian theory which is incorporated in natural selection, the modern evolutionary theory, but it does not deal with inheritance for instance (see also [Magalhães and Krause(2000)]). So, as some critics say, Williams does not achieve to the whole Darwinian theory. OK, you can (rightly) say, her objectives were of a different kind. The same can be said of the axiomatization of McKinsey, Sugar and Suppes of classical particle mechanics [Suppes(2002)], which was criticized since it does not involve interactions and collisions. The answer of the authors was that they did not intend to do it; clear and simple (see §4.2 [Krause and Arenhart(2016)]). This is one of the wings of the problems with axiomatization. In general, an informal theory is something not well stated, with no clear assumptions and concepts, so that it is not so easy to find its core and, worst, its limits. The most important thing for our discussion is that which concerns the results we get from the axiomatized version which in most cases were never even supposed to exist.

How should we understand such a situation? An interesting approach was advanced by Carnap in the 50's, and was called *explication*. The idea is roughly as follows: in order to deal scientifically with a concept that has vague borderlines, one must substitute the informal vague concept by a clearly defined concept. As Carnap [Carnap(1956), pp.7-8] puts it:

> The task of making more exact a vague or not quite exact concept used in everyday life or in an earlier stage of scientific or logical development, or rather of replacing it by a newly constructed, more exact concept, belongs among the most important tasks of logical analysis and logical construction. We call this the task of explicating, or of giving an **explication** for, the earlier concept; this earlier concept, or sometimes the term used for it, is called the **explicandum** and the new concept, or its term, is called the **explicatum** of the old one.

As example of explication, Carnap mentions the definitions by Frege and Russell of the term 'two'. The explicandum of course, is the informal notion of two, while the explicatum is the Russellian definition, for instance, of two as the class of two-membered classes. The idea, as it should be clear by now, is that the explicatum is a rigorous and well-defined version of the explicandum, which substitutes the explicandum for scientific purposes.

In [Carnap(1962), chap.1], Carnap develops the issue of explication a little further. It is clear that one can apply the procedure of explication to scientific concepts in general, empirical concepts included, not only mathematical concepts. Now, going in the same direction as in the analysis of Tarski and da Costa on the use of formal tools, which required some kind of adequacy with informal use, Carnap is also worried about the possibility of providing a correct explication. His answer is clear: given that the informal concept we start with (the explicandum) is inexact, there is no chance that we can check whether the explication is correct (see [Carnap(1962), p.4]). In fact, the question itself about the correctness of an explication ends up being senseless due to the lack of the possibility of providing for an answer to it.

Carnap advances not and adequacy criteria, but rather some requirements a good explication must satisfy. The conditions are four: i) similarity to the explicandum, ii) exactness, iii) fruitfulness, and iv) simplicity (see Carnap [Carnap(1962), pp.5-8]). *Exactness* means that the explicatum must be given explicit rules, perhaps in the form of a definition, which introduces it in a system of scientific concepts. *Simplicity* merely requires that the explication must be as simple as possible, after the other requirements are met (for further discussions on these requirements, see [Dutilh Novaes and Reck(2017)]).

The requirements that are more interesting for our purposes in this section are *Similarity* and *Fruitfulness*. Similarity means that the explicatum should apply to most of the same cases where the explicandum applies. However, Carnap does not demand perfect match (which we saw, is an impossible demand, given that the explicandum is inexact); he in fact allows for a difference. That is clearly relevant for us: the newly produced concept may depart from informal use. The example mentioned by Carnap, of the case of Fish versus Piscis is clear ([Carnap(1962), p.6]). Zoologists introduced a concept Piscis as an explicatum for the common notion Fish, which is "far removed from any prescientific language" (idem). So, the similarity required is not absolute. But why did scientists do that? As Carnap remarks, because that kind of explication was *fruitful*; it can be connected with other concepts and be employed to state laws. So, fruitfulness rules over a complete similarity. To be useful in the formulation of universal laws is seen as fruitfulness of an explication.

Due to these last two requirements, it not always easy to state when an explication fails. As Dutilh Novaes and Heck [Dutilh Novaes and Reck(2017), pp.204-205] put it, an explication may be inadequate due to *overgeneration* or *undergeneration*. In the first case, the extension of the explicatum has more instances than the extension of the explicandum, it applies to more things. An example is first order Peano arithmetic, which we shall discuss soon. There are non-standard models, where the domain of interpretation comprises more than the typical ω-sequences that represent the natural numbers (more on this soon). A case of undergeneration is the opposite: the explicatum does not capture some entities that are in the extension of the explicandum. As an example, Dutilh Novaes and Reck (idem) mention again first-order Peano arithmetic: due to Gödel's first incompleteness theorem, such a system cannot produce all the truth about the underlying structure.

Similarity and fruitfulness are in tension, as Dutilh Novaes and Heck [Dutilh Novaes and Reck(2017), p.205] observe. By evading similarity an explication may prove fruitful. We shall consider two cases in mathematics as exemplification.

Arithmetics The first case is that of arithmetics. It is well known from the history of this discipline that the XIXth century witnessed a 'return' to the foundations of mathematics (the so called Arithmetization of Analysis movement) with the aim of making concepts clear, such as that of real number, function (mapping) and so on. The first task was indeed to put arithmetics on a solid ground, since all Analysis depends on this theory. Dedekind and Peano provided axioms for arithmetics, but with the developments of the XXth century it was realized that their theory is what we today call *second-order arithmetics*. As we have said before, one of the most important developments of the XXth century was semantics, and later *semantic models* of Dedekind-Peano's arithmetics could be studied. The result was that all of them are isomorphic, that is, the theory is *categorical*. This looks good, for it seems that all 'realizations' (models) of the axioms differ only in what respects the nature of the involved objects, not concerning their mathematical properties.

But things are not so easy. With the rise of first-order logic, *first-order arithmetics* was also born. That is, the informal theory now grounded on a first-order system of logic. The differences are easy to explain. One of the postulates, the Postulate of Induction, says that if we have a property P which applies to natural numbers (the individuals we intend to deal with), then once the number ZERO (here represented by 0) has this property and from the fact that a number n whatever having the property it follows that the SUCESSOR of n has also this property, it results that *all* natural numbers have the property. The words in capital letters express primitive concepts. There are two ways of formally writing this axiom. In a second order logic, where we can quantify over properties and relations of individuals, it reads:

$$\forall P\Big(P(0) \wedge \forall n(P(n) \to P(s(n))) \to \forall n P(n)\Big). \tag{1}$$

Here, $s(n)$ stands for the sucessor of n. The domain is the set $\mathbb{N} = \{0, 1, 2, 3, \ldots\}$ of natural numbers. We can identify a property with a sub-set of this set (comprising precisely those natural numbers that have the property). So, the first quantifier is speaking of *all* subsets of the domain, that is (this must be acknowledged) $2^{\aleph_0}$ sets or properties, being $\aleph_0$ the cardinal of $\mathbb{N}$. This 'number' (a cardinal number) is strictly grater than $\aleph_0$ as proven by Cantor himself.

Now consider the first-order version of the postulate. In a first order setting, we cannot quantify over properties and relations, but only over individuals. Then the postulate needs to be transformed in a *schema* as follows: if P is a property that holds for natural numbers, then the following is a schema to generate Postulates of Induction, one for each predicate P (formula with just one free variable) we consider:

$$P(\mathbf{0}) \wedge \forall n(P(n) \to P(s(n))) \to \forall n P(n). \tag{2}$$

By hypothesis, our first-order language is recursive, that is, roughly speaking, generated by a countable list of primitive symbols. So, the most we can have is $\aleph_0$ formulas. As a consequence, the postulate (2) speaks of no more that $\aleph_0$ subsets of $\mathbb{N}$, and the resulting theory is essentially weaker than the former, formulated in second-order logic.

But this is not all. As a first-order theory, PA (Peano Arithmetics, according to our previous convention) is committed to the Compactness Theorem. In short, it says that if all finite subsets of a certain set Γ of formulas have a model (that is, there exists an interpretation that makes them true), then Γ itself has a model. What are the consequences?

The first consequence is the existence of non-standard models of PA [Mendelson(2010), pp.224-225]. Let us consider a new theory PA$'$ defined in a language including the language of Peano arithmetic together with a new constant symbol a. The axioms consist of the axioms of PA plus an infinite set of axioms which read: for each numeral n, add the axiom $a > n$. Thus, any finite subset of these axioms is satisfied by a model that is the standard model of arithmetic plus the constant a interpreted as any number larger than any numeral mentioned in the finite subset of PA$'$. Thus by the compactness theorem, there is a model satisfying all the axioms PA'. Since any model of PA$'$ is a model of PA, for a model of a set of axioms is also a model of any subset of that set of axioms, the extended model (of PA$'$) is also a model of PA. The element of this model corresponding to a cannot be a 'standard' natural number (that is $a \neq 0, 1, 2, 3, \ldots$), because it is larger than any standard natural number. The game may continue. Now, consider PA$'$ and add a new constant b ($b \neq a$) and repeat. You will find a sequence of models of PA one involving the precedents, and new 'natural numbers', different from anyone you have known before, is being presented.

Similar results can be obtained in considering the first-order theory of the real numbers, where *non standard* reals appear. Set theory is another field where we can see results which without axiomatization could be not even noted.

4 Explication

Set theory Set theory is attributed to Cantor. His theory is informal, that is, not based on the axiomatic method, and it is known to be inconsistent (for the history and consequences of the paradoxes, see [Fraenkel et al.(1973), chap.1]). Two solutions were proposed for the problem of the paradoxes of set theory, which basically arose due to the Principle of Abstraction, which roughly speaking says that given a property whatever, we can infer the existence of the set whose elements are precisely those individuals that satisfy the property. Bertrand Russell proposed Type Theory as a solution, by restricting the notion of property: the arguments of any property cannot have to the same *type* of the property itself. The second proposal was offered by Ernst Zermelo, who restricted the formation of such sets, imposing that the elements of the set of objects having the property must be taken from a set given already by the theory. Both solutions brought unexpected consequences.

Type theory, and here we shall make reference only to *simple* type theory as developed in the 1920s, provides an hierarchy of types for the universe. Firstly we have individuals, or objects of type 1; then we have properties and relations among individuals, which have type 2; then properties of properties and relations of relations, and so on. It results that there will be different mathematics for different type. There is a zero of each type beyond a certain type, an one, a two, and so on for any mathematical concept. That is, type theory has transformed completely the intuitive idea we had before, and further, constitutes in reality a completely different mathematical framework, in which the very notion of set can be put outside; as said Russell, it is a *no class theory*, that is, a theory where the notion of set (class, in Russell's terms) is absent. Zermelo's axioms also brought interesting results never thought before. For instance, as Mirimanoff has shown in 1919, it enables the existence of *circular sets*, that is, sequences of membership such as the x such that $x \in x_2 \in x_1 \in x_0 \in x$. Today, the theory of non-well-founded-sets has found lots of applications [Aczel(1988)].

Furthermore, it was realized that Zermelo's account was not the only way to approach Cantor's intuitive realm. Several and non-equivalent set theories were born; just to mention a few of them, the systems ZFC, NGB, NF, ML, KM, Church's theory, Ackerman's theory, and so on, all of them bringing us new insights and fields to be explored (again, for an overview see [Fraenkel et al.(1973)]). But this is not everything. Take some of these theories, like ZFC. The axiomatic version also has enlightened the role of the Axiom of Choice (AC) in mathematics.

This axiom was proven by Gödel (1938) and Paul Cohen (1963) to be *independent* of the remaining axioms of of ZF (supposed consistent). So, as it has occurred with the Axiom of the Parallels in Euclidian Geometry, mathematicians have realized that *non-Cantorian* (Cohen's expression) set theories could be developed, where some form of negation of AC is introduced; these theories have theorems which differ from standard set theories and permit the elaboration of alternative mathematics. The subject is very technical to be developed here, so we shall just mention it.

Models of set theory can be obtained also by a method termed *Boolean valued models* [Bell(1985)]. The starting concept is that of a *complete Boolean algebra. But Geisi Takeuti takes the lattice of quantum mechanics (an orthomodular lattice) instead that of a complete Boolean algebra and obtained quantum set theory [Takeuti(1981)], a mathematics also with properties which differ from standard mathematics; one of them is that Leibniz's Principle of the Identity of Indiscernibles is false.*

As we see, axiomatization has also an heuristic role not only in mathematics, but also in the empirical sciences.

In regarding the theories of empirical sciences, perhaps there is a way to control the proliferation of theories: empirical theories must answer to experience.

5 Summing up

There is much more that could be said about the problems concerning the use of formal methods. Our discussion was rather incomplete, but we hope to have brought the issues together in such a way that may prove useful for further exploration. In short: axiomatic and formal systems in general require at least a constructive point from which to depart. With such meagre resources one may elaborate the most sophisticated theories available. There is no real thread of circularity, but there is no possibility of getting everything out of nothing. In axiomatizing or in formalizing known theories, we create new frameworks which, we may attempt to prove, capture the informal content. That may be tried with a squeezing argument. However, that is rarely done, and, don't forget, it requires that the informal notion be first elaborated into a more well-behaved concept. In most of the cases the formal version does not clearly match the informal one; only by resemblance we can say it corresponds to the early theory. In this case, as we argued, the Carnapian notion of explication is useful. There is no proof that the axiomatic/formal version fits the informal version and, in several cases, the axiomatic/formal version presents unsuspected results which the informal theory does not enable us to realize. Typical cases are formal arithmetics and set theory. The question involving theories of empirical sciences is subtler, but similar discussion seems to be fruitful also in this field.

References

[Aczel(1988)] P. Aczel. Non-Well-Founded Sets. *CSLI Lecture Notes Number 14.* Stanford: CSLI Publications, 1988.

[Bell(1985)] J. L. Bell. Boolean-Valued Models and Independence Proofs in Set Theory. Oxford: Oxford Un. Press, 1985.

[Bueno(2014)] O. Bueno. Why identity is fundamental. American Philosophical Quarterly, *51(4):325–32, 2014.*

REFERENCES

[Carnap(1956)] R. Carnap. Meaning and Necessity. *Chicago: The Un. of Chicago Press, 2 edition, 1956.*

[Carnap(1962)] R. Carnap. Logical Foundations of Probability. *Chicago: The Un. of Chicago Press, 2 edition, 1962.*

[da Costa(1980)] N. C. A. da Costa. Ensaio sobre os Fundamentos da Lógica. *São Paulo: Hucitec-EdUSP, 1980.*

[Dutilh Novaes and Reck(2017)] C. Dutilh Novaes and E. Reck. Carnapian explications, formalisms as cognitive tools, and the paradox of adequate formalization. Synthese, *194:195–215, 2017.*

[Fraenkel et al.(1973)] A. A. Fraenkel, J. Bar-Hillel, and Levy. A. Foundations of Set Theory. *Amstredam: North-Holland, 1973.*

[Hilbert(1967)] D. Hilbert. On the infinite. *In* From Frege to Gödel. A Source Book in Mathematical Logic, 1897-1931, *pages 367–392. Cambridge, Mass.: Harvard University Press, 1967.*

[Hilbert and Ackerman(1950)] D. Hilbert and W. Ackerman. Principles of Mathematical Logic. *New York: Chelsea, 1950.*

[Krause and Arenhart(2015)] D. Krause and J. R. B. Arenhart. Is identity really so fundamental?, *2015. Available at: http://philsci-archive.pitt.edu/11295/1/IdentityFundamental.pdf, retrieved in 27/05/2016.*

[Krause and Arenhart(2016)] D. Krause and J. R. B. Arenhart. The Logical Foundations of Scientific Theories: Languages, Structures, and Models. *London: Routledge, 2016.*

[Kunen(2009)] K. Kunen. The Foundations of Mathematics. Studies in Logic, 19. *London: College Pu, 2009.*

[Lutz(2011)] S. Lutz. Artificial language philosophy of science. European Journal for Philosophy of Science, *2:181–203, 2011.*

[Magalhães and Krause(2000)] J. C. M. Magalhães and D. Krause. Suppes predicate for genetics and natural selection. J. Theor. Biology, *209:131–53, 2000.*

[Mendelson(2010)] E. Mendelson. Introduction to Mathematical Logic. *CRC Press, fifth edition, 2010.*

[Shapiro(2005)] S. Shapiro. Logical consequence, proof theory, and model theory. *In* S. Shapiro, editor, The Oxford handbook of the philosophy of mathematics and logic, *pages 651–670. Oxford: Oxford Un. Press, 2005.*

[Smith(2011)] P. Smith. *Squeezing arguments.* Analysis, *71(1):22–30, 2011.*

[Suppes(2002)] P. Suppes. Representation and Invariance of Scientific Structures. *Stanford, CSLI, 2002.*

[Susskind and Friedman(2014)] L. Susskind and A. Friedman. Quantum Mechanics: the Theoretical Minimum. *New York: Basic Books, 2014.*

[Takeuti(1981)] G. Takeuti. *Quantum set theory. In* Current Issues in Quantum Logic, *pages 303–22. Springer, 1981.*

[Tarski(1956)] A. Tarski. *On the concept of logical consequence. In* Logic, semantics, metamathematics. Papers from 1923 to 1938, *pages 409–420. Oxford: Clarendon Press, 1956. Transl. by J. H. Woodger.*

[Valencia(2004)] V. S. Valencia. *The algebra of logic. In D. M. Gabbay and J. Woods, editors,* The Handbook of the History of Logic: Vol.3 – The RIse of Modern Logic: From Leibniz to Frege, *pages 389–544. Elsevier, 2004.*

[Williams(1970)] M. B. Williams. *Deducing the consequences of evolution: a mathematical model.* J. Theor. Biology, *29:343–85, 1970.*

Explanation, Understanding, and Belief Revision*

Andrés Páez
Universidad de los Andes

Contents

1	Introduction	209
2	Three Theses about Explanation	210
3	The Objective Basis of Explanation	212
	3.1 Probability Values	212
	3.2 Epistemic Relativity	215
4	Potential Explanations	217
5	The Epistemic Value of Explanation	219
6	Conclusion	225

1 Introduction

For the longest time, philosophers of science avoided making reference to the notion of understanding in their accounts of explanation. Although Hempel, Salmon, and other philosophers who wrote about the subject in the 20th century recognized that understanding is one of the main goals of science, at the same time they feared that any mention of the epistemic states of the individuals involved would compromise the objectivity of explanation. Understanding is a pragmatic notion, they argued, and

---

*Previous versions of this paper were presented at the *Biennial Meeting of the Philosophy of Science Association* in Vancouver, at the *XVI Jornadas de Epistemología e Historia de la Ciencia* in Córdoba (Argentina), and at the *III Conference of the Latin American Association for Analytic Philosophy* in Buenos Aires. I am grateful to the audiences in these venues for their useful comments and questions.

although a subject worthy of psychological investigation, pragmatics should be kept at a safe distance from the universal, epistemological features of explanation. Although this attitude towards the notion of understanding has changed in the last decade[1], there are still many misgivings about using pragmatic notions in the analysis of one of the central concepts in the philosophy of science[2].

My main purpose of this paper is to defend the idea that there is a sense in which it is meaningful and useful to talk about understanding in an objective sense[3], and that to characterize this notion it is necessary to formulate an account of scientific explanation that makes reference to the doxastic states and epistemic goals of the participants in a cognitive enterprise. It is important to clarify at the outset that my goal is not to offer a general analysis of the notion of understanding, and that my approach is restricted to the understanding of singular facts in well-defined scientific contexts.

The essay is divided as follows. In the next section I introduce three theses about scientific explanation that will serve as the basis for the rest of the discussion. The first thesis, which is defended in sections 3 and 4, states that determining the potential explanations of a fact is essentially a non-pragmatic matter. This thesis is meant to allay the fears of those who see the introduction of pragmatic factors as the beginning of the road towards an unbounded relativism. Since the objective basis of explanation will be probabilistic, at the beginning of the paper I include a detailed discussion about the way in which probability will be used in my account of explanation. The second thesis, which is presented in section 5, states that it is possible to determine the epistemic value of most potential explanations of a fact, and that such value can be established in a non-arbitrary way despite being the result of the evaluation of individual researchers. Finally, towards the end of the essay I explain the third thesis, which establishes the criteria for the acceptance of an explanation in the corpus of beliefs of those researchers involved. These criteria are based on their joint assessment of the credibility and epistemic value of potential explanations.

2 Three Theses about Explanation

It has often been said that explanation is an interest-relative notion. Different inquiring agents impose different demands on the information they regard as explanatorily valuable. The interest-relativity of explanation has been accounted for in several ways: some authors have proposed a contrastive analysis of the explanandum (van Fraassen,

---

[1] See, for example, de Regt (2009), de Regt, Leonelli & Eigner (2009), Fey (2014), Grimm (2008), Khalifa (2012), Kvanvig (2009), and Strevens (2008, 2013).

[2] See, for example, Trout (2002, 2007) and Craver (2013) for more recent defenses of a purely ontic approach to explanation.

[3] Objective understanding in this sense will turn out to be the opposite of what de Regt (2009, p. 585) calls "the objectivist view of the relation between explanation and understanding," which he attributes to Hempel and Trout.

1980; Lipton, 2004) or a detailed description of the illocutionary context of an explanatory speech act (Achinstein, 1983). In my view, the interest-relativity of explanation has a much deeper origin. It derives from the interest-relativity of inquiry in general. Different agents use information for different purposes, and their acceptance of new information is directed by their cognitive and practical interests and goals. Far from being a superficial characteristic of inquiry, I believe that this is a fundamental trait of the acquisition of knowledge in general. The cost and effort that goes into obtaining new information makes the beliefs[4] that an inquiring agent has accepted a valuable asset that must be treated with care. Gratuitous losses must be prevented and the agent's acceptance of new information always involves the risk of bringing error into his system of beliefs. The risk must always be compensated by an epistemic incentive that outweighs the cost.

One of the biggest epistemic incentives of all is to obtain understanding of a fact.[5] But if understanding a given fact fulfills no purpose in the eyes of an inquiring agent, he will be more reluctant to incur the risks involved in accepting an explanation of it. On the other hand, if understanding a fact fulfills the cognitive interests and goals of the agent, but the information explains too much, it might be too good to be true. The acceptance of an explanation thus requires a delicate balance between two conflicting cognitive goals: the acquisition of valuable explanatory information and the avoidance of error.

The account of explanation that I present in this paper takes into account the difference between information and informational value, between the informational content of an explanation and the epistemic value of that content. When an agent seeks to expand his beliefs, his interest is restricted to information that promotes his cognitive goals or that is relevant to the problems he is trying to solve. In Catherine Elgin's words, "truth does not always enhance understanding. An irrelevant truth is epistemically inert" (1996, p. 124). I will argue that the goal of an inquiring agent is not just to find factually accurate explanations; it is to find explanations that are both factually accurate and epistemically valuable. This idea is captured by the following three theses:

1. Whether a piece of information is a *potential explanation* of the fact that P is mostly a non-pragmatic matter.

---

[4] In this paper beliefs should be understood as an agent's epistemic commitments, in the sense of Levi (1980). Some authors, such as Cohen (1989), use the term 'acceptance' for such attitudes, reserving the term 'belief' for involuntary epistemic states, akin to feelings. There is an extensive literature on the distinction between acceptance and belief (e.g., Engel, 2000; Cresto, 2009, among many others), but I cannot discuss the issue in this essay.

[5] Understanding laws and regularities is, of course, an equivalent or even greater epistemic incentive. The account presented here is restricted to the explanation of singular facts because the well-known objections against the explanation of laws require an entirely different analysis, one that most likely will not be probabilistic.

2. It is possible to determine the *objective epistemic value* of a subset of all the potential explanations of the fact that P.

3. In trying to understand the fact that P, an inquiring agent should only accept the potential explanations with positive objective epistemic value.

In the rest of the paper I discuss and defend each of these three theses.

3 The Objective Basis of Explanation

In this section and the next I defend the first of the three theses stated above, namely, that determining the potential explanations of a given fact is mostly a non-pragmatic matter. My basic contention is that an explanation of a singular fact should provide the information required to integrate the explanandum into an agent's cognitive system. An explanation should provide some of the factors that contributed to make P a fact, and some of the obstacles that could have, but did not prevent it from being one. Without such information, P will describe a brute fact, isolated from the rest of the agent's beliefs about the world. Probability sentences are the connecting tissue of an agent's corpus of beliefs. The influence of the preventing and contributing factors is captured by probability sentences of the form $p(P|Q) > p(P|\sim Q)$ and $p(P|Q) < p(P|\sim Q)$ that indicate that the fact that Q is statistically relevant to the explanandum[6].

The notion of statistical or probabilistic relevance has been used by many authors in the analysis of explanation. The best-known examples are Hempel's (1965) I-S model, Salmon's (1971, 1984) S-R model, Railton's (1978) D-N-P model, and Fetzer's (1974) causal-relevance model. All of these accounts consider precise probability values to be an essential part of an explanation. In contrast, I will argue that reference to probability values is largely unnecessary. Probability values have descriptive, predictive, and evidential value, but not explanatory value.

3.1 Probability Values

Probability values are thought to be important for two different reasons. If a statistical explanation is conceived of as an inductive argument, as it was in Hempel's original Inductive-Statistical model, the degree of expectation that a body of evidence confers upon a given event must be very high. Thus the value of the inductive probability must be kept in check to make sure it does not fall below a certain threshold as inquiry proceeds. On the other hand, if a statistical explanation is understood as an objective

---

[6]Many authors have used probabilities to model the epistemic states of researchers (e.g. Boutilier, 1995; van Fraassen, 1995; Halpern, 2003; van Benthem, 2003; Arló-Costa & Parikh, 2005). My account uses probability sentences to model the doxastic basis of an explanation, but an agent's epistemic states should not be understood to be probabilistic.

3.1 Probability Values

account of the stochastic process involved, as it is in Salmon's and Railton's models, it is crucial to avoid the attribution of false probability values to the probabilistic laws.

In response to criticism by Jeffrey (1971), Hempel (2001) gave up the high probability requirement, together with the claim that the explanans of an I-S explanation should show that the phenomenon described by the explanandum sentence was to be expected. Without this claim, however, the first reason to attribute any importance to probability values disappears. If the explanans is not supposed to justify our expectations that the explanandum will occur, there is no need to make sure that the value of the probability remains over a certain threshold.

Before we can evaluate the second reason why probability values are deemed to be explanatory, we must take a closer look at the logical structure of statistical explanations. One of the features of probability theory is that it does not have a weakening principle. A sound inductive argument that strongly supports its conclusion can be transformed into one that strongly undermines its conclusion with the insertion of additional true premises. An individual event can be referred to different reference classes, and the probability of the property associated with the event can vary considerably from one class to another. Hence, a body of evidence may confer a high degree of expectation upon a given event, while another body of evidence may confer a very low degree of expectation upon the same event. This is the problem that Hempel called the *ambiguity* of I-S explanation.

Hempel's partial solution to the problem is the requirement of maximal specificity. The requirement states that an acceptable statistical explanation should be based "on a statistical probability statement pertaining to the narrowest reference class of which, according to our total information, the particular occurrence under consideration is a member" (1965, p. 398). The requirement does not completely eliminate the ambiguity because the narrowest reference class can only be determined in the light of our current knowledge. It does not guarantee that there are no unknown statistical generalizations that can be used to construct a rival argument. In fact, Hempel claimed that *"the concept of statistical explanation for particular events is essentially relative to a given knowledge situation as represented by a set K of accepted sentences"* (p. 402, emphasis kept).

Salmon (1971) showed that the requirement of maximal specificity failed to rule out counterexamples in which irrelevant information finds its way into the explanation. But his main reason to reject Hempel's solution to the problem was his strong conviction that the appropriate reference class for a statistical explanation is one that is *objectively* homogeneous, not one that is *epistemically* homogeneous.

The notion of an objective homogenous reference class amounts to this: For any given reference class A, and for any given property C, there is, *in principle*, a partition of that class into two subclasses $A \wedge C$ and $A \wedge \sim C$. A property C is statistically relevant to a property B within A if and only if $p(B|A \wedge C) \neq p(B|A)$. Using von Mises's concept of place selection, Salmon defines a *homogeneous reference class* as follows:

> If every property $[C_1, C_2, C_3, \ldots, C_n]$ that determines a place selection is statistically irrelevant to B in A, I shall say that A is a homogeneous reference class for B. A reference class is homogeneous if there is no way, *even in principle*, to effect a statistically relevant partition without already knowing which elements have the attribute in question and which do not (1971, p. 43).

Salmon then replaces Hempel's requirement of maximal specificity for the *reference class rule*: "Choose the broadest homogeneous reference class to which the single event belongs" (p. 43). This characterization of statistical explanations is supposed to avoid any epistemic relativity because any statement of the form $p(G|F) = r$ that meets the homogeneity condition must be regarded as a fundamental statistical law of nature. Its reference class cannot be further specified, not because we do not know how to make a further relevant partition, but because *in principle* it is impossible to make a further relevant partition.

Salmon then defines a statistical explanation as follows. If we want to know why a member of the class A has the property B, the answer will be a S-R explanation that consists of: (i) the prior probability that a member of the class A will have the property $B : p(B|A) = r$, (ii) a partition into homogeneous cells with respect to the property in question: $A \wedge C_1$, $A \wedge C_2$, etc., (iii) the posterior probabilities of the property in cells of the partition $p(B|A \wedge C_1) = r_1, p(B|A \wedge C_2) = r_2$, etc., and (iv) a statement of the location of the individual in question in a particular cell of the partition: "a is a member of $A \wedge C_k$" (pp. 76-77).

Salmon explicitly requires the use of probability values in providing an explanation. The use of probability values stems from the fact that the S-R model is at bottom a covering-law model. Since any statement of the form $p(G|F) = r$ that meets the homogeneity condition must be regarded as a fundamental statistical law of nature, each of the probability sentences in the explanans of a S-R explanation is a law of nature. And since the factive condition on explanation demands that every element in an explanation must be true, the probability assigned to the explanandum by each of these probability sentences must be the right one.

To see how restrictive this requirement is, consider the following example provided by Humphreys:

> If a man dies from lung cancer, having been a heavy smoker, omitting from a probabilistic explanation any of the following minor relevant factors will result in a false probability claim: cosmic radiation from Alpha Centauri, particles from a chimney in Salem, Oregon, and a smoke-filled room he entered briefly at the Democratic convention eight years ago. It is good to be strict in matters of completeness, but not to the point of absurdity (1989, p. 111).

Humphreys argues that if one insists in providing the exact probability of the explanandum as part of the truth conditions of an explanation, it will be impossible to distinguish between a complete explanation and a true explanation. The omission of absurdly small probabilistically relevant factors, known or unknown, will result in a false explanation.

How can a true but incomplete statistical explanation be provided? Humphreys argues that instead of focusing on probability values, we should focus on causal relevance. An explanation should provide one or more of the factors that are causally relevant to an explanandum, and a factor is causally relevant if it changes the propensity for an outcome. His strategy has the advantage that it makes it possible to offer a true explanation of an event by providing a contributing or a counteracting cause even in cases where the other factors are not known and the true probability value cannot be calculated.

3.2 Epistemic Relativity

Although Humphreys' approach offers an appropriate formal basis for providing a statistical explanation, there is an obvious objection. As the many versions of Simpson's Paradox illustrate, one or more of the factors that the agent is unaware of can turn a contributing cause into a counteracting cause, or vice versa. Humphreys' response to this objection is puzzling. He says: "Of course, epistemically, we can never know for certain that such confounding factors do not exist, but that is an entirely separate matter, although regrettably relative frequentists have often failed to separate epistemic aspects of probabilistic causality from ontic aspects" (p. 114).

It seems to me that it is Humphreys who is guilty of not keeping epistemic and ontic matters in separate baskets. If Salmon's model is too demanding, as Humphreys maintains, it is because we can never know if we have met all the conditions that it imposes on explanation. But Humphreys' account suffers from a similar problem. In order for something to be a contributing or a counteracting cause in Humphreys' sense, there cannot be *any* further factor, known or unknown, that will invert the influence of these causes on the explanandum, or that will neutralize them altogether. Thus an agent who offers a causal statistical explanation will always have to relativize the explanation to a knowledge situation.

The accounts offered by Salmon and Humphreys avoid the epistemic relativity of statistical explanation by introducing a condition that effectively rules out the possibility that a bona fide statistical explanation will be defeated by a rival statistical claim. But the cost of avoiding the epistemic relativity of explanation is to render useless their accounts of explanation. It is hard to see how such a relativization can be eliminated if we want to provide a coherent picture of the role of explanation in inquiry. If we adopt the view that epistemic relativity is an unacceptable feature of explanation, we will be forced to conclude that there has never been a genuine scientific explanation in the history of science. Furthermore, we lose one of the main incentives for any scientific inquiry. Why would anyone want to incur the cost and effort involved in searching for explanations if the results cannot be assumed to be true in future decisions and deliberations? In Isaac Levi's words,

> If inquiry cannot be motivated by a concern to remove doubt, what is its rationale? If we cannot incorporate the solutions we come close to establishing into the evidence and background information for future investigations, why should we care that we come close? The truth of the well-established conjecture remains an open question and a legitimate issue for future investigation. Inquiry never settles anything and, hence, inquiry–even inquiry into a specific problem–never legitimately terminates because the matter is settled but only, so it seems, because the investigators are tired or bored or have run out of funds. No matter how minute a question might be, if inquiry into that question is free of costs, it should go on forever (1991, p. 2).

The reference to a specific epistemic context in the characterization of explanation is clearly a departure from tradition. Many philosophers have claimed that pragmatic elements have no place in the study of explanation. They recognize that there are interesting issues associated with the process of providing an explanation in an actual context, and their intention is not to belittle their importance. But the concept of explanation that they characterize is, in Hempel's words, "a concept which is abstracted, as it were, from the pragmatic one, and which does not require relativization with respect to questioning individuals any more than does the concept of mathematical proof" (1965, p. 426). The same general idea is defended by many other philosophers of science.

Michael Friedman has pointed out that there is a certain equivocation about the term 'pragmatic'. 'Pragmatic' can mean roughly the same as 'psychological', i.e., having to do with the thoughts, beliefs, attitudes, etc. of individuals. But 'pragmatic' can also be synonymous with 'subjective'. In the latter sense, a pragmatic notion must always be relativized to a particular individual. Friedman's claim is that "a concept can be pragmatic in the first sense without being pragmatic in the second." Further on he explains: "I don't see why there can't be an objective or rational sense of 'scientific understanding', a sense on which what is scientifically comprehensible is constant for a relatively large class of people" (1974, p. 8).

The traditional avoidance of any pragmatic element in a theory of explanation can thus be evaluated in two different ways. If one takes 'pragmatic' to mean the same as 'subjective', the insistence in providing a non-pragmatic analysis of explanation, i.e., an analysis that does not depend on the idiosyncrasies of the individuals involved, is perfectly justified. But if 'pragmatic' is interpreted in Friedman's first sense, there is no reason why an analysis of the concept of explanation should not make reference to the epistemic states of the individuals involved in a cognitive project.

I believe that we should take Friedman's suggestion seriously and explore the possibility of characterizing, in logically precise terms, a notion of explanation that is both objective and pragmatic, that does not depend on the idiosyncrasies of the individuals involved but that regards their epistemic states, their shared commitments, and their cognitive interests and goals as a fundamental part of the analysis. The concept of explanation will still be an "abstraction", in Hempel's sense, but an abstraction based on the decisions that take place when a group of inquiring agents rationally accept explanatory information. The resulting concept will be a hybrid, a combination of the formal, semantic, and pragmatic dimensions of explanation.

4 Potential Explanations

The epistemological framework for the account of explanation that I will present is Isaac Levi's version of the belief-doubt model first proposed by Peirce (1877)[7]. According to the belief-doubt model, an inquiring agent presupposes that everything he is currently committed to fully believing is true. This does not mean that truth or falsity is relative to what the agent believes. But the agent's *judgments* of truth are relative to what he believes. If the agent is concerned with establishing true explanations of phenomena, his decision to accept an explanation can only be made relative to the judgments of truth available to him.

To claim that an inquiring agent presupposes that everything he is currently committed to fully believing is true is not to say that he cannot change his mind. Certainty or full belief does not entail incorrigibility. Levi explains the claim thus: "To regard some proposition as certainly true and as settled is to rule out its falsity as a serious possibility for the time being. ...But from this it does not follow that good reasons will not become available in the future for a change of mind and for calling into question what is currently considered to be true" (1991, p. 3). Peirce puts it more graphically: "The scientific spirit requires a man to be at all times ready to dump his whole cartload of beliefs, the moment experience is against them" (1931, p. 55).

An inquiring agent has no doubt that all the sentences in his corpus of beliefs are true. Nonetheless, he does not regard all of the facts stated by these sentences as being equally well understood. The degree to which an agent understands the fact expressed by a sentence P will depend on how well integrated P is to the agent's cognitive system. It will not depend on how much support it has or on how epistemically entrenched

---

[7] Although my account of explanation uses Levi's belief revision theory as theoretical framework, it must be pointed out that Levi does not agree with my approach (personal communication). The main reason is that Levi believes that all explanations with probabilistic premises presuppose a D-N explanation stated in dispositional terms. Furthermore, Levi associates statistical explanations with the elimination of surprise and an increase in the expectation of the occurrence of the explanandum (Levi, 1988, 1995). The account of explanation that I present here does not entail those two consequences.

it is. On the one hand, if a sentence has been accepted in his corpus of beliefs, it is judged to be true and no further argument is necessary. On the other hand, poorly understood phenomena can be highly epistemically entrenched, and completely useless facts can be very well understood.

According to the belief-doubt model, an inquiring agent's judgments of truth are always relative to what he is currently committed to fully believing. Thus, an agent's decision to accept an explanation can only be made relative to the judgments of truth available to him. Naturally such decisions will lack any sort of *objectivity*. An agent who wants to claim objectivity for the explanations that he accepts must first make sure that the explanation is consistent with K, the set of beliefs that represents the shared agreement between the members of a community of experts. More technically, the states of belief of the individual experts can be partially ordered in a manner satisfying the requirements of a Boolean algebra. In consequence, it will be possible to form the meet of their individual states, i.e., the strongest common consequence of all their states of belief (Levi, 1991, p. 13).

Let P be a sentence in K. A set of sentences E is a *potential explanation* of the fact stated by P relative to K just in case the following conditions are fulfilled:

(i) $K \cup E$ is consistent.

(ii) $E \not\subset K$.

(iii) There is a sentence Q such that $Q \in E$.

(iv) Either $p(P|Q) > p(P|\sim Q) \in E$ or $p(P|Q) < p(P|\sim Q) \in E$.

(v) There is no $R \in K$ such that $p(P|Q\&R) = p(P|\sim Q\&R)$.

(vi) P and Q are logically independent.

(vii) Nothing else is an element of E.

The first condition states that a potential explanation must be consistent with the corpus of beliefs in which the explanandum is accepted. The second condition states that the potential explanation cannot be already accepted in K. The third condition says that the potential explanation must include a singular sentence Q that describes a potentially relevant factor. The fourth condition states that Q is positively or negatively relevant to the fact that P. The fifth condition guarantees that P and Q will not be spuriously correlated, as far as we know. Condition (vi) guarantees that P will not explain itself. It also prevents the inclusion of trivial cases in which $p(P|Q) = 1$ because $P \vdash Q$. A potential explanation is thus a set containing a singular sentence that describes a fact, and a probability sentence that states the potential statistical relevance of that fact to the explanandum.

Using this definition of a potential explanation, we can now characterize the notion of an *explanation space*. An explanation space can be understood as the set of sentences that contains all the potential explanations of P, regardless of whether the inquirers are aware of them or not.

(E_P) For every sentence P in K, there is a set $\{E_1, E_2, \ldots, E_k\}$ such that E_i is an element of the set iff it is a potential explanation of P. The set, denoted E_P, is the *explanation space* of P.

The explanation space will contain logically equivalent and empirically equivalent potential explanations. On the one hand, if $E_1 = \{Q, p(P|Q) > p(P|\sim Q)\}$ and $E_2 = \{R, p(P|R) > p(P|\tilde{R})\}$, where Q and R are logically equivalent, then E_1 and E_2 are logically equivalent potential explanations. If an agent accepts E_1, she is thereby committed to E_2. On the other hand, if Q and R contain coextensive singular terms or predicates that occupy the same places in Q and R, E_1 and E_2 will be empirically equivalent potential explanations. However, the explanatory value and the credibility of E_1 and E_2 will not be assessed in the same way unless the agents who assess them are aware that the singular terms or predicates are coextensive[8].

5 The Epistemic Value of Explanation

Consistency with K, the set of beliefs that represents the shared agreement between the members of a learning community, is not enough to guarantee the objectivity of an explanation. The objectivity of our conjectures lies, as Popper correctly points out, "in the fact that they can be intersubjectively tested" (1959, p. 44). The intersubjective test that an explanation must pass is the evaluation of its credibility and of its explanatory value in the eyes of the experts.

Suppose a group of inquirers–a community of experts in the field–wants to consider the adoption of an explanation. To do so, they must first adopt a belief state K representing the shared agreement between them. Such a belief state will be the strongest common consequence of all their states of belief. Obviously, such a state will contain more than just singular sentences representing facts and probability sentences. It will also include sentences that state which are the most relevant problems in the field, what type of experiments and observations are considered more reliable, in addition to basic methodological and reasoning principles.

Once the members of the community of experts have accepted a common corpus K, they must take it as the basis for establishing a set of potential explanations of the problem at hand, For example, suppose a group of inquirers are trying to establish why P. They must initially agree on a set of ground facts and low-level hypotheses.

---

[8] Condition (iv) also introduces an element of epistemic relativity because the non-existence of a screening off factor can only be guaranteed relative to K.

Statistical data and the chronology of the explanandum will be easy to agree upon. The explanation of some aspects of the phenomenon can be non-controversially accepted, while the explanation of others will be a matter of heated debate. After the inquirers have agreed on a common corpus of beliefs K, they can put together a *set of explanatory options*, denoted O_P, which will include all the factors consistent with K that might explain P and that have been identified by the inquirers. At this stage of inquiry it does not matter whether the potential explanations are uncontroversial or completely outlandish, as long as they are somehow relevant to the problem at hand and consistent with K, that is, if they fulfill the requirements to be in E_P.

It is possible for a group of agents to share the same information and yet disagree about the degree of belief or credal probability that they assign to the information in the set of explanatory options. Since the agents do not want to beg the question by assigning the highest marks to their favorite explanations, they must adopt a common credal probability measure. A common strategy to eliminate the conflict between different credal probability distributions is to represent the shared agreement as the weighted average of the distributions in conflict. The resulting credal probability function C determines the objective risk of error incurred in accepting a potential explanation in O_P. Let E_i be the conjunction of the elements of a potential explanation E_i in O_P, i.e., the conjunction of a singular sentence and a probability sentence. For every potential explanation E_i, the risk of error is $1 - C(E_i)$.

On the other hand, different inquirers will disagree in their assessment of the importance of the explanations contained in the set of explanatory options. Despite these differences, there must be a minimal objective criterion to measure the explanatory value of any potential explanation. That criterion is the new information carried by the potential explanation, which, following Levi, I identify with its logical strength. The set of potential expansions of a belief set K can be partially ordered by a classical consequence relation. The set is a Boolean algebra in which the minimum is K and the maximum is the inconsistent state. If a probability function M is defined over this set, and if the only element that has probability zero is the inconsistent state, potential expansions of K will strictly increase in probability with a decrease in logical strength. When the M-function is defined over the set of potential explanations of interest to the inquirer, we obtain a measure of the informational content of the potential explanations in O_P. The measure of the informational content of a potential explanation E_i, denoted $\text{Cont}(E_i)$, is $1 - M(E_i)$.

The informational content of a potential explanation is the first objective criterion that should be used in assessing the explanatory value of the elements of O_P. The evaluation of their explanatory value is subject to the following weak monotonicity requirement (WMR):

> (WMR) If a potential explanation E_1 in O_P carries at least as much information as another potential explanation E_2 in O_P, E_1 carries at least as much explanatory value as E_2.

Not all potential explanations of the fact that P are comparable in terms of logical content. Since the community of experts wants to consider all the explanations available to them, they might invoke further criteria in order to complete the quasi-ordering imposed by the weak monotonicity requirement. In order to assess the explanatory value of the remaining elements of O_P, they can evaluate if they have certain properties that are considered explanatorily virtuous.

There are several explanatory virtues mentioned in the philosophical literature. Friedman (1974) and Kitcher (1989), for example, argue that explanations improve our understanding through the unification of our knowledge. Explanations that reduce the number of independent assumptions we have to make about the world are to be preferred to those that do not. This suggests that the potential explanations in O_P could be ordered according to some set of rules that determines their unifying power.

The problem is that neither Friedman nor Kitcher have provided an account that can be applied to explanations generally. Friedman's original argument was intended as an account of the explanation of scientific laws. Friedman argued, for example, that the kinetic theory of gases is explanatory because it unified different laws and properties of gases that were previously disconnected. Friedman's only attempt to formalize and generalize the idea of explanation as unification was incisively criticized by Kitcher (1976) and Salmon (1989).

But Kitcher's account is no more helpful that Friedman's. According to Kitcher, the explanatory worth of candidates cannot be assessed individually. In his view, a successful explanation earns that name because it belongs to the explanatory store, a set that contains those derivations that collectively provide the best systematization of our beliefs. 'Science supplies us with explanations whose worth cannot be appreciated by considering them one-by-one but only by seeing how they form part of a systematic picture of the order of nature" (1989, p. 430). The idea that a virtuous explanation should have the potential to unify our beliefs is uncontroversial, but no one, to my knowledge, has provided a general account of explanation as unification that is not restricted to the case of scientific laws or scientific explanatory exemplars.

Mellor (1995) provides an account of explanatory value that is better suited for our purposes.

Mellor approaches explanation via his theory of causation. The theory requires every cause to raise the chances of its effects. That is, a fact C causes a fact E iff $ch_C(E) > ch_{\sim C}(E)$. When causes are used in the explanation of a given fact, Mellor argues that the explanans must necessitate its explanandum, or at least raise its probability as much as possible, thereby reducing its chance of not existing. Thus, he concludes, "the more C raises E's chance the better it explains it" (p. 77). If we were to accept Mellor's idea, it would be possible to order the potential explanations in O_P according to the difference between $ch_C(E)$ and $ch_{\sim C}(E)$.

The main problem with Mellor's proposal is that when we examine a genuinely stochastic process, the value of the chance that the cause confers on the explanandum will be irrelevant. As Jeffrey has convincingly argued, the information required to explain E is the same information used to explain $\sim E$, regardless of the value of the chance. Furthermore, if E is sometimes randomly caused by C and sometimes randomly caused by C^*, and $ch_C(E) > ch_{C*}(E)$, there is no reason to think that C is a better explanation than C^*.

Mellor will respond to the objection by claiming that chances measure possibilities. "The less possible $\sim E$ is, i.e. the less $ch(\sim E)$ is and hence the greatest $ch(E)$ is, the closer the fact E is to being necessary. This is the sense in which a cause C may explain E better or worse, depending on how close it comes to making E necessary, i.e. on how much it raises $ch(E)$" (p. 77). Independently of whether we can make sense of such concepts as *almost necessary* or *nearly impossible*, it is not clear how such notions would enhance our notion of explanation. Probabilities are important in statistical contexts because knowing that C raises the chance of E allows us to know what makes E possible, and because the chance that C gives E allows us to adjust our expectations of E's occurrence. But it seems to me that mixing chances and possibilities adds nothing to our understanding of why E is a fact.[9]

A third candidate for judging the epistemic value of an explanation is Whewell's (1837) notion of *consilience*. Consilience is intended to serve as a measure of how much a theory explains, and it can therefore be used to compare the explanatory value of two different hypotheses. One hypothesis has more explanatory value than another if the former explains more of the evidence than the latter. Thagard (1978) provides compelling evidence that this idea is often used by scientists in support of their theories. For example, Fresnel defended the wave theory of light by saying that it explained the facts of reflection and refraction at least as well as did the particle theory, and that there were other facts involving diffraction and polarization that only the wave theory could explain. Translated into my account, this means that if E_i raises the probability of more facts connected to the explanandum than E_j, then E_i is a better explanation than E_j.

The problem with consilience is that, once again, the account works well in the explanation of laws, but it will not work in the explanation of singular facts. Whether a given fact explains more aspects connected to the explanandum than another fact is hard to say. We would have to define what a fact "connected to the explanandum" is, and it is doubtful that a non-pragmatic formalization of this notion can be found. Besides, sometimes a theory can explain too much. Lavoisier accused the phlogiston theory of this particular crime.

Are there any other criteria that will allow us to assess the explanatory value of the potential explanations in O_P? We still have not examined the values that are usually mentioned in the context of theory choice: simplicity, accuracy, fruitfulness, and so

---

[9]In Páez (2013) I offer an exhaustive analysis of the relation between causation and explanation in Mellor's work.

5 The Epistemic Value of Explanation

on[10]. But such an analysis is unnecessary. If the criteria are such that the community of experts can agree on their importance and on how they should be applied in particular cases, they can be added to the belief state K that represents their shared agreement. The agents will then be able to complete, to some degree, the quasi-ordering generated by the monotonicity condition with respect to the M-function. But to expect a complete agreement in the way that all the agents engaged in common inquiry assess the explanatory value of different potential explanation is to expect a heterogeneous group of inquirers to agree on what aspects of reality they find interesting or useful.

If a common decision is required nonetheless, the community of experts can adopt the following compromise. The agents must first identify the elements of the set O_P that can be completely ordered because they are comparable in terms of strength or because they can be compared using the criteria to evaluate explanatory value that they have incorporated to K. The agents can then agree to disagree about the explanatory value of the remaining elements of O_P. Let O_P^* be a set of explanatory options such that $O_P^* \subseteq O_P$ and such that the M-value of each element of the set is determined. Combining the credal probability function C with the M-function defined over the elements of O_P^* we obtain a value that the community of experts can use to select the best explanation of P. I will call this result the *objective epistemic value* of a potential explanation[11]:

(OEV) $V(E_i) = \alpha C(E_i) + (1-\alpha)\text{Cont}(E_i)$

The agents' interest in valuable information should not outweigh the desideratum to avoid error; thus $\alpha \geq 0.5$. And since the information they seek should not be worthless, $1 > \alpha$.

Now, some researchers will be bolder than others in privileging content over credibility, while others will be more cautious and adopt the opposite attitude. Let q be a common boldness index, which is the average of their individual boldness indices. If $q = (1-\alpha)/\alpha$, we obtain the following affine transformation of OEV:

(OEV) $V(E_i) = C(E_i) - qM(E_i)$

The experts should reject a potential explanation in O_P^* if OEV is negative, remain uncommitted if it is 0, and accept it if it is positive. Any potential explanation in O_P^* with positive objective epistemic value is an *objective explanation* of P in K. The disjunction of all such objective explanations is *the* objective explanation of P in K:

(OE_P)   The objective explanation of P in K, denoted OE_P, is the disjunction of all the potential explanations in O_P^* with positive objective epistemic value.

One of the consequences of taking the functions C and M –which represent the average credibility and the agreed upon explanatory value, respectively– as a basis

---

[10]There is a vast literature on the epistemic and social values used in science. The compilations by Machamer and Wolters (2004) and Kinkaid, Dupré and Wylie (2007) offer a contemporary perspective on the topic.

[11]This strategy is similar to the one followed by Levi (1991) to characterize the maximization of the expected epistemic utility obtained by expanding a corpus of beliefs.

for the analysis of the potential explanations in O_P^* is that each individual agent was forced to sacrifice his personal evaluation of credibility and value in order to accept the verdict of the community of experts. Suppose an agent has accepted a potential explanation of P based on his individual assessment of its credibility and explanatory value. Now suppose that he submits his "subjective" explanation to the community of experts, and the explanation is judged to be maximally credible and maximally valuable by the community, thus becoming an objective explanation. Does the agent understand *more* now that his explanation has been certified by others? It seems to me that he does not. But if the agent does not obtain more understanding from this recognition, why should anyone seek objectivity for an explanation that he or she already believes?

Part of the answer is that the belief-doubt model is not a recipe for dogmatism. A seldom-noted fact about inquiry is that most newly suggested explanatory hypotheses do not survive the test of intersubjective scrutiny. If the agent is aware of this fact– and he should be if he is a responsible inquirer-it would be imprudent for him to give his full assent to an explanatory hypothesis that contradicts firmly established theories and findings without obtaining at least a partial intersubjective assessment of its merit. An agent does not need to fully believe that an explanation is true to obtain the understanding that the explanation provides. Any inquirer can explore the consequences of a hypothesis by assuming, for the sake of argument, that it is true. If the hypothesis is judged to have positive objective epistemic value by a community of experts, the inquirer will then be fully justified in giving it his full assent.

But the question remains. If the agent does not obtain new understanding in the approval that he receives from his peers, why should he seek their approval? What prevents an agent from individually assessing the credibility and explanatory value of a potential explanation, and deciding to fully believe it if his individual understanding is thereby increased? In other words, why should objectivity matter? The answer is that objectivity itself is a property of information that some agents find valuable and some do not. An agent who decides to be a member of a learning community does so because he is convinced that his beliefs will be more valuable if they are objective. Other agents will find that objectivity adds no value to their corpus of beliefs. Just as there is a difference between objective and subjective explanation, there is an analogous distinction between objective and subjective understanding. The latter is the type of understanding that Hempel (1965) correctly believed should be shunned at all costs from an account of scientific explanation. But the reason it should be shunned is not that it is an inferior type of understanding. The reason is that the members of a scientific community are among the many agents who find objectivity valuable. Therefore, an account of scientific explanation should avoid any reference to an evaluative process in which the agent shows no concern for the views of others.

6 Conclusion

The belief-doubt model provides an adequate basis for an account of explanation that takes into consideration the epistemic value of the information that we acquire through inquiry. By including the shared commitments and the cognitive interests and goals of the individuals engaged in a cognitive enterprise, we obtain a notion of explanation that is objective by any reasonable standard of objectivity, and that clarifies the connection between explanation and understanding. The main reason why I have adopted the belief-doubt model is that an account of explanation that takes into consideration the epistemic value of the information that we acquire through inquiry leads to a natural resolution of the conflict between the purely pragmatic approach to explanation defended by Achinstein and van Fraassen, for example, and the more common approach in which pragmatic considerations are not assigned any serious role.

References

[Achinstein(1983)] P. Achinstein. *The nature of explanation.* New York: Oxford University Press, 1983.

[Aliseda(2006)] A. Aliseda. *Abductive reasoning.* Dordrecht: Springer, 2006.

[Arló-Costa and Parikh(2005)] H. Arló-Costa and R. Parikh. Conditional probability and defeasible inference. *Journal of Philosophical Logic*, 34:97–119, 2005.

[Boutilier (1995)] C. Boutilier. On the revision of probabilistic belief states. *Notre Dame Journal of Formal Logic*, 36:158–183, 1995.

[Boutilier and Becher(1995)] C. Boutilier and V. Becher. Abduction as belief revision. *Artificial Intelligence*, 77:43–94, 1995.

[Cohen(1989)] L. J. Cohen. Belief and acceptance. *Mind*, 98:367–389, 1989.

[Craver(2013)] C. Craver. The ontic conception of scientific explanation. In A. Hütteman & M. Kaiser (Eds.), *Explanation in the biological and historical sciences.* Dordrecht: Springer, 2013.

[Cresto(2010)] E. Cresto. Belief and contextual acceptance. *Synthese*, 177:41–66, 2010.

[de Regt, Leonelli, and Eigner (2009)] H. de Regt, S. Leonelli, and K. (Eds.) Eigner. *Scientific understanding: Philosophical perspectives.* Pittsburgh: University of Pittsburgh Press, 2009.

[Elgin(1996)] C. Z. Elgin. *Considered judgment*. Princeton: Princeton University Press, 1996.

[Engel(2000)] P. Engel. (Ed.) *Believing and accepting*. Dordrecht: Springer, 2000.

[Faye(2014)] J. Faye. *The nature of scientific thinking: On interpretation, explanation, and understanding*. New York: Palgrave Macmillan, 2014.

[Fetzer(1974)] J. H. Fetzer. A single case propensity theory of explanation. *Synthese*, 28:171–198, 1974.

[Friedman(1974)] M. Friedman. Explanation and scientific understanding. *Journal of Philosophy*, 71:5–19, 1974.

[Grimm(2008)] S. R. Grimm. Explanatory inquiry and the need for explanation. *British Journal for the Philosophy of Science*, 59:481–497, 2008.

[Halpern(2003)] J. Y. Halpern. *Reasoning about uncertainty*. Cambridge: MIT Press, 2003.

[Hempel(1965)] C. G. Hempel. *Aspects of scientific explanation*. New York: The Free Press, 1965.

[Hempel(2001)] C. G. Hempel: Postscript 1976. More recent ideas on the problem of statistical explanation. In J. H. Fetzer (Ed.), *The philosophy of Carl G. Hempel. Studies in science, explanation, and rationality*. New York: Oxford University Press, 2001.

[Humphreys(1989)] P. Humphreys. *The chances of explanation. Causal explanation in the social, medical, and physical sciences*. Princeton: Princeton University Press, 1989.

[Jeffrey(1971)] R. Jeffrey. Statistical explanation vs. statistical inference. In W. C. Salmon (Ed.), *Statistical explanation and statistical relevance*. Pittsburgh: Pittsburgh University Press, 1971.

[Khalifa(2012)] K. Khalifa. Inaugurating understanding or repackaging explanation? *Philosophy of Science*, 79:15–37, 2012.

[Kinkaid, Dupré, and Wylie (2007)] H. Kinkaid, J. Dupré, and A. Wylie. (Eds.) *Value-free science? Ideals and illusions*. New York: Oxford, 2007.

[Kitcher(1976)] P. Kitcher. Explanation, conjunction, and unification. *Journal of Philosophy*, 73:207–212, 1976.

[Kitcher(1989)] P. Kitcher. Explanatory unification and the causal structure of the world. En P. Kitcher & W. Salmon (Eds.), Scientific explanation. Minnesota studies in the philosophy of science, Volume XIII. Minneapolis: University of Minnesota Press, 1989.

[Kvanvig(2009)] J. L. Kvanvig. The value of understanding. In A. Haddock, A. Millar & D. Pritchard (Eds.), Epistemic value, Oxford: Oxford Unviersity Press, 95–111, 2009.

[Levi(1988)] I. Levi. Four themes in statistical explanation. In W. L. Harper & B. Skyrms (Eds.), Causation in decision, belief change and statistics, 1988.

[Levi(1991)] I. Levi. *The fixation of belief and its undoing*. New York: Cambridge University Press, 1991.

[Levi(1995)] I. Levi. *Review of The Facts of Causation* by D.H. Mellor. *Theoria*, 61: 283–288, 1995.

[Lipton(2004)] P. Lipton. *Inference to the best explanation*. London: Routledge, 2nd edition, 2004.

[Machamer and Wolters(2004)] P. Machamer and G. Wolters. (Eds). *Science, values, and objectivity*. Pittsburgh: University of Pittsburgh Press, 2004.

[Mellor(1995)] H. D. Mellor. *The facts of causation*. London: Routledge, 1995.

[Páez(2009)] A. Páez. Artificial explanations: The epistemological interpretation of explanation in AI. *Synthese*, 170:131–146, 2009.

[Páez(2013)] A. Páez. Probability-lowering causes and the connotations of causation. *Ideas y Valores*, 151:43–55, 2013.

[Peirce(1877)] C. S. Peirce. The fixation of belief. *Popular Science Monthly*, 12: 1–15, 1877.

[Peirce(1931)] C. S. Peirce. Lessons from the history of science. In C. Hartshorne, & P. Weiss (Eds.), Collected Papers of Charles S. Peirce, Cambridge: Belknap Press, 1931.

[Popper(1959)] K. Popper. *The logic of scientific discovery*. London: Hutchinson, 1959.

[Railton(1978)] P. Railton. A deductive-nomological model of probabilistic explanation. *Philosophy of Science*, 45:206–226, 1978.

[Salmon(1971)] W. C. Salmon. Statistical explanation. In W. C. Salmon (Ed.), *Statistical explanation and statistical relevance*. Pittsburgh: Pittsburgh University Press, 1971.

[Salmon(1984)] W. C. Salmon. *Scientific explanation and the causal structure of the world*. Princeton: Princeton University Press, 1984.

[Salmon(1989)] W. C. Salmon. *Four decades of scientific explanation*. Minneapolis: University of Minnesota Press, 1989.

[Strevens(2008)] M. Strevens. *Depth: An account of scientifc explanation*. Cambridge: Harvard University Press, 2008.

[Strevens(2013)] M. Strevens. No understanding without explanation. *Studies in History and Philosophy of Science*, 44:510–515, 2013.

[Thagard(1978)] P. Thagard. The best explanation. criteria for theory choice. *Journal of Philosophy*, 75:76–92, 1978.

[Trout(2002)] J. D. Trout. Scientific explanation and the sense of understanding. *Philosophy of Science*, 69:212–233, 2002.

[Trout(2007)] J. D. Trout. The psychology of explanation. *Philosophy Compass*, 2: 564–596, 2007.

[van Benthem(2003)] J. van Benthem. Conditional probability meets update logic. *Journal of Philosophical Logic*, 12:409–421, 2003.

[van Fraassen(1980)] B. C. van Fraassen. *The scientific image*. Oxford: Clarendon Press, 1980.

[van Fraassen(1995)] B. C. van Fraassen. Fine-grained opinion, probability, and the logic of full belief. *Journal of Philosophical Logic*, 24:349–377, 1995.

[Whewell(1837)] W. Whewell. *History of the inductive sciences*. London: Parker, 1837.

Logic and Language

Yablo's Paradox and omega-Paradoxes

Eduardo Alejandro Barrio
IIF-Conicet
University of Buenos Aires

Abstract

In this paper, I analyze the paradox of Yablo: a semantic antinomy that involves an infinite sequence of sentences each of them claims that all linguistic items occurring later in the series are not truth. At least on a superficial level, the paradox does not seem to implicate any circularity. And this would mean that the set of sentences of Yablo shows that circularity is not a necessary condition to paradoxes. I start by describing and examining the main results about the option of formalizing the Yablo Paradox in arithmetic. As it is known, although it is natural to assume that there is a correct representation of that paradox in first-order arithmetic, there are some technical results that give rise to doubts about this approach. In particular, one of the aims of this work is to show that the formalization of this paradox in first order arithmetic is omega-inconsistent, even though the set of this sentences is consistent and has a (non-standard) model. Then, the plan is to draw the philosophical consequences of this result. Further, I am going to take into account Priest's point according to which this representation is also circular according to the same standard that the Liar sentence. All these reasons justify the necessity to look for alternative formalizations to the paradox. So, another proposal of this paper is also to consider different versions of Yablo's paradox that do not use first order arithmetic. Then, I will show some problems connected with such formulations. I will argue that there are reasonable doubts to adopt any of these formalizations as modeling the set of sentences of Yablo.

KEYWORDS: Yablo's Paradox - Truth - Circularity - Finitism.

1 Informal Yablo' s Paradox

Stephen Yablo (1985, 1993) formulated his paradox as an infinite set of sentences that is linearly ordered. Each of them claims that all sentences occurring later in the series are not true. In particular, the paradox consists in imagining a denumerably infinite sequence of sentences $S_1, S_2, S_3, \ldots$, each of them claims that all sentences occurring later in the series are not truth:

(S_1) For all $k > 0, S_k$ is untrue
(S_2) For all $k > 1, S_k$ is untrue
(S_3) For all $k > 2, S_k$ is untrue
(S_4) For all $k > 3, S_k$ is untrue

$$\vdots$$

On a superficial level, the series doesn't seem to involve any kind of self-reference. According to Yablo, this means that this sequence generates a liar-like paradox without any kind of circularity involved: no sentence in the list seems to refer to itself, and opposed to liar cycles, no sentence seems to refer to sentences above it in the list.

Informally, Yablo describes what happens with his sequence in the following way: suppose, for *reductio*, that some particular sentence in the sequence is indeed true. For example, let's take S_1 as true. S_1 says that for all $k > 1$, S_k is not true. Accordingly, we may conclude the following:

S_2 is not true,

and

For every $k > 2, S_k$ is not true.

The latter, however, entails that what S_2 says is in fact the case. This contradicts the first assumption: S_2 turns out to be true. Thus, S_1 must be false after all. The proof can be generalized, then, for each number k, we can prove that S_k is false. Thus, for all $k > 1$, S_k is not true. Therefore, S_1 is true, which we have just shown to be impossible.

As is well known, from the previous informal proof, Yablo intends to extract significant consequences about the concept of *truth*. Since the proof doesn't seem to appeal to self-referential expressions, Yablo's infinite sequence of sentences seems to show that the feature mentioned above is not necessary for the existence of a semantic paradox. Of course, the notion of *self-referentiality* is hard to characterize, and its link to the concept of *circularity* is not at all helpful to this task. Maybe for that, this issue has been the focus of a fascinating discussion. For example, Roy Sorensen (1998) and Otavio Bueno & Mark Colyvan (2003) have argued that the list produces a semantic paradox without circularity. Graham Priest (1997) and JC Beall (2001) have instead argued the paradox involves a fixed-point construction, and as a result of this, the list is basically circular. And Roy Cook (2006 / 2013) claims that the arithmetic variant of the Yablo's list is circular, but a slight modification of the original Yablo's construction allows us to generate a truly non-circular paradox. To show some details of this discussion, it is necessary to present a formal version of Yablo's sequence in first order arithmetic. This is going to be the main proposal of next section. I also discuss some conceptual problems linked with formalization.

2 Yablo Paradox in Arithmetic

a) Consistency and non-standard models

Formalizing the sentences that appear in the Yablo Paradox with a truth predicate T, one gets that for all natural numbers n, S_n is the sentence $\forall k > n, \neg T(S_k)$. Since the sentences on the right-hand side are the truth conditions for the sentences named in the Yablo sequence, one could formulate the list of Yablo's sentences by the set of biconditionals:

$$\{S_n \leftrightarrow \forall k > n, \neg T(S_k) : n \in \omega\}$$

In this way, one could try to formalize the proof of the Yablo Paradox increasing first-order arithmetic. Being that some non-logical expressions that appear in the sequence are part of informal arithmetic (numerals, order relations), this option seems natural. Of course, since T is also a non-logical expression, one needs some kind of truth-theoretical principle. According to this, one can consider

the Local Disquotational Principle (LDP):

$$T(S_n) \leftrightarrow S_n : n \in \omega.$$

Then, paradox of Yablo could be the following. Assume for reductio:
- 1. $T(S_n)$
- 2. S_n 1, (LDP) for S_n^*.
- 3. $\forall k > n, \neg T(S_k)$ 2, eq.
- 4. $\forall k > n+1, \neg T(S_k)$ 3, Arithmetic.
- 5. S_{n+1} 4, eq
- 6. $T(S_{n+1})$ 5, (LDP) for $S_{n+1}*$
- 7. $\neg T(S_{n+1})$ 3, Arithmetic
- 8. $\bot$ 6 and 7.
- 9. $\neg T(S_n)$ 7, $I\neg$

But n is arbitrary,

- 10. $\forall n, \neg T(S_n)$,   9, Universal Generalization $(UG)^{**}$.
- 11. $\forall n > 1, \neg T(S_n)$ 10, Arith.
- 12. S_1 11, Eq.
- 13. $T(S_1)$ 12, $(LDP)^*$
- 14. $\neg T(S_1)$ 10, Universal Elimination (UE)
- 15. $\bot$ 12 and 13.

Of course, there are some problems with this proof. Firstly, the demonstration uses UG in step 10 (line marked $(**)$) and (LDP) in 2, 6 and 13 (lines marked $(*)$). Then, it's natural to suppose that n and k are variables. Otherwise, it is not possible to apply UG. Nevertheless, as Priest correctly (1997: 237) focus on, the application of the (LDP) would not be possible in this case. This principle only applies to sentences, not to formulae with free variables in.

So, it is necessary to be more accurate. To state the paradox of Yablo in a more rigorous way, we would like to define a function on the natural numbers assigning to each number i the sentence S_i. But, as Halbach & Zhang (2016) focus on, since the definition of S_j presupposes the sentences S_k with $k > i$ this cannot be done recursively. However, as they also show, such a function can be defined using the Second Recursion Theorem.[1] Alternatively, the Gödel diagonal lemma can be employed and this method has become the standard way of presenting Yablo's paradox in arithmetic (Priest's (1997)). In fact, Priest proposes constructing the Yablo list adding the predicate name Y to first order arithmetic. Then, one needs a different disquotational principle connected with the predicate $Y(x)$. So,

The Local Yablo Disquotational Principle (LYDP):

$$Tr(Y(n)) \leftrightarrow Y(1) : n \in \omega$$

Simply put, according Priest, Yablo's Paradox consists of an ω-sequence of formulas: $Y(1), Y(2), Y(3), \ldots Y(n)$. In other words, this is just the infinite

---

[1] Cantini 2009, p. 987, fn. 242.

2 Yablo Paradox in Arithmetic

sequence[2]:

$$Y(1) \leftrightarrow (\forall x)(x > 1 \rightarrow \neg T(< Y(\cdot(x)) >, n))$$
$$Y(2) \leftrightarrow (\forall x)(x > 2 \rightarrow \neg T(< Y(\cdot(x)) >, n))$$
$$Y(3) \leftrightarrow (\forall x)(x > 3 \rightarrow \neg T(< Y(\cdot(x)) >, n))$$
$$\vdots$$
$$Y(n) \leftrightarrow (\forall x)(x > n \rightarrow \neg T(< Y(\cdot(x)) >, n))$$
$$\vdots$$

But even adopting this formulation, problems do not finish. As Priest also note, the application of UG in line 10 (step marked $(**)$) is wrong: only in case n is a constant, a numeral for an (unknown) particular natural number, $LYDP$ can be used. But, Priest concludes that in this case UG could not be applied correctly. UG could only be applied in the case the sentence Y_n were arbitrary. But, this is not the case.

In sum, we expected that adding the list of the Yablo biconditionals and the $LYDP$ to first order arithmetic yields an inconsistency. But different proposals to get a contradiction fail. Moreover, surprisingly, it can be shown that this theory is consistent, although ω-inconsistent. The last result should be evident: the set of numerical instances of $\{Y(n) \leftrightarrow \forall x > n, \neg T(\langle Y(\cdot(x))\rangle) : n \in \omega\}$ must be consistent. If it were not, by Compactness, this should mean that there is a proof of a contradiction from some finite subset of the Yablo sentences. Nevertheless, as Hardy claims,

> If we restrict ourselves to a finite collection of the Yablo sentences, then no paradox arises. The upshot of this is that there is no first-order derivation of a contradiction from Yablo's premises (The Yablo List) and the Tarski biconditionals. (1996: 197)

And Ketland adds:

> Each finite subset of Yablo biconditionals is satisfiable. By the Compactness Theorem, the whole set is satisfiable (2005: 165, note 1)

This model-theoretical result has an important consequence in proof theory. As Ketland shows, adding the $LYDP$ and the list of Yablo's biconditionals to first

---
[2]Dot notation was proposed by Feferman to allow quantification into formulae containing quotation terms. The dots above the variables indicate that these symbols are bound from outside in the usual way, where a function replaces the variables by the corresponding numerals.

order arithmetic produce a consistent theory. No contradiction could be obtained in this theory. Even more, Ketland shows that "with an appropriate definition of the extension of 'true", it is possible to satisfy this combination on any non-standard model of arithmetic" (2005: 165). Nevertheless, as Ketland marks out, this theory is ω-inconsistent. Formally, if a set of formulas is ω-inconsistent, it can be proved that even when each natural number fulfills the condition $Y(x)$, there is a number that doesn't. Of course, we are unable to prove that there is a specific number that doesn't fulfill the condition, but we can prove that there is one that doesn't fulfill it. Moreover, as it's well known, Gödel's results imply that First-order arithmetic is ω-incomplete, if this theory is consistent: there are cases where it can be proved, case by case, that each number satisfies some condition $Y(x)$, but it can't be proved that all numbers satisfy $Y(x)$. Then, assuming that this theory is consistent, Gödel's sentence can't be proved. Nevertheless, compared with a theory that results ω-inconsistent, it seems that ω-incompleteness in a theory of arithmetic is a regrettable weakness, but ω-inconsistency is a very bad news (nor as bad as outright inconsistency, of course, but still bad enough.) A ω-inconsistent theory can prove each of $Y(n)$) and yet also prove $\neg \forall x Y(x)$ is just not going to be an acceptable candidate for expressing arithmetic.

In order to appreciate the point with more detail, let $A(k)$ be $= Tr(Y(k))$ and T_{YA} is the first order theory that include first order arithmetic, all instance of Yablo's biconditionals and the $LYDP$. Then, Barrio (2010) shows that T_{YA} is ω-inconsistent in the following way. Assume that:

1. $T_{YA} \vdash Tr(Y(1))$
2. $T_{YA} \vdash Y(1)$ by 1 and the $LYDP$
3. $T_{YA} \vdash \forall k > 1, \neg Tr(Y((k)))$ by 2 and Yablo's biconditionals
4. $T_{YA} \vdash \forall k > 2, \neg Tr(Y((k)))$ by first order PA
5. $T_{YA} \vdash \neg Tr(Y(2))$ by 2
6. $T_{YA} \vdash Tr(Y(2))$ by 3
7. $T_{YA} \vdash \bot$ by 5 and 6 and Logic
8. $T_{YA} \vdash Tr(Y(1))$ Logic
9. $T_{YA} \vdash \neg Tr(Y(2)), T_{YA} \vdash \neg Tr(Y(3))$, $T_{YA} \vdash \neg Tr(Y(4))$, etc
10. $T_{YA} \vdash (Y(1))$ by 8 and $LYDP$
11. $T_{YA} \vdash \exists k > 1, Tr(Y((k)))$ by 10 and Yablo's biconditionals
12. $T_{YA} \vdash \exists k > 1, Tr(Y((k)))$ by 11 and first order PA

9 and 12 imply that T_{YA} is ω-inconsistent.

This point is disconcerting: against what we would have thought, it is possible to find a model for T_{YA}, after all. However, this model cannot have as its domain the set of the standard natural numbers. For this reason, even though Yablo's sequence is consistent, it has not at the same time standard model. Thus, even though the theory that includes the list of Yablo's sentences is not strictly paradoxical (since it has non-standard models), it cannot be interpreted using a structure in which the numbers appearing in each of the biconditionals are standard natural numbers. Because of this, even if the sequence is consistent, it turns out to be ω-inconsistent. Thus, Yablo's Paradox is not strictly a paradox but actually a ω-paradox. Based on these results, Leitgeb (2001), Barrio (2010), Barrio & Picollo (2013) and Barrio & DaRé (2017) have argued against theories of truth that only have non-standard models. In particular, ω-inconsistency causes a dramatic deviation in the theory's intended ontology of first-order arithmetic. In order to be able to express the concept of *arithmetic truth*, the theory has to abandon the possibility of speaking about *standard natural numbers*. For this reason, ω-consistency is an additional requirement that should be satisfied every time we are concerned with expressing truth: not only do we want the addition of Yablo's sequence to first order arithmetic to result in a consistent theory, but we also want it to result be capable of conserving the ontology of the intended interpretation of this arithmetic theory, for it seems plausible to maintain that no theory of truth should imply a substantive answer to what numbers are or to what the ontology of arithmetic is. Barrio & Picollo (2013) has extended these criticisms showing other negative consequences that share the theories of truth that only have non-standard models as the revision theories of truth FS and $T\#$. And in the same direction, Barrio & DaRé (2017) analyze as these results impact on the non-classical theory of naïve truth based on Lukasiewicz infinitely-valued logic: $PALT$.

Of course, the second-order arithmetic with standard semantics avoids the existence of non-standard models. So, as Barrio (2010) has shown, adding Yablo's sequence to this theory produces a theory of truth that doesn't have a model. However, Picollo [2011] has shown that even in higher-order cases, the theory is consistent: one is not able to derive a contradiction from the set of Yablo sentences. Barrio & Picollo (2013) have generalized this result to several omega-inconsistent theories of truth based on second order arithmetic.

In sum, contrary to what happens with the Liar Paradox, the set of Yablo sentences formalized in first -order arithmetic is consistent and satisfiable, even though is ω-inconsistent and only has non-standard models. In the second-order case with standard semantics, one has only standard models. So, adding the set of Yablo sentences to the second-order arithmetic produces a theory that doesn't have a model. But there is not a finitary proof of a contradiction.

One important issue is to evaluate if it is enough to formalize a paradox to get a model-theoretical result. In particular, in Yablo's case formalized in second order, one does not get a contradiction using proof-theoretical resources. The topic is connected with the general question about what a paradox is[3]. What is usual is to ask the one can get a contradiction using some intuitive resources involving a naive notion. In this case, it could be natural to get a contradiction from some intuitive principles linked with the notion of truth and the the list of Yablo's sentences. But, this not the case. The only that one can get is an unsatisfiability result. As we analyze later, this could imply a problem: proofs provide epistemic warrants that involve human beings that use a naive notion (in this case, the naive concept of truth). But in any case, a model-theoretical result is what we can get using arithmetic.

b) The Charge of Circularity

Perhaps the results about ω-inconsistency given above are not so bad as to reject the formulation in the arithmetic of the paradox of Yablo. But, there are also other reasons to suspect. In this section, I briefly review the charge of circularity. I will present three arguments. In all of them, one attempts to show that there is no way of reformulating Yablo's construction that does not involve circularity implicitly.

The Argument of Existence of the Sequence

An important point that has been discussed is how one *knows* that the Yablo list exists. Yablo seems to assume the existence of the list in order to show that that list generates a paradox. Nonetheless, as Priest claims:

He [Yablo] asks us to imagine a certain sequence. How can one be sure that there is such a sequence? (We can imagine all sorts of things that do not exist.) As he presents things, the answer is not at all obvious. In fact, we can be sure that it exists because it can be defined in terms of $Y(x)$: the n-th member of the sequence is exactly the predicate $Y(x)$ with "x" replaced by $\langle Y(x) \rangle$. (Priest, 1997: 238, notation was changed to match that the one used by me)

Nevertheless, in this case, the fixed-point construction required to generate the sequence of Yablo involves an implicit circularity. So, from Priest's perspective, the list of Yablo's sentences itself is circular:

...the paradox concerns a predicate $Y(x)$ of the form $(\forall k > x)(\neg T(\langle Y(x) \rangle, k))$, and the fact that $Y(x)$'$(\forall k > x)(\neg T(\langle Y(x) \rangle, k))$' shows that we have a fixed point, $Y(x)$ here of exactly the same self-referential kind as in the liar paradox. In a nutshell, $Y(x)$ is the predicate 'no number greater than x satisfies this predicate'. The circularity is now manifest. (Priest (1997): 238)).

---
[3] For details, I recommend Barrio, E (2014) "Introducción" in Barrio, E (2014).

2 Yablo Paradox in Arithmetic

Specifically, moving onto Cook's terminology[4], the predicate $Y(x)$ is a weak predicate fixed point of the predicate: '$\forall k > x(\neg T(\langle Y(z) \rangle, k)$'. In other words, each member of the list $Y(1), Y(2), Y(3), \ldots Y(n)$ is implied by what Ketland [2005] calls *the Uniform Fixed-Point Yablo Principle*(UFPYP):

$$\forall x(Y(x) \leftrightarrow \forall k > x, \neg T(\langle Y(dot(x)) \rangle, k))$$

The point of Priest is that the UFPYP involves circularity, because it provides a definition for the predicate $Y(x)$ in terms of itself. And since this principle guarantees the existence of the sequence, the list itself involves circularity. As Ketland pointed out:

> To stress, it is a *theorem of mathematical logic that the Yablo list exists*. This is a direct and well-understood construction. Priest does not 'presuppose the existence of the list, in order to establish that to derive a contradiction from the latter, a fixed-point construction is required'." (2004: 169)

Now, consider the stronger Uniform Yablo Disquotation principle (**UYPYP**):

$$\forall x(T(Y(dot(x))) \leftrightarrow Y(x))$$

It is important to note that adding the **UFPYP** and the Uniform Yablo Disquotation principle to PA yields an inconsistency. Nevertheless, in that case, the infinity of the list of Yablo biconditionals would not play any important role in the paradox. So, this did not appear to be acceptable.

To summarize, the list of Yablo, formulated within arithmetic, seems to involve a kind of circularity. And this type of circularity involved is not distinct from the sort found in the arithmetic Liar. As Roy Cook points out:

> "if the existence of fixed points is enough for a statement or predicate to be circular, then the Yablo paradox is circular." (2013, p. 96).

It is important to note that this point affects both first-order and second-order arithmetic formulations. That is, this result has a broader scope than the objection of the ω-inconsistency.

Nevertheless, this objection is not without its doubts. As Cook also emphasizes (2013), this sort of circularity (fixed point) is not a plausible cause of the paradox. And what is more important, he argues that the circularity involved in both is too broad to be relevant. His argument takes into account that every unary predicate (in a strong enough language) is a weak fixed point of some binary

---
[4] Cook (2013)

predicate, and every statement is a weak sentential fixed point of some unary predicate. Hence, according to him, this mathematical fact seems to throw serious doubts on the prospects of explaining the roots of paradoxes concerning the presence of (this sort of the weak fixed point) circularity. Cook points out that the sort of circularity found in both the Liar paradox and the Yablo paradox seems to be an innocuous type of circularity since this sort of circularity is endemic throughout arithmetic.

Of course, accepting Cook's point would be just a start. To avoid objection, we should give a different circularity criterion that applies to the liar's sentence and does not apply to Yablo's sequence.Identifying this point is an interesting challenge for those who would like to formalize Yablo's paradox within arithmetic.

The Epistemically Circular Argument

Turning to the second argument, Beall (2001) has offered new reasons in support of the circularity of the sequence. He focuses on our knowledge of the meaning of the predicate $Y(x)$. Then, he claims that we have no way coming to know what the predicate $Y(x)$ means without employing a circular fixed-point principle. From his position, the Yablo sequence is epistemically circular because "everyone, I think, will agree: we have not fixed the reference of 'Yablo's paradox" via demonstration. Nobody, I should think, has seen a denumerable paradoxical sequence of sentences, at least in the sense of 'see' involved in uncontroversial cases of demonstration" (Beall 2001: 179). But, for Beall, any such description is circular. So, any entity that can only be referred to by a circular description must itself be circular. Then, Beall concludes, Yablo's paradox is circular.

That argument does not seem to be too forceful. Beall's point depends on the idea that the only way we can know that the Yablo sequence exists in PA is to apply the $UFPYP$.But this is not the case: because theorems of PA are enumerable using an enumeration of valid proofs, in order to construct the Yablo sequence, it would be enough to run through an enumeration of valid proofs until we get one whose final line is:

$$\forall x(\phi(x) \leftrightarrow k > x, \neg T(\langle\phi((x))\rangle, k))$$

Then, one can apply countably many instances of UE to arrive at the ω-sequence of Yablo biconditionals. There is nothing circular in the process of carrying out proofs, enumerating them, or surveying the resulting enumeration. Thus, there is nothing circular in (this way of obtaining) the Yablo paradox

The Argument against ω-Rule

Moving onto the following argument on the charge of circularity, Priest intends to support that resource to the ω-rule does not help prevent circularity. He claims:

One might suggest the following. We leave the deduction as just laid out, but construe the n in the reductio part of the argument as schematic, standing for any natural number. This give us an infinity of proofs, one of $\neg T(S_n)$, for each n. We may then obtain the conclusion $\forall n \neg T(S_n)$ by an application of the ω-rule:

$\alpha(0), \alpha(1), \ldots$

$\forall n \alpha(n)$

The rest of the argument is as before. Construing the argument in this way, we do not have to talk of satisfaction. There is no predicate involved, a fortiori no fixed point predicate. We therefore have a paradox without circularity. (Priest, (1997): 238-239)

But, Priest adds:

As a matter of fact, we did not apply the ω-rule [in his earlier sketch of the derivation of a contradiction], and could not have. The reason we know that $\neg T(S_n)$ is provable for all n is that we have a uniform proof, i.e. a proof for variable n. Moreover, no finite reasoner ever really applies the ω-rule. The only way that they can know there is such a proof of each $\alpha(i)$ is because they have a uniform method of constructing such proofs. And it is this finite information that grounds the conclusion the $\forall n \alpha(n)$. (1997: 239)

Priest's position against the use of ω-rule in Yablo's Paradox is the following: being that no finite human being ever really applies the ω-rule (or any infinitary analogue such as our infinitary variant of conjunction introduction above), then the only way we can know that the Yablo Paradox truly is paradoxical is through a proof depending on fixed points of the sort described above.

In contrast, Selmer Bringsjord & Bram van Heuveln defend:

The point (...) is that in light of such arguments, Priest is in no position to simply assume's [that we are finite reasoners i.e. Turing machines] ... and hence he hasn't derailed the infinitary version of Yablo's paradox. (2003: 65)

And they add:

... also argued ... specifically that logicians who work with infinitary systems routinely and genuinely use the ω-rule. Again, the claim isn't that such arguments are sound, and that therefore some human persons, contra Priest, genuinely use the ω-rule. The claim is a simple, undeniable one: if any of these arguments are sound, then we can really use the ω-rule, and the infinitary reasoning we gave above would appear to be above reproach. (2003: 67)

One can not just assume without argument that we are finite reasoner. Neither that a finite reasoner is a Turing machine. There are many arguments that show that we are not Turing machines. Hypercomputers, unlike Turing machines and their equivalents (and lesser systems), can make essential use of the ω-rule. And this is just a brute mathematical fact. Further, there are no compelling reasons for thinking that the performance of supertasks is a logical impossibility.

Besides, according to Cook (2013), Priest's objection "(...) relies on the idea that we might restrict the notion of truth (...) to natural languages or finitary languages (or both). The motivation for such a restriction, one assumes, would be the observation that all language users that we have come into contact with (and, importantly, all language users that matter) speak finitary languages that do not allow for the construction of the truly non-circular paradox sketched above." However, the main point of Cook is that restricting our account of truth (and our development of a view on the semantic paradoxes) to languages that we are able to speak looks worryingly provincial. Cook changes the focus of the discussion. There are two different problems:

ONTOLOGICAL: Is there any infinite sequence that represents the truth predicate of some infinitary language and does not be circular?

EPISTEMOLOGICAL: Could a being with our epistemic capabilities knowing that infinite sequence by means no circular?

According to Cook, the main discussion about Yablo's paradox is linked to the ontological problem. Firstly, logic is modeling truth preservation and not all systems of logic are complete. Secondly, logic is used to describe mathematical structures. Infinitary languages, whose models are structures under study, might raise conceptual problems as Yablo's construction seems to show. I agree with this point. The notion of truth is so complex that it could be beyond the limits of human epistemic capacities. And Cook does not care if the paradox that he presents is or not a good formalization of Yablo's paradox. But, if one is interested on offering a formal version of that paradox, the infinite resources give rise to doubts about the formalization of the original paradox.

Final remarks using arithmetic:

In sum, I sympathize with the response of Cook to Priest's point: the circularity involved in PA is too broad to be relevant. The sort of circularity found the Yablo paradox formulated in PA overgeneralizes: all arithmetic predicates turn out to be circular. But this does not be the case. So, one can use PA to formulate the Yablo paradox avoiding the risk of circularity. In my opinion, the problems associated with the ω-inconsistency in first order arithmetic are evidence to consider that one has a good non-circular representation of the list of Yablo's sentences. Moreover, second-order arithmetic with standard semantics avoids the existence of non-standard models. So, adding Yablo's sequence to this theory produces a theory of truth that doesn't have a model. I think that if a theory of truth that be ω-inconsistent is a bad thing, having a unsatisfiable theory is really bad. In this case, one shows that adding Yablo's list to arithmetic produces serious problems.

3.- Yablo's Paradox without arithmetic

The last section shows some problems involved with the formulation of Yablo's Paradox using arithmetic. And it could also be added that Yablo's paradox involves circularity because its proof relies on the Gödel diagonal lemma, the recursion theorem, or the like. As I've pointed out before, Graham Priest has instead argued that the paradox involves a fixed-point construction and as a result of this the list is circular. But, as we have analyzed, this point, nevertheless, can be weighted in various ways.

Using infinitely resources:

One option was developed by Roy Cook. Cook's idea is to show that there are non-circular paradoxes, but Yablo's construction in PA is not one of them. Consequently, Cook introduces an infinitary language Lp. This language only allows conjunctions (possibly infinite) of predications of falsity to sentences names. In other words, every sentence is of the form $\wedge_{i \in I} F(S_i)$, where $\{S_i : i \in I\}$ is a (possibly infinite) class of sentence names and where F is the falsity predicate. He also uses a denotation function δ in order to provide the denotation of sentence names in Lp. If C is the collection of every sentence names in Lp, for each of the sentence name in C, δ is denotation function such that $\delta : C \to$ {fórmulas de Lp}. For example, the liar sentence can be formulated in this system as $\delta(S1) : F(S1)$. Now the Cook-Yablo sequence consists of the set $\{S_n\}_{n \in \omega}$ under the denotation function $\delta(S_n) = \wedge\{F(S_m) : m \in \omega \; m > n\}$. This sequence can be expressed as "the unwinding" of the Liar sentence:

$$\delta(S_1) = F(S_2) \wedge F(S_3) \wedge F(S_4) \wedge \ldots \quad (1)$$

$$\delta(S_2) = F(S_3) \wedge F(S_4) \wedge F(S_5) \wedge \ldots \quad (2)$$

$$\delta(S_3) = F(S_4) \wedge F(S_5) \wedge F(S_6) \wedge \ldots \quad (3)$$

The logic of L_p is the infinitary system D. Proofs within D admit possibly transfinite sequences of expressions, where each expression is either a finite or infinite conjunction of instances of the falsity predicate applied to sentence names or an instance of the truth predicate applied to a sentence name. System D has some introduction and elimination rules. It's really important to note that Conjunction Introduction is in some applications an infinitary rule that plays the same function that ω-rule in formal arithmetic.

In L_p, the Yablo sequence is the set $\{\langle S_i, \wedge k > i F(S_k)\rangle i \geq 1\}$ (where i ranges over the integers). Cook shows that Yablo paradox is in the context of D provably inconsistent. The proof has $\omega^2 + 3$ steps. Semantically, the paradoxicality is apparent in the fact that no valuation can be found for these sentences if F is really interpreted as the falsity predicate. It is interesting to note that Cook's formalization of the Yablo Paradox in L_p is a genuinely non-circular paradox. Cook shows the absence of weak fixed points in L_p. This is evidence for the non-circularity of its construction. Interestingly, Cook defines an operation of "unwinding" which transforms any set of formulas with an assignment of denotations to the sentence names into another such set which (i) does not involve any (direct or indirect) self-reference, but which (ii) shares important semantic properties with the "original". Cook's goal was to define the simplest framework in which Yablo's construction could be somewhat generalized.

Cook asks for the circularity of his construction. Then, he defines a notion of fixed point for D, and proves the following:

Theorem 2.4.3: *Given any denotation function δ such that $\delta(Yn) = \wedge\{F(Ym) : m \in m > n\}$, there is no $\kappa \wedge \omega$ such that $\delta(Y_\kappa)$ is a weak fixed point in D of $\langle\{Y_n\}n \in \omega, \delta\rangle$.*

According to him, the absence of a fixed point is evidence for the non-circularity in Lp. The proof that none of the sentences of Lp involves in the construction are fixed points shows that there are non-circular constructions that are paradoxical. Of course, a way to deny such achievement as a genuine paradox is to reject infinitary systems. In particular, it is possible to deny the validity of the rules such as the omega rule or a version of it, the Conjunction Introduction Rule of Cook's system. Graham Priest, for example, claims that nobody can actually use an infinitary rule, and that when we talk "as if" we performed one, it is actually finite information which is justifying us (Priest, 1997). In Cook's systems, proofs may be infinitely long. This fact could disturb us. In sum, adopting infinitary resources allows avoiding the charge of circularity. But the price seems high. The approach is not formalizing a paradox that involves languages used by human beings. Maybe this is even enough to show that there are non-circular paradoxes. And obviously this proposal should be compare with the second-order arithmetic case. Remember that in second order arithmetic case, one gets a model-theoretical result. Yablo's sequence of sentences and second-order arithmetic have not a model. But if one uses only finite resources, adding this set of sentences to second-order arithmetic does not produce a inconsistent theory. If one uses Lp, one can get an inconsistency because one is adopting non-finite apparatus of proof. Finally, Theorem 2.4.3 shows that one can avoid circularity considering as a weak fixed point. This is not the case for second order arithmetic. So, if circularity were linked with having a (weak) fixed point, Lp would be a better option than Yablo's paradox formulated on second order arithmetic.

Using first order principles:

Another option to formulate Yablo's paradox without arithmetic is proposed by Halbach & Zhang (2016). In this version, infinitary resources are not involved. Here, the proof of Yablo's paradox requires a binary predicate, a ternary satisfaction predicate, and some sets of intuitive assumptions. Directly, they work in a language of predicate logic with identity. The language contains a binary predicate symbol $<$ and a ternary predicate symbol Sat (x, y, z). For each formula ϕ in the language there is a closed term ϕ in the language. As they claim, this can be achieved by adding countably many new constants $c_1, c_2, \ldots$ to the language and then fixing some 1-1-mapping between the set of constants and the set of formulae in the language with all constants. Although this mapping does not play any role, it only helps to motivate the version of the Local Yablo Disquotation principle as governing the intuitive notion of truth. As it's usual, 'ϕ' is the name for ϕ (It is a constant in this context without first-order Arithmetic).

Now, using a proof very close to Yablo's paradox, Halbach & Zhang show that the theory T_{YA}^* given by the following schema and two axioms is inconsistent:

$$(SYD*) \forall x \forall y (Sat(\phi(x,y), x, y) \leftrightarrow \phi(x, y))$$
$$(SER) \forall x \exists y\, x > y$$
$$(TRANS) \forall x \forall y \forall z (x > y \rightarrow (y < z \rightarrow x < z))$$

As they focus on, the first axiom schema is a variant of Yablo's Disquotation T-schema. The predicate Sat stands for is satisfied or is true for. It is to be read: For all x and y, the formula '$\phi(x, y)$' is satisfied by x and y iff '$\phi(x, y)$'. An instance in English would be the following sentence: 'is bigger than' is satisfied by objects x and y i x is bigger than y. the variables x and y are fixed; the first two variables in alphabetic order could be used. Generally, $Sat(z, v, w)$ expresses that z is true if the free variable x in z is assigned v and y are assigned w. (SER) and $(TRANS)$ lay down that $<$ is a serial and transitive relation[5].

---

[5]This axioms are connected with a result showed by Ketland (2005). In that paper is showed for an arithmetical setting that only a serial and transitive relation is needed to obtain the paradox.

For the proof, they stand $\forall z > y\phi$ for $\forall z(y < z \to \phi)$ and 'ψ' for '$\forall z > y \neg Sat(x,x,z)$'. Then, one can reasons:

- $(1) T^*_{YA} \vdash \forall x \forall y (Sat(`\psi', x, y) \leftrightarrow \forall z > y \neg Sat(x,x,z))$
- $(2) T^*_{YA} \vdash \forall y (Sat(`\psi', `\psi', y) \leftrightarrow \forall z > y \neg Sat(`\psi', `\psi', z))$.

The first line is obtained by instantiating $\phi(x,y)$ with ψ in $(SYD*)$. The second step is obtained by universal instantiation. Now one assumes $Sat(\psi,\psi,a)$ for arbitrary a and argue in T^*_{YA} as follows:

- $(3) \forall z > a \neg Sat(`\psi', `\psi', z)$ from (2)
- $(4) \vdash \nu > a \neg Sat(`\psi', `\psi', \nu)$ by (SER)
- $(5) \vdash \nu > a(\forall z > \nu \neg Sat(`\psi', `\psi', z) \land \neg Sat(`\psi', `\psi', \nu))$ by $(TRANS)$ and (3)
- $(6) \vdash \nu > a(Sat(\psi,\psi,\nu) \land \neg Sat(\psi,\psi,\nu))$ by (2)
- $(7) \bot$

Then, one can see that the assumption $Sat(`\psi', `\psi', a)$ leads to a contradiction in T^*_{YA}. Thus one can conclude that $\neg Sat(`\psi', `\psi', a)$ for any a:

- $(8) T^*_{YA} \vdash z Sat(`\psi', `\psi', z)$
- $(9) T^*_{YA} \vdash z > a \neg Sat(`\psi', `\psi', z)$ for arbitrary a
- $(10) T^*_{YA} \vdash Sat(`\psi', `\psi', a)$ by (2)

But as shown above, $Sat(`\psi', `\psi', a)$ is refutable in T^*_{YA}. This is concludes the proof of the paradox.

Unlike the previous first order strategies, one gets a contradiction in T^*_{YA}. But, as Halbach & Zhang notes, the schema $(SYD*)$ is already inconsistent by itself. It can be seen by choosing $\neg Sat(x,x,x)$ as $\varphi(x,y)$ in the schema and then instantiating both universal quantifiers ?x and ?y with $\neg Sat(x,x,x)$[6]. So, this version of the paradox does not allow to get a new paradox: the additional assumptions (SER) and (TRANS) connected with the idea of an infinite sequence of sentences with certain properties does not play any fundamental role. So, it's necessary to adapt the last reasoning to capture Yablo's Paradox. It's crucial to offer some weaker version of (SYD*): a consistent principle that can be shown to be inconsistent with (SER) and (TRANS) by an argument similar to the one given above. So, they propose a solution to this problem adapting Visser's paradox[7].

The starting point of Visser's paradox is the following quote of Kripke:

---

[6] The inconsistency follows in the style of the proof of Russell's paradox.
[7] Visser (1989)

2 Yablo Paradox in Arithmetic

"One surprise to me was the fact that the orthodox approach by no means obviously guarantees groundedness in the intuitive sense mentioned above ... Even if unrestricted truth-definitions are in question, standard theorems easily allow us to construct a descending chain of first order languages $L_0, L_1, L_2, \ldots$ such that L_i contains a truth predicate for L_{i+1} · I don't know whether such a chain can engender ungrounded sentences, or even quite how to state the problem here: some substantial questions in this area are yet to be solved." (Kripke [1975], pp. 697-698.)

Visser formulated his paradox in an arithmetical setting with infinitely many truth predicates indexed by natural numbers. According to him, "the orthodox approach" is simply too indefinite to be able to claim whether it does or does not exclude a descending hierarchy. So, He shows that a new paradox may appear in descending hierarchies. To see the point, let L be the language of arithmetic and let: $L_i = L + T_1, \ldots, T_{1+1}$. Then, Visser considers the set of T-sentences T_n'ϕ' $\leftrightarrow \phi$ where 'ϕ' is a sentence containing only truth predicates with index $k > n$. Now, he proves that this set is ω-inconsistent over arithmetic.

One important property of this construction is the following: Visser's hierarchy is like Tarski's hierarchy of the truth predicate. But the indexing of the levels of the hierarchy is reversed. So, the truth predicate applies to sentences with truth predicates T_n with $n > 0$ and so on. So, it is an infinitely descending hierarchy of languages, that is, the hierarchy is ill-founded. To see the point, consider:

$T+ = Th(N)+$ Tarski style axioms for T_0 as truth predicate for L_1 for T, as truth predicate for L_2, etc. So, e.g., if '$L, (x)$' represents the arithmetical predicate 'to be a Godel number of a sentence of L_1. then:

$T+ \vdash \forall x \forall y (L_1(x) \wedge L_1(y)) \to (T_0(\text{conj}(x,y)) \leftrightarrow T_0(x) \wedge T_0(y))$. As in Yablo's Pardox, by a simple compactness argument one sees that $T+$ is consistent: the reason is that, for any finite set of axioms of $T+$, this set can be interpreted in a finite ascending hierarchy (taking, for example, the empty set as interpretations of the truth predicates not occurring in our set of axioms.). Besides, as well in Yablo's paradox, the theory has a model.

Let $T = PA + \{T_n$'ϕ' $\leftrightarrow \phi$ where 'ϕ' is a sentence containing only truth predicates with index $k > n\}$. T has no standard model. The moral here obviously is: even if strong $T+$ is consistent in the usual sense, already weak T is 'semantically inconsistent' in the sense that it excludes the standard model.

Halbach & Zhang adapt this construction to formulate Yablo's paradox in first order without arithmetic. Of course, they have to avoid to use an arithmetic framework. Then, in contrast to Visser, the new version should not use infinitely many truth or satisfaction predicates. Instead, they use a quaternary satisfaction predicate $\text{Sat}_x(y, z, w)$ whose level index x is a quantifiable variable. The axioms have to be chosen in such a way that $\text{Sat}_x(y, z, w)$ is a satisfaction predicate for formulae where all quantifiers over indices are restricted to objects v with $v > y$. Thus, if $v > w$, the predicate

$\text{Sat}_v(y, z, w)$ is 'lower' than $\text{Sat}_w(y, z, w)$ in the hierarchy. The effect of using that quantifiable variable as a level index is that one does not only obtain a ω-inconsistency as Visser did but rather an outright inconsistency.

The idea is simple: an occurrence of a variable is in index position iff it occurs in the 'hierarchy level' position. In $\text{Sat}_x(y, z, w)$, for instance, exactly x occurs in index position.

So, the axioms for $\text{Sat}_x(y, z, w)$ are now formulated in such a way that $\text{Sat}_x(y, z, w)$ applies only to formulae y about level v of the hierarchy with $v > x$. More precisely one uses the following axiom instead of (SYD*) :

(SYD**) $\qquad \forall x \forall y (\text{Sat}_y(\phi(x,y), x, y) \leftrightarrow \phi(x,y))$

All occurrences of variables in index position in $\phi(x,y)$ must be bound. All occurrences of quantifiers $\forall v$ or v (for some variable v) in $\phi(x,y)$ that bind some occurrence of the variable v in index position must be restricted by $v > y$. For instance, $?w?v > y\text{Sat}_y(w, x, y)$ would be an admissible instance of $\phi(x,y)$. Now, Halbach & Zhang show that:

> "The proof (presented above) still goes through mutatis mutandis, if the formula $\forall z > y \neg \text{Sat}_z(x, x, z)$ is chosen as $\phi(x,y)$. It shows that SER and TRANS together with (SYD**) are inconsistent. In contrast to (SYD*), the schema (SYD**) by itself is consistent. As long as $>$ is well-founded, that is, for any set of objects there is always a $<$-maximal element, models of (SYD**) can be defined by induction on $>$. For the proof of consistency it suffices to assume that $>$ denotes the empty relation. In this case the restricted quantifiers $v > y$ and $\forall v > y$ in the instances $\phi(x,y)$ of (SYD**) are idling."[8]

In sum, adapting Visser's Paradox, Halbach & Zhang get an inconsistent first-order theory without using an arithmetic setting. The theory T^{**}_{YA}

(SYD**) $\forall x \forall y (\text{Sat}_y(\phi(x,y), x, y) \leftrightarrow \phi(x,y))$

(SER) $\forall x \exists y x < y$

(TRANS) $\forall x \forall y \forall z (x < y \rightarrow (y < z \rightarrow x < z))$

allows formulating a structure that is inconsistent and has not a model at all. Nevertheless, I have different doubts about this achievement. It is not clear to me that this theory formalizes Yablo's Paradox. Firstly, it is important to note that Haibach-Zhang's Paradox is not about the concept of truth. It is about the concept of satisfaction. Secondly,

---
[8] Halbach & Zhang (2016) p 4

it is clear at all how to read intuitively the list of Halbach-Zhang's sentences. One possible reading could be:

$$\forall x \forall y (\text{Sat}_z(\forall z < y \neg \text{Sat}_z, x, x, z), x, y) \leftrightarrow (\forall z < y) \neg \text{Sat}_z, x, x, z)$$

Finally, informal Yablo's paradox formulated in natural language is constituted by a list of sentences. But, Halbach-Zhang's Paradox is constituted by an infinite list of conditions or predications. Another point is connected to the assumption that $>$ is ill-founded in the proof. That is, one needs a nonempty set of objects that does not have a $<$-maximal object. According to Haibach and Zhang, whether $>$ is ill-founded because it is circular or because it is an infinite linear ordering without end point doesn't matter. Nevertheless, it may be the case that there are conceptual connections between circularity and ill-foundedness. In this case, Halbach-Zhang's paradox would not have achieved its objective: get a semantic paradox without circularity.

3 Conclusions

In this paper, I have shown that there are some conceptual problems connected with usual formalizations of Yablo's Paradox. I have argue that using first-order arithmetic could not be a good idea: omega-inconsistency and the idea of considering circularity as a fixed point involve different problems that make us doubt about this approach. I have also argue that some options that avoid to use arithmetic present important difficulties. In particular, it is not clear that these type of approaches are formalizing the original version of Yablo's Paradox.

References

[Barrio(2010)] E.A Barrio. Theories of truth without standard models and Yablo's sequences. *Studia Logica*, 96:375–391, 2010.

[Barrio(2014)] E.A. Barrio. *Paradojas, Paradojas y más Paradojas*. Londres, College PU, 2014.

[Barrio and Picollo(2013)] Eduardo Barrio and Lavinia Picollo. Notes on ω-inconsistent theories of truth in second-order languages. *The Review of Symbolic Logic*, 6(4):733–741, 2013.

[Barrio and Ré(2017)] Eduardo Barrio and Bruno Da Ré. Truth without standard models: some conceptual problems reloaded. *Journal of Applied Non-Classical Logics*, 2017. https://doi.org/10.1080/11663081.2017.1397326.

[Beall(2001)] JC Beall. Is Yablo's paradox non circular? *Analysis*, 61:176–187, 2001.

[Bringsjord and Heuveln(2003)] S. Bringsjord and Van Heuveln. The mental eye defense of an infinitized version of Yablo's paradox. *Analysis*, 63:61–70, 2003.

[Bueno()] M. Bueno, O.and Colyvan. Yablo's Paradox rides again: a reply to Ketland. Manuscript.

[Bueno and Colyvan(2003)] O. Bueno and M. Colyvan. Paradox without satisfaction. *Analysis*, 63:152–156, 2003.

[Cook(2006)] R. T. Cook. There are non-circular paradoxes (but Yablo's Isn't One of Them!). *The Monist*, 89:118–149, 2006.

[Cook(2013)] R. T. Cook. *The Yablo Paradox: An Essay on Circularity*. Oxford, Oxford UP, 2013.

[Forster(1996)] T. Forster. The significance of Yablo's Paradox without self-reference, 1996. Manuscript.

[Halbach and Zhang(2016)] V. Halbach and S Zhang. Yablo without Gödel Analysis. *https://doi.org/10.1093/analys/anw062*, 2016.

[Hardy(1996)] J. Hardy. Is Yablo's paradox liar-like? *Analysis*, 55:197–198, 1996.

[Ketland(2004)] J. Ketland. Bueno and Colyvan on Yablo's Paradox. *Analysis*, 64: 165–172, 2004.

[Ketland(2005)] J. Ketland. Yablo's Paradox and omega-inconsistency. *Synthese*, 145:295–307, 2005.

[Leitgeb(2001)] H. Leitgeb. Theories of Truth which have no Standard Models. *Studia Logica*, 68:69–87, 2001.

[Leitgeb(2002)] H. Leitgeb. What is a Self-Referential Sentence? Critical Remarks on the Alleged (Non-)Circularity of Yablo's Paradox. *Logique and Analyse*, 177: 3–14, 2002.

[Picollo(2011, 2012)] L. Picollo. The Yablo Paradox in second-order languages: Consistency and Unsatisfiability. *Studia Logica*, 101 (3):601–617, 2011, 2012.

[Priest(1997)] G. Priest. Yablo's Paradox. *Analysis*, 57:236–242, 1997.

[Sorensen(1998)] R. Sorensen. Yablo's paradox and kindred infinite liars. *Mind*, 107: 137–155, 1998.

REFERENCES

[Yablo(1985)] S. Yablo. Truth and reflexion. *Journal of Philosophical Logic*, 14: 297–349, 1985.

[Yablo(1993)] S. Yablo. Paradox without self-reference. *Analysis*, 53:251–252, 1993.

[Yablo(2004)] S. Yablo. Circularity and paradox. *Self-Reference, Bolander, Hendricks & Pedersen, (Eds.) CSLI Publications, Stanford*, pages 139–157, 2004.

Kripke's Interpretation of the Theories of Frege and Russell in Comparison with Mill's Views

Luis Fernández Moreno
Complutense University of Madrid

Abstract

In this paper I will mainly deal with Frege's and Russell's theories of singular terms, especially proper names, as well as of general terms. My aim is to elucidate whether Kripke's interpretation of those theories and of what they have in common is acceptable as well as to analyze the similarities and differences of those theories with regard to Mill's views alleged by Kripke.

Contents

1 Introduction   253

2 Frege's theory of proper names and general terms   254

3 Russell's theory of proper names and general terms   262

1 Introduction

In a passage of Kripke (1980), in which he puts forward Mill's division of names or terms into connotative and non-connotative in (1843), Kripke asserts:

> "He [Mill] says of 'singular' names that they are connotative if they are definite descriptions but non-connotative if they are proper names. On the other hand, Mill says that *all* 'general' names are connotative; such a predicate as 'human being' is defined as the conjunction of certain properties which give necessary and sufficient conditions for humanity –rationality, animality, and certain physical features. The modern logical tradition, as represented by Frege and Russell, *seems to hold* that Mill was wrong about singular names, but right about general names." (1980, p. 127; last emphasis added).

A few pages later Kripke does not introduce the qualification "seems to hold", and he claims:

> [...] "The modern logical tradition, as represented by Frege and Russell, disputed Mill on the issue of singular names, but endorsed him on that of general names. Thus *all* terms, both singular and general, have a 'connotation' or Fregean sense [...]" (Kripke 1980, p. 134).

Kripke equates Frege's theory of proper names with Russell's theory of ordinary names (see Kripke 2013, p. 8). In this regard Kripke talks of "the Frege-Russell view" or "the Frege-Russell analysis" (2011b, p. 53) or "the Frege-Russell doctrine" (2013, p. 11), and he asserts:

> "In [...] [(1980)] I constantly refer to the Frege-Russell doctrine, mostly having in mind what they would hold in common about (ordinary) proper names of historical figures, or names of fictional characters, etc." (2013, p. 11, n. 13).

However, since Kripke does not make any detailed considerations on Frege's and Russell's views of general terms, it has to be assumed that he is extending to this sort of terms Frege's and Russell's theories of (ordinary) proper names, which are usually regarded as versions of the classical descriptivist theory.

In this paper I will deal with Frege's and Russell's theories of singular terms, especially proper names, as well as of general terms. My aim is to elucidate whether Kripke's interpretation of those theories and of what they have in common is acceptable[1] as well as to establish the similarities and differences of those theories with regard to Mill's theory alleged by Kripke.

2 Frege's theory of proper names and general terms

Frege distinguishes two types of expressions, saturated (or complete) and unsaturated (or incomplete), and to this linguistic division corresponds the ontological division into saturated entities or *objects* and unsaturated entities or *functions*. Functions are referred to by means of unsaturated expressions to which Frege also alludes as "expressions of function" (1892b, p. 22, n. 7), "function-names" (1983, p. 259) and "function-signs" (ibid). On the other hand, objects are referred to by saturated expressions, denominated by Frege *proper names*. Thus, Frege understands the notion of proper name in a broad sense, which includes not only proper names in the usual sense, i.e., ordinary proper names, but also definite descriptions and even sentences. The consideration of sentences as a type of proper name –in the Fregean wide use of

---

[1] I will not enter into the discussion of Kripke's interpretation of Frege's semantics put forward in (2008) according to which Frege, like Russell (see section 3.), has a doctrine of (direct) acquaintance.

the term– that is, as a type of saturated expressions can be justified because unsaturated expressions can be completed by saturated ones to form other saturated expressions. So, e.g., elementary sentences are saturated expressions that result from the completion by proper names, in the usual sense[2], or by definite descriptions of predicates or general terms[3].

In a parallel way, this time at the ontological level, functions can be completed or saturated by objects, which will be the arguments of the function, to give rise to other objects, which are the values of the function for such arguments. Among those values, and therefore among the objects, are the truth values, denominated by Frege the *True* and the *False*, which according to him are the entities referred to by sentences. The type of functions Frege paid more attention to are the *concepts*, i.e., one-argument functions whose value is a truth value; when the value of a concept for an object is the True the object *falls under* the concept and this is a *property* of the object.

Frege used different expressions to denominate the general terms that designate concepts, among which there are "concept-expressions" (1892a, p. 41), "concept-words" (1983, p. 128), "concept-signs" (1983, p. 191) and "concept-names" (1983, p. 217). I will allude to them as *conceptual terms*. This is the type of general terms Frege took more into consideration, and I will be the only ones we will deal with. However, regarding these terms I will have to take into account their sense as well as their reference, both semantic dimensions similar to those of connotation and denotation in Mill's theory.

Frege introduced the notion of sense in (1892a) to explain the different cognitive value of the (true) identity sentences of the sort "$a = b$" with respect to those of the type "$a = a$" –"a" and "b" substitute proper names–, since the identity sentences of the first sort, e.g., "the Morning Star is (identical with) the Evening Star", unlike those of the second sort, e.g., "the Morning Star is (identical with) the Morning Star" can increase our knowledge. This happens when the names "a" and "b" have the same reference, which is required for the identity sentence to be true, but a different sense –and the speaker does not know in advance that they have the same reference. Thus, the sense of a proper name is a dimension of the name that contributes to the *cognitive value* of the sentences in which it appears; the notion of sense is therefore an *epistemic* notion.

---

[2] In the following when I use without qualifications the expression "proper name" in the framework of Frege's theory I will only take into account proper names in the usual sense, i.e., ordinary proper names.

[3] It should be distinguished between a general term and the monadic predicate formed by it, e.g., linking the general name with an indefinite article, thus building an indefinite description, and the predicative copula. Nonetheless, in this section I will interchangeably talk of predicates or general terms, thus using the last notion in a wide sense. The "conceptual terms" to which I will allude below are monadic predicates or in a broad sense the main components of those predicates, especially general terms - usually common names - and indefinite descriptions; Frege seems to have expressed himself also in this ambiguous way (see Frege 1892b). However, the unsaturated expression that may be completed by means of a proper name or a definite description is in strict sense the monadic predicate that contains a general term.

However, the notion of sense is also a *semantic notion*, since it is the mode or way in which the expression presents its referent. Thus, using the examples of proper names already mentioned, given by Frege, the expressions "the Morning Star" and "the Evening Star"[4] have the same referent, since they both refer to the planet Venus, but they present their referent in a different mode, let us say, as the first heavenly body seen at dawn and as the last heavenly body to fade at dusk, respectively. The sense of an expression, in its semantic role, i.e., as mode of presentation of its referent, can be characterized as a condition to identify its referent; the referent of an expression will be the entity that satisfies its sense. Thus, it is understandable that in Frege's theory the sense of an expression *determines* its reference, i.e., if two expressions have the same sense they will have the same reference –if they have reference– and if two expressions have a different reference they will have a different sense. Of course, there can be expressions with the same reference and a different sense, like "the Morning Star" and "the Evening Star".

In the following, I will attend to further details of the theory of the reference of proper names and of conceptual terms put forward by Frege, but also of his theory concerning their sense, since, as already said, the sense of an expression determines its reference. However, in this regard it is to be pointed out beforehand that Frege mainly dealt with the sense of proper names, making hardly any assertions concerning the sense of conceptual terms; the writing intended to be his main contribution to that subject is an incomplete manuscript, published posthumously, i.e., Frege (1982-1985). Nonetheless, it has to be assumed that Frege's theory of sense concerning conceptual terms will be similar to that concerning proper names.

As already indicated, the reference of a proper name is an object, and this reference is determined, and therefore mediated, by the sense of the name. In a famous footnote of (1892a) Frege affirms:

> "In the case of an actual proper name such as 'Aristotle' opinions as to the sense may differ. It might, for instance, be taken to be the following: the pupil of Plato and teacher of Alexander the Great. Anybody who does this will attach another sense to the sentence 'Aristotle was born in Stagira' than will somebody who takes as the sense of the name: the teacher of Alexander the Great who was born in Stagira. So long as the reference remains the same, such variations of sense may be tolerated [...]" (Frege 1892a, p. 27, n. 2)[5].

---

[4] As Dummett asserts, "these expressions, though typographically complex, may be claimed as logically simple, for we cannot be expected to determine their sense just for knowing the senses of the constituents: for one thing, they both refer to a planet, not to a star at all" (Dummett 1981, pp. 96-97).

[5] In Frege's framework the senses of sentences are thoughts. Thus, the sentence "Aristotle was born in Stagira" will express different thoughts for the two speakers he takes into consideration, since Frege endorses the principle of compositionality of sense (and also of reference), according to which the sense (reference) of a complex expression, like a sentence is determined by the senses (references) of its constituent

In this passage Frege admits that different competent speakers in the use of a proper name can associate with it different senses as they can associate with the proper name a different *definite description*; therefore, there is not, as a rule, a unique definite description that expresses (and thus makes explicit) the sense of a proper name[6].

If we generalize what is said by Frege in the quoted passage concerning the proper name "Aristotle", Frege's position seems to be that the sense of a proper name (for a speaker) would be expressed by a *definite description*, and the justification for this claim would be that the sense of a proper name is *identical* with the sense of a definite description; the conclusion to be drawn is that the reference of a proper name is determined by (the sense of) a definite description, that is, by the definite description that expresses the sense of the proper name. For this reason, it is understandable that the main thesis on the reference of proper names usually attributed to Frege is the following: the reference of a proper name is determined by *one* definite description –leaving aside the mentioned caveat that this description can be different for different speakers. This is regarded as the characteristic thesis of the *classic version* of the descriptivist theory for proper names.

As already said, this conclusion follows from two theses. Firstly, the reference of a proper name is determined by its sense. Secondly, the sense of a proper name is identical with the sense of a definite description –and therefore the sense of the name is expressed by a definite description. The question to be posed is whether Frege has really sustained that second thesis.

There are authors, Michael Dummett being one of the main ones, who have rejected that this thesis might be attributed to Frege. Dummett points out that the sense of a proper name provides a criterion to identify or recognize an object as the referent of the name, and although Frege admits that the sense of a proper name *can* be identical with that of a definite description, Dummett claims that Frege made no assertions from which it would follow that the sense of a proper name is *always* the sense of a definite description. Dummett claims that when Frege intended to say what the sense of a proper name is he was led in a natural way to point out a definite description, but from here it cannot be inferred that he maintained that the sense of a proper name is always identical with the sense of a definite description (see Dummett 1981, pp. 97 f.). He also asserts that for speakers to understand a proper name, i.e., to grasp its sense, they must have the ability to recognize the object that constitutes the referent

---

expressions and the way in which they are combined. On the other hand, since the definite descriptions "the pupil of Plato and teacher of Alexander the Great" and "the teacher of Alexander the Great who was born in Stagira" have the same reference, the sentence "Aristotle was born in Stagira" will have for both speakers the same reference, i.e., the same truth-value, and in this case, we assume, the True, although for the second speaker, not for the first one, that sentence will be a priori true and hence not informative.

[6]The same thesis is supported by the example put forward by Frege in (1918) concerning the name "Dr. Gustav Lauben". Frege alludes to two speakers, Herbert Gauner y Leo Peter, who associate with this name a different sense; one of them would be expressed by the definite description "the only individual born on the 13th of September of 1875 in N.N.", while the other sense would correspond to a definite description similar to "the only doctor who lives in such-and-such house" (1918, pp. 65 f.).

of the name in the presence of that object, but it is not required of them to be able to specify in a verbalizable way how they recognize the object in question, since the understanding of a name, at least in the case of some proper names, can simply consist in the association of the name with such ability (Dummett 1978, p. 129)[7].

Obviously, if the sense of a proper name can consist in an ability or in a non-linguistic criterion to recognize or identify the referent of the name, the equation of the sense of a proper name with the sense of a definite description would have to be rejected and also the subsequent claim that the sense of a proper name would always be expressed by a definite description.

However, Dummett's proposal seems to be subject at least to one important *limitation*, that is, non-linguistic criteria to recognize objects as the referents of names cannot be regarded as the senses of *all* names, as Dummett concedes, since there are objects with which we have never been –and will probably never be– in perceptive contact with, and therefore we can only identify them *linguistically*. For this reason, although Frege has never *explicitly* sustained that the sense of each name is *always* identical with the sense of a definite description and hence could be expressed by it, in the framework of Frege's position the only way to make it explicit in a *general* way and to communicate the sense of a proper name seems to require the resort to some definite description. From here it would follow that the reference of a proper name is determined by a *definite description*. This is at least what Frege would have to sustain concerning the reference of proper names that designate objects the speaker has never been in perceptive contact with and therefore which he can only identify or recognize linguistically.

The thesis that the sense of a proper name is identical with the sense of a definite description involves the assertion that the name is definable by means of the description in question. In this regard it is noteworthy that Kripke claims that according to Frege –and Russell– the notion of sense is identified with that of "defining properties" (1980, p. 127) and with that "given by a particular conjunction of properties" (1980, p. 135)[8]. In this assertion the notion of properties is not understood the way Frege does, since in Frege's theory properties belong to the sphere of reference: the properties of an object are the concepts under which the object falls, and concepts are the referents of conceptual terms. As Kripke recognizes, he is using the term "properties" in a wide sense (1980, p.137). In this respect one could say that the sense of a proper name in Frege's theory is given by the property or by the conjunction of properties expressed by the definite description that expresses its sense, which constitutes the *definiens* of

---

[7]There are other authors, like Geach and Searle, who have presented a similar interpretation to Dummett's on Frege's notion of sense; see Geach (1980) and Searle (1969), pp. 173-174. Another position on the Fregean notion of sense that rejects the descriptivist conception of Fregean senses is the one put forward by the "Neo-Fregeans", who interpret Frege's senses as non-descriptive ways of presentations or "de re senses"; this view involves that there cannot be sense without reference; see in this regard, for instance, McDowell (1977) y (1984) as well as Evans (1982).

[8]Concerning the attribution to Russell of that notion of sense (1980, p. 134) see section 3.

the proper name in question[9].

However, since each definite description is formed by a *conceptual term*, simple or complex, we are led to deal with Frege's theory of conceptual terms.

As already said, Frege hardly made any assertions concerning the sense of conceptual terms, although it is to be assumed that his theory on this matter would be similar –although as I will point out in the following *only* in part– to that regarding the sense of proper names. The sense of a conceptual term determines the concept that constitutes the referent of the term by providing the condition to be fulfilled by such concept and also by the objects belonging to the class that constitutes the *extension* of the concept; this is the class of the objects that fall under the concept. Thus, the sense of a conceptual term supplies a criterion to identify or recognize such objects and hence the corresponding class of objects. In the case of proper names I have already indicated that the most plausible interpretation of Frege's position, at least regarding names of objects with which we have not had perceptive contact, is that the reference of a proper name is determined by a definite description that expresses the sense of the name and therefore can be regarded as providing the definition of the name.

Concerning conceptual terms, we could sustain a similar thesis at least regarding those terms that refer to concepts under which non-sensorially perceptible objects fall and, in fact, the concepts of such kind were those Frege was mostly interested in, i.e., logical or logical-mathematical concepts. In any case, the similarity exists regarding the conceptual terms that designate *complex* concepts, i.e., concepts composed by other concepts. The reference of a conceptual term whose reference is a complex concept is determined by a *definition* which contains the conceptual terms that compose that concept, in such a way that the conjunction of those conceptual terms constitutes the *definiens* of the term[10] and expresses its sense. The concept in question is in the *subordination* relation[11] to each of its component concepts and those concepts are *marks* of the former one. Thus, when an object falls under a complex concept, it falls also under the concepts that constitute their marks; and this concept as well as the concepts by means of which it has been defined, are properties of the object[12].

---

[9] In (2013) Kripke characterizes the theory of Frege on proper names - as well as the theory of Russell on ordinary proper names - in a somewhat different way: "[t]o each proper name [...] there corresponds a criterion or property picking out which individual is supposed to be named by the name" (2013, p. 8) and "with each proper name one associates some predicate that is supposed to be uniquely instantiated" (2013, p. 9). In the case of Frege, the predicate in question will be the one contained in the definite description that expresses the sense of the name and the property (in the mentioned broad sense), the one expressed by that predicate. Concerning Russell's view in this regard see note 30 infra.

[10] An example put forward by Frege (see 1892b, p. 202) is the following. The concept of being a positive whole number inferior to 10 is defined by means of the concepts of being a positive number, being a whole number and being inferior to 10.

[11] A concept is subordinate to another concept if every object that falls under the first concept falls also under the second concept.

[12] Concerning the distinction between the notions of mark and property in Frege see, e.g., Frege (1884), section 53, as well as (1892b).

Nevertheless, not every conceptual term designates a complex concept; therefore, not every conceptual term can be defined. Taking into account what Frege claimed on many occasions concerning *simple* entities and the expressions that designate them, there will be conceptual terms whose clarification cannot be carried out by means of a definition, but through elucidations (see, for instance, 1983, p. 224). In the elucidation of a term we make understandable what it is understood by it –that is, its sense and therefore its reference– resorting to the use of expressions of the natural language, despite their uses not being sufficiently precise and being subject to fluctuations, counting in this regard on the good disposition of the audience to understand.

According to our preceding considerations, Kripke's interpretation of Frege's notion of sense mentioned above is appropriate concerning conceptual terms that refer to complex concepts, but it is questionable concerning conceptual terms that designate simple concepts and hence conceptual terms that are not definable. Kripke seems to extend Frege's theory of proper names to Frege's theory of general terms (and thus of all conceptual terms), but that extension is not justified.

Little more can be directly said concerning the sense of conceptual terms. Something else derives from connecting the sense of a conceptual term with its reference, since Frege made explicit some remarks on the reference of conceptual terms. The two most important claims in this regard are the following. One of them is that there are some conceptual terms *without* reference. However, Frege affirms in this concern:

> "These are not such as, say, contain a contradiction - for there is nothing at all wrong in a concept's being empty - but such as have vague boundaries. It must be determinate for every object whether it falls under a concept or not; a concept-word which does not meet this requirement on its reference is referenceless." (Frege 1892-1895, p. 133)[13].

Therefore, Frege's thesis is that there are conceptual terms without reference, these being the conceptual terms that are vague, that is, whose sense is *vague* and hence those whose sense would determine a concept whose extension has "vague boundaries". Using the notion of property as Frege understands it, instead of in the wide sense used by Kripke, elementary sentences, which contain a monadic predicate or conceptual term, will express, as a rule, the attribution of a property to an object, but in the case of vague conceptual terms there is, according to Frege, no property to be attributed, and hence neither correctly nor incorrectly. However, taking into account the compositionality principle of reference, this implies that the definite descriptions composed by conceptual terms of that sort will lack in reference; thus, the field is reduced for the descriptions that would determine the reference of proper names.

---

[13] In this passage I have translated the German words "Bedeutung" and "bedutungslos" by "reference" and "referenceless", instead of following the proposal of the translators of rendering them in English as "meaning" and "meaningless".

Frege states the other important thesis concerning the reference of conceptual terms in the following way:

> "[W]hat two concept-words refer to is the same if and only if the extensions of the corresponding concepts coincide." (1892-1895, p. 133)[14].

Put in another way, as far as the reference of conceptual terms is concerned, two concepts are identical if and only if they have the same extension, that is, if every object that falls under one of the concepts also falls under the other, and vice versa. However, from here it follows a not very intuitive claim, that is, that conceptual terms as "creature with a heart" and "creature with a kidney" refer to the same concept, although those terms will have a different sense[15]. Of course, this is consistent with the thesis of the determination of reference by sense, from which it follows –with the exception of the cases in which according to Frege there are conceptual terms without reference– that conceptual terms with the same sense will refer to the same concept, i.e., to concepts with the same extension, as well as those with different extensions, that is, different concepts are referred to by conceptual terms that express different senses.

According to our aforementioned considerations, the comparison between Frege's and Mill's theories could be summarized in the following theses. Firstly, their theories coincide concerning definite descriptions, since in Mill's theory they are singular but connotative terms and accordingly they have, as in Frege's theory, two semantic dimensions; thus, regarding definite descriptions there would be no conflict between Frege and Mill. Secondly, there would be a disagreement concerning proper names, since in Mill's theory these terms are non-connotative, and consequently they only have a semantic dimension, namely their denotation, while in Frege's theory proper names have two semantic dimensions. For this reason in Kripke's passages quoted at the beginning of this chapter, where it is said that "[according to Frege] Mill was wrong about singular names" (1980, p. 127) and it is asserted that "Frege [...] disputed Mill on the issue of singular names" (1980, p. 134), by "singular name" it should be understood *proper name*. Thirdly, in Frege's theory all grammatically well-formed expressions, like conceptual terms and, in general, general terms have sense and, but for the abovementioned exception, also reference, and according to Mill general terms are connotative terms and therefore they also possess two semantic dimensions, connotation and denotation. And in the same way as in Mill's theory the connotation of

---

[14] In this passage I have translated the German verb "bedeuten" by "refer", although the translators render it in English as "mean". See the preceding note.

[15] According to Frege two senses are identical if anybody who grasps them might not hold that there corresponds to them a different reference. However, a competent speaker of English can understand the expressions - i.e., grasp the senses of the expressions - "creature with a heart" and "creature with a kidney", but if he did not know that they have the same reference, by the mere understanding of them he might hold that they have a different reference; hence those expressions do not have the same sense. The mentioned criterion for the identity of senses is an application to senses in general of the criterion for the identity of thoughts put forward by Frege in (1892a), p. 32.

a general term determines its denotation, in Frege's theory the sense of a general term determines its reference. Thus, Kripke's claim can be justified that "[according to Frege] Mill was [...] right about general names" (1980, p. 127) and that "Frege [...] endorsed him [Mill] on [...] [issue of] of general names" (1980, p. 134). Although the notions of connotation and of sense are not quite identical, they are very similar, and Kripke seems to be identifying them, although writing "connotation" in quotes; given that identification, Frege would have agreed with Mill's thesis that general terms are connotative. Thus Frege could have subscribed, in Kripke's words, the claim that "*all* terms, both singular and general, have a 'connotation' or Fregean sense", although the Fregean sense of some general terms, and especially of the conceptual terms that are not definable, is in Frege's theory more complicated than Kripke seems to assume.

3 Russell's theory of proper names and general terms

Russell's theory on proper names or at least the theory on these terms traditionally associated with him is paradigmatically exposed in the writings published between 1905, the date when "On denoting" came out, and 1919, the publication date of *Introduction to Mathematical Philosophy*. Russell's theory on proper names includes a theory of *logically proper names*, and another of *ordinary proper names* (i.e., proper names in the usual sense). According to Russell, the latter are "really" definite descriptions (1912, p. 29; 1911, p. 206), or really stand for (definite) descriptions (1911, p. 215)[16], or they are "abbreviations for [definite] descriptions" (1918-19, p. 201), or "truncated or telescoped" (1918-19, p. 243) definite descriptions. I will allude to this thesis as the claim that ordinary proper names are *abbreviated definite descriptions* (see Russell 1969, p. 125).

Russell's theory of proper names and of general terms is put forward within an epistemological framework. In this regard, it becomes necessary to mention two sorts of epistemological distinctions; on the one hand, the distinction between *knowledge of things* and *knowledge of truths*, and on the other hand, between two sorts of the first type of knowledge, *knowledge by acquaintance* and *knowledge by description*. One of the features on which this distinction between two types of knowledge of things is based is that the knowledge by acquaintance is independent of the knowledge of truths, while the knowledge by description always involves some knowledge of truths.

We know an entity by *acquaintance* when we have a direct and immediate cognitive relation with it, without the mediation of any process of inference or any knowledge of truths; it is a sort of indubitable knowledge, which excludes error. According to Russell, we know by acquaintance two sorts of entities. On the one hand, a sort of entities called by Russell *particulars* and whose prototypical example are the

---

[16]Russell sometimes qualifies the last two claims with the words "usually" and "as a rule", but in later formulations Russell does not mention such caveats (see, e.g., 1918-19, p. 201).

sense-data[17]. On the other hand, we know by acquaintance certain *universals*, which paradigmatically include the universals instantiated in particulars from which we abstract them, although Russell makes an important concession, namely, that among universals "there seems to be no principle by which we can decide which can be known by acquaintance" (1912, p. 62-63).

On the opposite, we know an entity by *description* when we know it as the entity that satisfies a definite description, i.e., as the entity that instantiates the universal designated by the general term by means of which the definite description has been composed. According to Russell, physical objects, including human beings –perhaps with the exception of oneself– are known by description. Thus, we do not know a table by acquaintance, but by description; a table is "the physical object which causes such-and-such sense-data" (1912, p. 26). However, although this description describes the table resorting to sense-data, our knowledge of the table requires the knowledge of truths that connect the table with entities we know by acquaintance; we have to know that "such-and-such sense-data are caused by a physical object" (ibid). In this way, all knowledge by description involves knowledge by acquaintance as well as knowledge of truths. However, since all knowledge of truths requires acquaintance with universals, all our knowledge, of things as well as of truths, has its foundation in the knowledge by acquaintance. Russell claims that the knowledge by description is reducible to the knowledge by acquaintance and, additionally, that every proposition we can understand must be composed, in the ultimate analysis, of expressions that designate entities we know by acquaintance (see 1911, p. 209, and 1912, p. 32), i.e., of expressions that designate particulars as well as universals known by acquaintance –besides, if necessary, of logical signs[18]– and therefore, that every proposition we can understand is analyzable on the basis of such sorts of expressions[19].

---

[17] In (1912) Russell also includes among the particulars our mental states as well as the memories of sense-data and of our mental states; likewise, Russell considered probable that everyone be acquainted with oneself. However, in the following, as a rule, I will concentrate on the prototypical case of particulars, that is, on sense-data. It is noteworthy to point out that Russell occasionally uses the term "sense- data" in a wide sense, which includes all sorts of particulars (see 1911, p. 203).

[18] However, Russell claims that the logical signs contribute to the logical form of a proposition, but they are not constituents of the proposition (see 1918-19, p. 239). In the following, I will not take the logical signs into account, although I will mention them when we present Russell's theory of descriptions.

[19] The use of the term "proposition" by Russell is not univocal. In some of his writings - especially in some of his first writings and paradigmatically in (1903) - Russell regards propositions as compound extra-linguistic entities whose structure corresponds to that of the sentences that express them and that constitute the meaning of those sentences - and also the objects of belief expressed by belief sentences, in which the proposition in question corresponds to the sentence that constitutes the subordinate clause. Nonetheless, in other writings, especially later writings, Russell understands by "proposition" simply a declarative sentence (1918-19, p. 185), although in the same writing (1918-19) as well as in (1911), (1912) there are assertions according to which propositions seem to be regarded as composed of extra-linguistic entities. However, these assertions, which are traces of the first meaning of the term "proposition" already mentioned, could be understood as claiming that there are extra-linguistic entities which constitute the meaning of the expressions (by which propositions are formed). In general, I will follow the last use of the

The epistemological theses mentioned and especially the distinction between knowledge by acquaintance and knowledge by description, are at the basis of Russell's theory of the analysis of language, which runs parallel with a theory of the analysis of the world, at least such as his theory of the analysis of language is put forward in the writings of the said period.

The basic entities by means of which the rest of entities have to be analyzed are those entities we know by acquaintance; on the other hand, the entities we are acquainted with constitute the meaning of the most basic linguistic expressions by means of which the rest of linguistic expressions must be analysed[20]. We can put this thesis also in the following way. The ultimate foundation of our ontology is constituted by the entities we know by acquaintance and if the most basic and simple entities of our ontology have to provide the meaning of the simplest linguistic expressions, these will have to be expressions that designate particulars or universals –in the latter case, in strict sense, only universals we are acquainted with. The expressions that designate universals are general terms[21], while the expressions that designate particulars are the proper names "in the logical sense" (1918-19, p. 201) – "in the narrow logical sense" (1918-19, p. 201), "in the sense of logic" (ibid), or "in the proper sense of the word" (ibid) – or as it is usually said, the *logically proper names*, which according to Russell are the demonstrative pronouns like "this" and "that", but only as long as we use them to refer to something we know by acquaintance, i.e., to our own sense-data. Other sorts of expressions, like ordinary proper names as well as definite descriptions, are not proper constituents of the propositions in whose grammatical form they appear, which is shown in the reducibility of such propositions to propositions not containing such expressions. In particular, Russell claims that ordinary proper names are, as already said, abbreviated definite descriptions. Under that assumption, the analysis of the former is reduced to the analysis of the latter; in this regard, Russell puts forward his famous theory of descriptions and, in particular, of definite descriptions[22].

---

term "proposition" by Russell, although in order to conform to that procedure I will have to reformulate some of Russell's assertions, as we have already begun to do.

[20] Russell identifies the meaning of those expressions with their designation; concerning propositions he assumes a compositionality principle similar to Frege's compositionality principles for sense and for reference which Frege explicitly endorsed. However, Russell rejects Frege's notion of sense and hence the distinction between sense and reference; see Russell (1905). In this writing, using "meaning" for "sense" and "denotation" for "reference", Russell claims that Frege's theory leads to "an inextricable tangle" (1905, p. 50). However, Russell's objections are rather obscure and they might result from a misunderstanding of Frege's theory.

[21] Russell's way of expression concerning the terms that designate universals and, in particular, universals we know by acquaintance, is not uniform. Thus in (1911) he mentions as an example of a universal we know by acquaintance "the universal yellow", where "yellow" is a general term, while in (1912) he gives as examples of universals we are acquainted with yellow as well as yellowness; however, while the linguistic expression designating the former is a general term, the one designating the latter should be regarded is an abstract term. In the following, I will assume the view that universals are designated by general terms.

[22] Russell exposes his theory of descriptions mainly in (1905), (1910), (1918-19) and (1919), as well as in Russell/Whitehead (1910).

Russell sustains that, although the propositions that grammatically contain definite descriptions are meaningful sentences, definite descriptions do not possess meaning by themselves or in isolation and they are not authentic constituents of the proposition in whose grammatical form they appear. No proposition in which a definite description appears will contain[23], once it has been analyzed, the description in question, which will have disappeared; thus, in the proposition wholly analyzed there will not be a component that corresponds to the description as a whole. According to Russell's theory of descriptions, every assertion made by means of a proposition in which a definite description appears contains the assertion of the existence and uniqueness of the individual that satisfies the definite description. Thus, the assertion made by a proposition of the type "The F is G" –like the proposition "The present King of France is bald'– is asserted in a more explicit way by the proposition "There is a unique individual (i.e., at least an individual and at most an individual) that is F and that individual is G[24]", where "F" and "G" represent general terms, and still more explicitly by the symbolic formula "$\exists x(Fx \land \forall y(Fy \to y = x) \land Gx)$". When there is only one individual who satisfies the definite description, the description has *denotation*[25], but the individual denoted by the definite description is not the contribution of the description to the meaning of the propositions in which it appears. This is mainly due to two reasons. Firstly, we can understand the proposition, although we are not acquainted with the denoted individual, and secondly we can understand propositions that contain descriptions lacking in denotation. The contribution of a definite description to the meaning of the propositions containing it is not a constituent corresponding to the description as a whole, but such contribution consists in that of the component signs of the definite description according to the analysis of the definite descriptions proposed by Russell.

If an ordinary proper name is really an abbreviated definite description, the entity denoted by the proper name will be the one denoted by the definite description, and we find here again the thesis of the *classic version* of the descriptivist theory of reference for proper names, to wit, the claim that the reference (in this case, the denotation) of a name is determined by a definite description. However, Russell conceded, as Frege did, that different speakers can associate with a proper name *different* definite descriptions, and furthermore that the same speaker can associate with it different definite descriptions at different times (see note 26 *infra*).

---

[23] The assertions that a definite description appears in a proposition or that a proposition contains a definite description concern only the grammatical form of the proposition, not the authentic, logical form of the proposition.

[24] Or rather by the three propositions "there is at least one individual that is F", "there is at most one individual that is G", and "whoever is F is G" (see e.g. Russell 1919, p. 177).

[25] As Russell points out: "if 'C' is a [...] [definite description], it may happen that there is one entity x (there cannot be more than one) for which the proposition 'x is identical with C' is true [...] We may say that the entity x is the denotation of the phrase 'C'. Thus Scott, e.g., is the denotation of 'The author of Waverley' " (Russell 1905, p. 51).

As already indicated, Russell sustains that ordinary proper names are really abbreviated definite descriptions and –with the already mentioned exception of the logical signs– the components of a definite description should be in the last analysis expressions designating entities we are acquainted with. Something similar should be applicable to the general terms that designate universals with which we are not acquainted. A proposition which contains a general term that designates a universal we are not acquainted with will be analysable through a proposition –or propositions– which contain general terms that designate universals we are acquainted with. Thus, following the analysis by Russell of the ordinary proper names and hence of the definite descriptions, the general terms of the first type will be contextually analyzable on the basis of general terms of the second type, although Russell did not show the form of that analysis[26].

The comparison between Russell's and Mill's theories could be summarized in the following theses. Firstly, singular names include in Mill's theory proper names and definite descriptions. Russell's theory on definite descriptions disagrees with Mill's theory, since for Russell definite descriptions are not a sort of singular terms, but a type of quantifier phrases. Secondly, Russell's logically proper names only possess a semantic dimension, that of designation[27], and the same happens with proper names in Mill's theory, where they, as every non-connotative terms, only have the dimension of denotation; thus, as it is often said, Russellian logically proper names are Milllian. However, the type of entities designated by logically proper names is different from those denoted by proper names in Mill's theory. Thirdly, according to Russell ordinary proper names have a descriptive content (at least at the level of thought), since they

---

[26] Sometimes Russell puts forward the core of his view on ordinary proper names in a more nuanced way than the one mentioned before; thus, he asserts: "[...] the thought in the mind of a person using a proper name correctly can generally only be expressed explicitly if we replace the proper name by a description. Moreover, the description required to express the thought will vary from different people, of for the same person at different times. The only constant thing (as long as the name is rightly used) is the object to which the name applies. But as far as this remains constant, the particular description involved usually makes no difference to the truth or falsehood of the proposition in which the name appears." (1912, pp. 29-30, 1911, p. 206). From this passage it follows that at the level of thought the contribution of a proper name is given by a (definite) description, while at the level of the truth-conditions of a proposition the contribution of the name is the object to which it applies; thus, it is only at the first level that an ordinary proper name abbreviates a definite description. This two-level view of Russell's theory has been put forward by Sainsbury; see his (2002, pp. 86 ff.) and (2005, pp. 25-26). It is plausible that a similar view would also be applicable to the general terms designating universals we are not acquainted with.

[27] Russell employed the verb "to denote" – in the passage quoted in the preceding note, also "to apply" – and the corresponding substantives, like "denotation", to express the relation between an ordinary proper name or a definite description and the unique entity that satisfies the latter - since an ordinary proper name is at least at the level of thought an abbreviated definite description. In order to express the relationship between a logically proper name and the corresponding particular Russell mainly used verbs like "to name" and "to designate", and the corresponding substantives, like "designation". However, since in Mill's theory there is no difference between denotation and designation, in the comparison between Mill's and Russell's theories I will not put much weight on the distinction between the different semantical relations mentioned by Russell.

3 Russell's theory of proper names and general terms   267

are equivalent to definite descriptions they abbreviate, whilst in Mill's theory proper names and definite descriptions have different semantic features, since definite descriptions are connotative terms while proper names are non-connotative. It is in this regard that in Kripke's words, Russell would have held "that Mill was wrong about singular names" (1980, p. 127) and "Russell [...] disputed Mill on the issue of singular names" (1980, p. 134), where by "singular names" are understood (ordinary) proper names – let us bear in mind that in Kripke's passages he is speaking about "Frege and Russell". However, as already said, Russell's disagreement with Mill's view on singular terms concerns not only proper names but also definite descriptions which, in Mill's theory but not in Russell's, are singular terms.

Fourthly, as already mentioned, Russell does not provide a criterion to determine which are the universals we are acquainted with and hence nor to distinguish between general terms that stand for universals we are acquainted with and those that stand for universals we do not know by acquaintance. Sometimes he uses what apparently is a neutral expression applicable to definite descriptions and therefore to the general terms by which descriptions are formed, when he asserts that definite descriptions express "properties" (Russell 1911, p. 206 and 1912, p. 29), but by "properties" it will probably be understood just universals. Nonetheless, it could be alleged that there is a similarity between general terms in Russell's and Mill's theories, since in Mill's theory general terms, as connotative terms, express attributes or properties[28]. Another way to sustain the mentioned similarity, and the one intended by Kripke –since he does not enter into details on Russell's view of general terms–, results from extending Russell's view of ordinary proper names to general terms; in that case general terms would also have a descriptive content and would express attributes or properties, like general terms in Mill's theory. Thus, Kripke claims that Russell could have held, using his words quoted above, "that Mill [...] was right about general names" (1980, p. 127) and that Russell could have "endorsed him [i.e., Mill] on [...] [the issue] of general names" (1980, p. 134)[29].

However, Kripke also asserts that Russell would accept that " all terms, both singular and general, have a 'connotation' or Fregean sense" (1980, p. 134). In this regard, since Russell rejected Frege's notion of sense (see note 20 *supra*), Kripke recognizes that he is departing from Russell's literal claims, but Kripke justifies his procedure in the following way:

"Though we won't put things the way Russell does, we could describe

---

[28] The similarity would be stronger between general terms that stand for universals we are not acquainted with in Russell's theory and general terms in Mill's theory, since it could be claimed that the former have (at least at the level of thought) a descriptive content, and general terms and definite descriptions share in Mills' view the feature of being connotative terms.

[29] If we take into account Russell's two-level view mentioned in note 26, the contribution of a logically proper name, of an ordinary proper name and of a general term of every sort to the truth-conditions of the propositions in which they appear is to stand for entities, as it happens in Mill's theory concerning proper names and general terms.

Russell as saying that names, as they are ordinarily called, *do* have sense. They have sense in a strong way, namely, we should be able to give a definite description such that the referent of the name, by definition, is the object satisfying the description [...] In reporting Russell's views, we thus deviate from him [...] we regard descriptions, and their abbreviations, as having sense." (Kripke 1980, p. 27, n. 4).

Kripke seems to be taking as starting point the theory of Frege concerning the sense of proper names, according to which the sense of a name is expressed by a definite description. However, as said in section 2., Kripke attributes to Russell the same notion of sense he ascribes to Frege, whereas in Kripke's interpretation this notion is identified with that of "defining properties" (1980, p. 127) and with that "given by a particular conjunction of properties" (1980, p. 135). Since the "meaning" of an ordinary proper name in Russell's theory is provided (at least at the level of thought) by the definite description it abbreviates, the name is defined by that definite description, and its "meaning" is given by the property or the conjunction of properties expressed by the description, which also delivers the "meaning" of the description. Thus Kripke claims that we are justified in departing from Russell's literal view in such a way that "we regard descriptions, and their abbreviations [ordinary proper names] as having sense" (1980, p. 27, n. 4). Those considerations are at the basis of Kripke's attribution to Russell of the thesis that ordinary proper names have "a 'connotation' or Fregean sense'" (1980, p. 134)[30].

However, the same procedure cannot be applied in order to justify Kripke's attribution to Russell of the corresponding thesis concerning all general terms, since general terms that designate universals we are acquainted with are not abbreviations of definite descriptions. Besides, according to Frege, the sense of all general terms cannot be given by definite descriptions. Nonetheless, as already said, Kripke seems to be assuming the extension of Russell's theory of (ordinary) proper names to general terms, and this would make it possible for Kripke to attribute to Russell the claim that general terms have "a 'connotation' or Fregean sense" (ibid)[31].

References

[Black(1952)] M. Black. English translation: On sense and reference. *P. Geach and M. Black (eds.), Translations from the Philosophical Writings of Gottlob Frege, Oxford, Blackwell*, pages 56–78, 1952.

---

[30] In note 9 I indicated how Kripke (2013) characterizes in a somewhat different way the theory of Frege on proper names as well as the theory of Russell on ordinary proper names. In the case of Russell the predicate in question will be the one contained in the definite description that the proper name abbreviates (at least at the level of thought), and the property, the one expressed by that predicate.

[31] This paper has been supported by the Spanish Ministry of Economy and Competitiveness in the framework of the research project FFI2014-52244-P.

[Dummett(1978)] M. Dummett. *Truth and Other Enigmas.* Cambridge, Mass.: Harvard University Press, 1978.

[Dummett(1981)] M. Dummett. *Frege. Philosophy of Language.* London: Duckworth, 2nd ed., rev. and extended; 1st ed., 1973, 1981.

[Evans(1982)] G. Evans. *The Varieties of Reference.* J. McDowell (ed.). Oxford: Clarendon Press., 1982.

[Frege(1884)] G. Frege. *Die Grundlagen der Arithmetik.* Breslau: M. & H. Marcus. Reprinted in G. Frege, Hamburg, Meiner, 1884.

[Frege(1892a)] G. Frege. Über Sinn und Bedeutung. *Zeitschrift für Philosophie und philosophische Kritik*, 100:25–50, 1892a.

[Frege(1892b)] G. Frege. Über Begriff und Gegenstand. *Vierteljahresschrift für wissenschaftliche Philosophie*, 16:192–205, 1892b.

[Frege(1918)] G. Frege. Der Gedanke. *Beiträge zur Philosophie des deutschen Idealismus*, 2:58–77, 1918.

[Frege(1979a)] G. Frege. 1892-1895): "Ausführungen über sinn und bedeutung", in G. Frege (1983), pp. 128-136. *Posthumous Writings, Oxford, Blackwell*, pages 118–125, 1979a. English translation by P. Long and R. White: "Comments on sense and meaning" , in H. Hermes, F. Kambartel and F. Kaulbach (eds.).

[Frege(1979b)] G. Frege. (1983): *Nachgelassene Schriften. H. Hermes, F. Kambartel and F. Kaulbach (eds.).* Hamburg: Felix Meiner, 2nd edition, rev. and extended; 1st edition, 1969. (English translation of the 1st edition by P. Long and R. White, in H. Hermes, F. Kambartel and F. Kaulbach (eds.). Posthumous Writings, Oxford, Blackwell, 1979b.

[Geach(1980)] P. Geach. Some problems about the sense and reference of proper names. *in F.J. Pelletier and C. G. Normore (eds.), New Essays in Philosophy of Language, Guelph, Cannadian Association for Publishing in Philosophy*, pages 83–98, 1980.

[Kripke(1980)] S. Kripke. *Naming and Necessity.* Oxford: Blackwell. (Revised and enlarged edition, first published in D. Davidson and G. Harman (eds.), Semantics of Natural Language, Dordrecht, Reidel, 1972), 1980.

[Kripke(2008)] S. Kripke. Frege's theory of sense and reference: Some exegetical notes. *Theoria, Reprinted in Kripke (2001a)*, 74:254–291, 2008.

[Kripke(2011a)] S. Kripke. *Philosophical Troubles. Collected Papers, vol. I.* New York. Oxford University Press, 2011a.

[Kripke(2011b)] S. Kripke. Vacuous names and fictional entities. *In Kripke (2011a)*, pages 52–74, 2011b.

[Kripke(2013)] S. Kripke. *Reference and Existence* . Oxford: Oxford University Press, 2013.

[McDowell(1977)] J. McDowell. On the sense and reference of a proper name. *Mind*, 76:159–185, 1977.

[McDowell(1984)] J. McDowell. De re senses. *Philosophical Quarterly*, 34:283–294, 1984.

[Mill(1843)] J.S. Mill. *A System of Logic Ratiocinative and Inductive*. Reprinted in J.M. Robson (ed.), Toronto, University of Toronto Press, 1973, 1843.

[Russell(1903)] B. Russell. *The Principles of Mathematics*. New York: Norton. Reprinted in B. Russell, Cambridge, Cambridge University Press, 2nd edition, 1935, 1903.

[Russell(1905)] B. Russell. On denoting. *Mind. Reprinted in B. Russell (1956)*, 14: 39–56, 1905.

[Russell(1910-11)] B. Russell. Knowledge by acquaintance and knowledge by description. *Proceedings of the Aristotelian Society. Reprinted in B. Russell, A Free Man's Worship and Other Essays, London, George Allen & Unwin, 1976*, 11:200–221, 1910-11.

[Russell(1912)] B. Russell. *The Problems of Philosophy*. London: Williams Norgate. Reprinted in B. Russell, Oxford, Oxford University Press, 1959, 1912.

[Russell(1918-19)] B. Russell. The philosophy of logical atomism . *The Monist. Reprinted in B. Russell (1956)*, 28 and 29:175–282, 1918-19.

[Russell(1919)] B. Russell. *Introduction to Mathematical Philosophy*. London: George Allen & Unwin, 1919.

[Russell(1956)] B. Russell. Logic and Knowledge. *Essays 1901–1950, R.Ch. Marsh (ed.). London: Unwin*, 1956.

[Russell(1969)] B. Russell. *My Philosophical Development*. London: Unwin, 1969.

[Russell and Whitehead(1910)] B. Russell and A. Whitehead. *Principia Mathematica, vol. I*. Cambridge: Cambridge University Press, 1910.

[Sainsbury(2002)] R.M. Sainsbury. Russell on names and communication. *In R.M. Sainsbury, Departing from Frege. Essays in the Philosophy of Language London/New York, Routledge*, pages 85–101, 2002.

[Sainsbury(2005)] R.M. Sainsbury. *Reference without Referents*. Oxford: Clarendon Press, 2005.

[Searle(1969)] J. Searle. *Speech Acts. An Essay in the Philosophy of Language* . Cambridge: Cambridge University Press, 1969.

Logics of programs as a fuelling force for semantics

Francisco Hernández-Quiroz
Facultad de Ciencias
Universidad Nacional Autónoma de México
Ciudad Universitaria, D.F. 04510, MEXICO
e-mail: fhq@ciencias.unam.mx

Abstract

Programming Language Semantics is a field with a staggering variety of approaches that do not necessarily share aims and can be in conflict with each other sometimes.

Some of the reasons for these differences have been discussed in the past. Turner [Turner(2007)] has commented on the Platonist/formalism divide as one possible source. He suggests this distinction partially overlaps with the distinction between operational and denotational semantics (but it also runs through and against it).

Nevertheless, the origin of part of the variety can be found elsewhere. This paper explores some motivation behind a different approach: axiomatic semantics and logics of programs—and how this approach projects semantic problems against a very different background. By looking at attempts at logics of programs, the role of semantics in programming languages can be better understood as an effort to find convergent and complementary explanations of programming languages (full abstraction and domain theory in logical form are just two examples of this project) lead by a desire to produce more reliable software rather than by a purely theoretical intent as is so often seen in denotational semantics.

keywords: semantics, logics of programs, domain theory in logical form.

1 Introduction

As White has pointed out [White(2004)], programming language semantics and the Davidson-Dummett program of a formal semantics ("a theory of meaning") share some aims. And program language semantics also shares with its philosophical counterpart the presence of competing, divergent views about semantics.

While it cannot be said that programming language semantics is a real jungle, it is fair to call it a diverse ecosystem. Axiomatic, operational and denotational semantics (of various kinds) populate this realm. This variety demands an explanation, and it will probably not be simple.

Nevertheless, some reasons for this diversity have been advanced [White(2004)]. The traditional divide between formalism and Platonism in mathematics has been put forward as one of the sources of the different semantic approaches. Roughly speaking, denotational semantics is said to lean towards Platonism, while operational semantics towards formalism, both with some caveats [Turner(2007)].

According to Turner, operational semantics is translational, while denotational semantics refers to the mathematical objects behind a programming language syntactical constructions. In this author's view, Turner's observation about operational semantics applies only to Landin's original proposal in the 1960's, but not to the more widespread approach of structural operational semantics of Plotkin [Plotkin()], where the meaning of program constructs is given by inference rules dealing with the effect of the execution of a command or the evaluation of an expression. No translation is involved.

Nevertheless, these are not the only reasons for diversity. Another major force behind the flourishing of semantic approaches in the 70 was the search for techniques for improving software reliability. This endeavour fuelled the development of axiomatic semantics (already in the scene since the end of the 60 [Floyd(1967), Hoare(1969)]). Axiomatic semantics defines the meaning of programs by means of an external language (as did initial attempts at operational semantics), but this aim is not reached through translation, and thus axiomatic semantics is set on very different grounds.

2 Software crisis and verification of programs

At the end of the sixties, computers had become very complex and powerful compared to their earlier predecessors and their users had also become accustomed to the idea that this trend would not stop soon. Programmers had begun to find increasingly difficult to build reliable software for these machines. Awareness of this situation led to the adoption of the leitmotif of "software crisis" [Naur and B. Randell(1969)] by many leading computer scientist, Dijkstra among them [Dijkstra(1972)].

The software crisis was a strong stimulus for the adoption of more formal tools for developing computer software. It was expected that formal methods would help to produce more reliable software. This, in its turn, would help overcome or at least soften the software crisis. Two very influential books emerged after a few years of effort in this direction: Dijkstra's *A Discipline of Program* [Dijkstra(1976)] and Gries's *Science of Computer Programming* [Gries(1981)].

3 Verification of programs and axiomatic semantics

Axiomatic semantics does not try necessarily to provide a definition of the meaning of a program, but to offer formal tools to verify software. Verification means here a mathematical proof that a program (written in an ideal or in a "real" computer language) meets the requirements it was designed for. So, the most common product of an axiomatic semantics is a "logic of programs" for conducting such proofs. Related approaches are formal specification and derivation of programs, which also rely on a particular axiomatic semantics.

Program verification is then a series of techniques for proving programs "right". It developed partially in opposition to the idea of testing programs for errors in order to produce reliable software [Dijkstra(1976)]. In program verification, software properties (v.g., intended behaviour, termination or not termination) are expressed ("specified") in a formal language, generally a logical language, and then a formal method (sometimes automatic or computer assisted) is applied to prove that a program meets the requirement. This author has been unable to find a definite original source for the expression "program verification", but related terms can already be found in Floyd seminal work [Floyd(1967)].

Program derivation is akin to program verification. In this case, intended properties of a program are specified also in a formal language (generally a type of logic), but instead of having a program tested against these properties, the program itself is derived by means of logical rules from the specification. This is also known as "correctness by construction". Of course, a suitable axiomatic semantics is needed for this approach. Again a leading early example of this approach is Dijkstra [Dijkstra(1976)].

In addition to a formal language for specifying properties of a program, an axiomatic semantics consists also of a set of inference rules (axioms) to prove that a program has the intended properties or to guide the derivation of a program from a specification. The earliest example is Floyd-Hoare logic [Floyd(1967), Hoare(1969)], based partially on unpublished ideas by Perlis and Gorn (see [Jones(2003)] for a very illuminating historical account).

Right since the beginning, Floyd intended to produce "an adequate basis for formal definitions of programs" [Floyd(1967), p. 19] and to use this basis to prove correctness, termination and equivalence of programs.

Floyd-Hoare logic works with triplets $\{\alpha\}P\{\beta\}$, where α and β are formulas in a suitable logical language (generally a subset of first-order calculus) and P is a program in a typical imperative language. The meaning of $\{\alpha\}P\{\beta\}$ is intuitively the following:

If the computer is in a state satisfying the *precondition* α, then a successful (i.e. terminating) execution of program P leads to a state satisfying *postcondition* β.

The formal meaning of the previous assertion can be defined implicitly via a set of inference rules ("axioms" in Hoare's words) or explicitly via a traditional Tarski-based semantics for the logical language of pre- and post- conditions. Interpretations à la Tarsky are extended in order to cover the memory states of an abstract machine executing the program. Because of this general scheme, some further logics of programs were called "predicate transformers" [Dijkstra(1976)].

A close follow-up to Floyd and Hoare's attempt is Dijkstra's "weakest precondition" logic: program meaning is defined as the weakest statement that is true in any state that leads to a successful execution of the program, that is one that is both terminating and ends in a state satisfying the postcondition. Weakest precondition logic was intended as a tool for deriving correct programs from a specification (in this case, the postcondition).

A totally different approach can be seen in logics like Hennessy-Milner's: programs (or processes in their terminology) are paired with formulae by means of a *satisfaction relation*: $P \models \alpha$. In this case, the formula α plays the role of a specification and the correctness proof goes along model-theoretic lines rather than deductive ones. Temporal logic was used in a similar way for *model checking* finite-state systems [Pnueli(1977)].

4 Two problems of axiomatic semantics

Two charges can be levelled easily against axiomatic semantics:

- Is it really a semantics or was it just a fashionable name for a software engineering technique at its inception time? What is the ontology behind it?

- Axiomatic semantics are themselves quite diverse and this diversity can be sometimes a major weakness. There are different systems based on the programming language they target or the logical language they are built upon. They also differ on the properties of programs they can prove. As a consequence, it is not so easy to evade the accusation of being ad hoc systems that do not reflect any intrinsic features of a given programming language.

We will deal with both charges in the next two subsections.

4.1 Semantics without an explicit ontology

If we agree with Turner that operational semantics gives an explanation of the meaning of a program via a translational device and that denotational semantics gives a mathematical ontology for the program constructs, what does axiomatic semantics give us?

Let us take, for instance, the case of Floyd-Hoare logic. Given a program P, let us say that its meaning is defined by

$$[\![P]\!] = \{(\alpha, \beta) \mid \{\alpha\} P \{\beta\} \text{ is true.}\}$$

This approach is miles away from operational or denotational semantics. But it is not alien to Davidson's program for meaning: the meaning of an assertion is the conditions on which the assertion is true [Davidson(1984)], following Tarski's successful definition of truth for predicate calculus [Tarski(1936)]. In fact, the set of pairs of pre- and post-conditions met by programs is made of the formulae true before and after the execution of a program. Truth in this case comes from the initial and final states in which a computer executes the program, and these states then play the role of the conditions for truth.

4.2 Ad hoc character and lack of abstraction

But, why an ad hoc character would be detrimental to axiomatic semantics? Let us say we have two very different program logics for a language: they can express different sets of properties without an obvious relation between them (for instance, inclusion); they also possess different inference rules with disparate deductive power. How can they claim to give *the* semantics of the language in the way a denotational model would? Here it is when the ad hoc character of the logics comes to the fore: they were designed with distinct purposes and these purposes shaped the resulting logics in ways that render them incomparable.

This is in stark contrast to denotational semantics, one of whose aims is to give an abstract model of a programming language. A denotational semantics is considered to be adequate when it meets a mathematical requirement about its canonical (as opposed to ad hoc) character. Full abstraction [Milner(1977)], i.e. a form of strong equivalence with a given operational semantics, normally takes this role [Ong(1995)].

5 Domain Theory in Logical Form

There have been efforts to eliminate this ad hoc appearance by grounding logics of programs to a more abstract foundation. An obvious candidate for this foundational role is denotational semantics. One of the most ambitious attempts in this direction is Abramsky's Domain Theory in Logical Form [Abramsky(1991)], which intended to derive logics of programs directly from denotational semantics.

Domain theory in logical form is based on the idea that "given a denotational description of a computational situation in our meta-language, we can turn the handle to obtain a logic for that situation", as Abramsky himself explains [Abramsky(1991), p. 1]. This is done by using Stone's duality [Stone(1936)]. The denotational side of the semantics takes away the ad hoc flavour of the resulting logic and hence brings together the best of two worlds (at least in theory).

Let P be a program in any given programming language and α a formula expressing a property of this program in a given logical language. This situation can be stated by $P \models \alpha$.

In a denotational semantics, the meaning of P, in symbols $[\![P]\!]$ is a "point" in a suitable mathematical space $\mathcal{D}$. The meaning of a formula like α is the set $[\![\alpha]\!] \subseteq \mathcal{D}$. Then

$$P \models \alpha \quad \text{means} \quad [\![P]\!] \in [\![\alpha]\!].$$

In axiomatic semantics, the relation $\models$ is axiomatized so that the meaning of a program is the set of properties (expressed by formulas) that it satisfies:

$$[\![P]\!] = \{\phi \mid P \models \phi\}.$$

This logic is a lattice of properties of programs. Given this

$$P \models \alpha \quad \text{means} \quad \alpha \in [\![P]\!].$$

Stone's representation theorem for Boolean algebras proved that every Boolean algebra B can be represented as a field of sets. Operations of meet, join and complement are represented by intersection, union and set complement. Stone's method started building a topological space *Spec B* and then proving that B is isomorphic to the closed-open sets of *Spec B*. They are said to be *Stone duals* of each other.

This very method can be applied to denotational semantics by means of a metalanguage that assigns domains to types in programming languages and elements in these domains to programs. On the other hand, a logical interpretation of the metalanguage assigns propositional theories to domains and determines what properties are satisfied by programs by means of an axiomatization. These two interpretation of types and terms are Stone duals and so they determine each other up to isomorphism and the axiomatic and denotational interpretation of $P \models \alpha$ are equivalent.

As the initial ingredient in this process we have a programming language. By giving it a denotational semantics, we can "turn the handle" of domain theory in logical form and obtain a program logic that is guaranteed to be uniquely determined by the semantics. Ad hoc-ness has disappeared from the scene!

In a way analogous to full abstraction's unification of two semantic paradigms (the denotational one and the operational one), domain theory in logical form shows that the opposition between the very practical, software engineering-based approach of axiomatic semantics is not totally unrelated from the extremely abstract style of denotational semantics and that the entities populating both universes have not so obvious (though not less real) connections.

6 Conclusions

Axiomatic semantics arose from two diverging concerns: to have a formal definition of the meaning of a program (as in operational or denotational semantics) *and* to have a tool for helping to produce reliable software. This was achieved in the way of program correctness: the property of meeting a specification expressed by formulae in a logical language. Proof-theoretic techniques and model checking are standard methods in logics of programs.

The case of axiomatic semantics for the role of a mathematically well-grounded semantic model for a programming language may seem not as strong as those of operational or denotational semantics. But the way axiomatic semantics can fill this role is not very different from theories of meaning found in different contexts, such as Davidson's.

Still logics of programs are open to the charge of being ad hoc, weakening its generality as an abstract model for programming languages. A possible solution is not to design logics of programs from scratch, but to derive them from a denotational model. Abramsky's Domain Theory in Logical Form provides a method for this. By relating axiomatic and denotational semantics by means of a strong form of equivalence, Domain Theory in Logical Form performs a role similar to that of full abstraction in operational and denotational semantics.

Nevertheless, Domain Theory in Logical Form, as it stands, is not the whole solution to the problem of the ad hoc character of axiomatic semantics. To begin with, it is able to produce only a certain type of propositional program logics and not some very complex program logics based on predicate calculus like Dijkstra's weakest precondition calculus.

The question of how to frame an equivalence between axiomatic semantics and other semantic approaches remains open. Hopefully further work can take this question up from the point reached by Domain Theory in Logical Form.

References

[Abramsky(1991)] S. Abramsky. Domain theory in logical form. *Annals of Pure and Applied Logic*, 51:1–77, 1991.

[Bjorner and Jones(1978)] D. Bjorner and C.B. Jones. *The Vienna Development Method: The Meta-Language*. Lecture Notes in Computer Science 61. Berlin, Heidelberg, New York: Springer, 1978.

[Davidson(1984)] D. Davidson. *Inquiries into Truth and Interpretation*. Oxford: Oxford University Press, 1984.

[Dijkstra(1972)] E. Dijkstra. The humble programmer. *Communications of the ACM*, *15*, 10:859–8, 1972. ACM Turing Lecture 1972.

[Dijkstra(1976)] E. Dijkstra. *A Discipline of Programming*. Prentice Hall, 1976.

[Floyd(1967)] R.W. Floyd. Assigning meanings to programs. *Proceedings of the American Mathematical Society Symposia on Applied Mathematics*, 19:19–31, 1967.

[Gries(1981)] D. Gries. *The Science of Programming*. Springer Verlag, 1981.

[Hoare(1969)] C.A.R. Hoare. An axiomatic basis for computer programming. *Communications of the ACM*, 12(10) 576:580,583, 1969.

[Jones(2003)] C.B Jones. The early search for tractable ways of reasoning about programs. *Annals of the History of Computing*, 25(5):139–143, 2003.

[Milner(1977)] R. Milner. Fully abstract models of typed lambda-calculi. *Theoretical Computer Science*, 4:1–22, 1977.

[Naur and B. Randell(1969)] P. Naur and (eds.) B. Randell. *Software Engineering: Report of a conference sponsored by the NATO Science Committee, Garmisch, Germany, 7–11 Oct. 1968.* Brussels, Scientific Affairs Division, NATO, 1969.

[Ong(1995)] C.-H.L. Ong. Correspondence between operational and denotational semantics: the full abstraction problem for PCF. In *Handbook of Logic in Computer Science*, volume 4. Oxford University Press, 1995.

[Plotkin()] G.D. Plotkin. The origins of structural operational semantics. *The Journal of Logic and Algebraic Programming*, 60-61:3–15.

[Pnueli(1977)] A. Pnueli. *The Temporal Logic of Programs*. Foundations of Computer Science, FOCS, 1977.

[Spivey(1992)] J.M. Spivey. *The Z Notation: A reference manual*. Prentice Hall International Series in Computer Science, 2nd edition, 1992.

[Stone(1936)] M.H. Stone. The theory of representations for Boolean algebras. *Transactions of the American Mathematical Society*, pages 37–111, 1936.

[Tarski(1936)] A. Tarski. The Semantic Conception of Truth and the Foundations of Semantics. *Philosophy and Phenomenological Research*, 4(3):341–376, 1936.

[Turner(2007)] Raymond Turner. Understanding Programming Languages. *Minds & Machines*, 17:203–216, 2007.

[White(2004)] Graham White. The Philosophy of Computer Languages. In *The Blackwell Guide to the Philosophy of Computing and Information*, pages 237–248. Blackwell, 2004.

¿Puede un personaje literario estar durmiendo en una playa? Ficción, referencia y verdad

Eleonora Orlando
Universidad de Buenos Aires / CONICET

Índice

1. **Introducción: un dilema para el análisis semántico del discurso de ficción** 282
2. **Los nombres de ficción como designadores rígidos**   **288**
    2.1. Crítica de algunas posiciones estándar 288
    2.2. La idea del 'cambio de contexto' y los mundos posibles ficticios . . . 291
3. **Conclusión**   **295**

En este ensayo me ocupo del análisis semántico de los enunciados que contienen nombres de ficción. En la primera parte, introduzco el problema semántico fundamental planteado por ese tipo de enunciados, a saber, ¿cómo es posible explicar la intuición de que dan lugar a usos lingüísticos significativos y, en algún sentido, verdaderos, aun cuando uno no está hablando acerca de nada o nadie real? Asimismo, describo las principales respuestas ofrecidas: por un lado, las posiciones realistas según las cuales esos usos expresan proposiciones singulares, constituidas en parte por objetos ficticios referidos por los nombres en cuestión; por otro lado, el antirrealismo, según el cual deben ser parafraseados en términos de enunciados que no contienen nombres de ficción en posiciones referenciales. En la segunda parte, defiendo una versión de realismo abstractista, en la línea de Kripke (2011, 2013 basado en su 1973) y Thomasson (1999). De acuerdo con ella, los nombres de ficción son expresiones genuinamente referenciales y sus referentes son entidades abstractas, los personajes literarios, de un tipo peculiar. Más específicamente, mi defensa está basada en el ofrecimiento de una explicación, dentro de ese marco, de cómo los usos ficticios de enunciados que contienen nombres de ficción pueden ser entendidos como afirmaciones verdaderas (o falsas) con respecto a mundos posibles internos al marco de la narración correspondiente, esto es, mundos puramente concebibles.

1. Introducción: un dilema para el análisis semántico del discurso de ficción

Como ha sido ampliamente reconocido, los nombres de ficción pueden aparecer en distintos tipos de usos lingüísticos. Por un lado, están los usos *fictivos*, como el de

(1) Ulises duerme en la playa de Ítaca,

en el contexto de la creación de la *Odisea*, reeditado luego por cada lector de la obra, y los usos *parafictivos*, como por ejemplo el de

(2) Ulises es un hombre, no un dios,

para hacer una aclaración acerca de los hechos narrados en la *Odisea* en el contexto de una clase de literatura griega. En ambos casos, la intención del hablante es hacer referencia a hechos de una historia de ficción, adscribirle a un personaje literario características o propiedades que le son atribuidas en el marco de la narración correspondiente, por lo que los llamaré globalmente 'usos ficticios'. Por otro, están los usos *metafictivos*, tales como el de

(3) Ulises es un famoso personaje literario,

en el contexto de la escritura o la lectura de un ensayo crítico, en donde la intención del hablante es hacer referencia a hechos reales, adscribirle a un personaje literario propiedades que de hecho le pertenecen. Pueden distinguirse de ellos los usos *fácticos*, en los cuales, si bien también se hace referencia al mundo real, se cree (falsamente) que se está hablando de una persona y no de un personaje, como sería el caso de un uso de (1) bajo la impresión de que Ulises existió realmente. Nótese entonces que, desde cierto punto de vista, el uso fictivo de (1) lo hace intuitivamente verdadero, a diferencia de lo que ocurre con un enunciado como

(4) Ulises duerme en una playa del Caribe,

mientras que el uso fáctico lo vuelve falso; asimismo, (2) resulta intuitivamente verdadero en el uso parafictivo, pero falso en el fáctico[1]. Finalmente, (3) es intuitivamente falso en los usos fictivo y parafictivo, pero verdadero en el metafictivo[2].

---

[1] Nótese que los usos metafictivos de (1) y (2), es decir, aquellos realizados con la intención de adscribirle a un personaje literario características que posee en el mundo real, son muy poco probables (dado que las características que se adscriben a Ulises por medio de (1) y (2) son claramente características que el personaje posee en la historia de ficción pero no en el mundo real).

[2] Véase, por ejemplo, Bonomi (2008) para una distinción de usos semejante a la aquí presentada. Kripke (2013, basado en su 1973) distingue originalmente entre dos posibles lecturas de los enunciados que contienen nombres de ficción: por un lado, los usos externos o fácticos (*out and out*) se realizan para describir el mundo real; por otro, los internos o ficticios (*according to the story*) son aquéllos mediante los cuales se describe lo que ocurre en una historia de ficción.

1 Introducción: un dilema para el análisis semántico del discurso de ficción

Ahora bien, el problema que presentan todos estos usos es que es necesario explicar la intuición de que son significativos y, eventualmente, tienen valor de verdad, aun cuando uno parece no estar hablando acerca de nada ni de nadie. Como señala Simpson (1964), este problema, que se encuentra explícitamente planteado ya en el *Teéteto* de Platón, ha conducido a una ampliación del compromiso ontológico basada en razones gramaticales y semánticas, tal como se pone de manifiesto en el siguiente argumento[3]:

(i) (1), (2), (3) y (4) son oraciones de la forma gramatical sujeto-predicado.

(ii) 'Ulises' es el sujeto gramatical de (1), (2), (3) y (4).

(iii) (1), (2), (3) y (4) son significativas.

(iv) Si (1), (2), (3) y (4) son significativas, entonces son verdaderas o falsas (dado el principio de bivalencia).

(v) Toda oración de la forma sujeto-predicado es verdadera si y solo si el objeto nombrado por el sujeto tiene la propiedad expresada por el predicado, y es falsa si y solo si el objeto nombrado por el sujeto no tiene la propiedad expresada por el predicado.

Por lo tanto,

(vi) Ulises es un objeto (de algún tipo).

Un razonamiento como el anterior parece subyacer a las diferentes *propuestas realistas* acerca del discurso de ficción. La propuesta original de Alexis von Meinong (1904), retomada por los neo-meinongianos (Parsons 1980, 1982, Priest 2005), ha sido comprometerse ontológicamente con *objetos concretos inexistentes* que pudieran oficiar de referentes de los nombres de ficción (así como de otras expresiones vacías como 'el cuadrado redondo'). Dado que está claro que Ulises no es ningún objeto existente, para Meinong y los neo-meinongianos, se trata de un objeto inexistente; es decir, a partir del presupuesto de que ser no es lo mismo que existir, se concluye que se trata de un objeto que tiene algún tipo de ser que no es la existencia. Más recientemente, Lewis (1978), en el marco de su realismo modal, ha propuesto considerar a las entidades de las historias de ficción no como objetos inexistentes sino como *objetos existentes pero meramente posibles*. Finalmente, los abstractistas (Kripke 2011, 2013, basado en el manuscrito de 1973, Salmon 1998, 2002, Thomasson 1999, Predelli 2002) consideran que se trata de *objetos existentes y actuales, pero abstractos*. Este incremento ontológico involucra la ventaja semántica de considerar a los nombres de ficción como expresiones genuinamente referenciales, al igual que el resto de

---

[3] Véase Simpson (1964: 57-61) para la propuesta de argumentos semejantes a éste en relación con usos aquí llamados 'fictivos' y 'metafictivos' (tales como los de existenciales negativos del tipo de "Ulises no existe").

los nombres. Los enunciados que componen expresan, por tanto, proposiciones singulares, esto es, constituidas en parte por particulares, a saber, los objetos *sui generis* en cuestión. Las oraciones anteriores tienen entonces la misma estructura gramatical y semántica que cualquier oración que tenga un nombre como sujeto.

Por otro lado, quienes no son partidarios de ampliar el compromiso ontológico sobre la base de razones gramaticales y semánticas, empezando por filósofos clásicos como Frege y Russell, han optado por bloquear la conclusión anterior, de diferentes maneras. De la teoría de Frege (1892) se sigue la negación de la premisa (iv); de este modo, ningún uso de (1), (2), (3) o (4) es verdadero o falso puesto que todos ellos carecen de valor de verdad. La posición de Russell (1905), por otro lado, involucra la negación de la premisa (i): (1), (2), (3) y (4) no tienen *realmente* la forma sujeto-predicado, por lo que no es preciso comprometerse con ningún objeto presuntamente designado por sus respectivos sujetos gramaticales. Lo que tienen en común ambos enfoques es la adopción de lo que puede llamarse '*una estrategia descriptivista*', según la cual los nombres de ficción son semánticamente equivalentes a términos de naturaleza descriptiva. De este modo, una oración como (1) es entendida, a grandes rasgos, en términos de otra, descriptiva y por lo tanto, según Russell, veladamente cuantificada, tal como

(5) El protagonista de la *Odisea* duerme en la playa de Ítaca.

En nuestros días, la posición antirrealista de Walton (1990) también involucra un rechazo de la premisa (i), aunque basado en razones distintas relacionadas con su concepción de lo que es una obra de arte en general: desde su perspectiva, los enunciados anteriores, si bien tienen la forma sujeto-predicado, no expresan ninguna proposición y deben ser parafraseados en términos de oraciones tales como, en relación con (1),

(6) Existe un juego de *make-believe* originado por la *Odisea* en cuyo marco afirmar (1) es hacer una afirmación verdadera,

que expresan, claramente, proposiciones generales. Se renuncia por tanto a una teoría general de la referencia para los nombres: los nombres de ficción, a diferencia de los restantes, no refieren a particulares, esto es, no son expresiones genuinamente referenciales[4].

Hasta aquí podría decirse entonces que el análisis semántico del discurso de ficción nos enfrenta con el siguiente dilema: o bien *se interpreta, a la manera realista, que los enunciados que contienen nombres de ficción expresan proposiciones singulares al precio de un controvertido incremento en la ontología; o bien se los parafrasea, a la manera antirrealista, en términos de enunciados que expresan proposiciones generales al precio de sacrificar su carácter prima facie singular*. En otras palabras, las opciones parecen ser preservar la simplicidad de la teoría semántica y complicar la

---

[4] Para otra posición antirrealista actual, véanse Sainsbury, 2005, 2010a, 2010b.

metafísica con la introducción de objetos *sui generis* o bien, a la inversa, revisar la semántica y preservar un compromiso ontológico austero. En este trabajo optaré por la primera opción, sobre la base de dos razones principales.

En cuanto a la primera razón, el revisionismo semántico involucrado por el antirrealismo tiene un alto costo teórico: como se destacó más arriba, implica considerar que los nombres de ficción no son expresiones genuinamente referenciales, esto es, no son genuinos nombres. El filósofo antirrealista parece motivar su posición en el análisis de los usos fictivos: muchos de ellos piensan, en la línea de Frege, que tales usos no involucran afirmaciones puesto que el innegable componente de simulación (*pretense*) que los caracteriza elimina la fuerza asertórica; de este modo, no pueden ser ni verdaderos ni falsos[5]. Desde esta perspectiva, no hay lugar para la verdad y la falsedad en el ámbito de la ficción, en donde puede hablarse a lo sumo de *fidelidad o infidelidad* a una narración. Por consiguiente, dado que las nociones de referencia y verdad son ajenas al ámbito de la ficción, no habrá razón para pensar que los usos fictivos introducen un compromiso con la existencia de objetos particulares de algún tipo peculiar. Ahora bien, por más que se considere que los usos fictivos no tienen fuerza asertórica y por tanto no pueden dar lugar a verdad o falsedad, es necesario explicar cuál es el contenido proposicional de los enunciados involucrados: dado que no puede tratarse de proposiciones singulares (puesto que no hay objeto particular que las constituya en cada caso), como se señaló en el parágrafo anterior, la opción principal del antirrealista ha sido considerar que expresan, por medio de sus respectivas paráfrasis, proposiciones generales. Una opción alternativa ha sido considerar que expresan proposiciones *gappy*, esto es, proposiciones singulares que no tienen *nada* en el lugar del objeto, con lo cual se les hizo preciso asociar con cada enunciado que contiene un nombre de ficción *algún otro contenido proposicional*, pragmáticamente transmitido, que por lo general también es concebido en términos descriptivos[6]. De este modo, la tesis de que los usos fictivos no tienen fuerza asertórica involucra, de una manera u otra, la adopción de lo que he llamado 'una estrategia descriptivista': los usos fictivos se interpretan en términos de enunciados descriptivos asociados con ellos –a veces propuestos como paráfrasis que ofrecen sus respectivos contenidos semánticos, y otras veces como aportando contenidos pragmáticos que son fundamentales para su interpretación. Por otro lado, la tesis de la ausencia de fuerza asertórica no parece

---

[5]Véase, a modo de ejemplo, el siguiente fragmento de Frege: "En la forma de una oración asertórica expresamos la aceptación de la verdad. Para esto no necesitamos la palabra 'verdadero'. E incluso cuando la usamos la fuerza asertórica no reside en ella, sino en la forma de la oración asertórica, y cuando ésta pierde su fuerza asertórica, la palabra 'verdadero' no puede restablecérsela. Esto sucede cuando no hablamos en serio. Así como el tronar en el teatro es solamente tronar aparente y la lucha en el teatro es solamente lucha aparente, así también la aserción en el teatro es solamente aserción aparente. Se trata solamente de escenificación, de ficción. [...]" (Frege 1918-1919: 203). Para un análisis de la posición fregeana, véase mi 2014b.

[6]El recurso a las proposiciones *gappy* fue inicialmente introducido por Braun (2005). Véanse también, por ejemplo, Taylor (2000), Lo Guercio (2016), Predelli (2017).

poder extenderse a los restantes usos, el parafictivo y el metafictivo: en términos de los ejemplos anteriores, tanto en el uso parafictivo de (2) como en el metafictivo de (3) es claro que el hablante está *afirmando algo verdadero*. Por consiguiente, la única manera de evitar el compromiso ontológico con particulares ficticios parece ser, en estos casos, recurrir a una paráfrasis que evite la aparición de nombres de ficción en posiciones referenciales.

De este modo, el antirrealista parece recurrir a una estrategia descriptivista en *todos* los casos, esto es, para explicar todos los usos de los enunciados que contienen nombres de ficción. La adopción de tal estrategia es problemática. En primer lugar, la defensa de una concepción descriptivista del significado (aun cuando solo se trate de un subconjunto de los nombres, los de ficción), como es el caso de las teorías clásicas de Frege y Russell, debe enfrentar los poderosos argumentos de Kripke (1972) en contra del descriptivismo. Además, en el caso de que esté acompañada por una propuesta de paráfrasis, ¿por qué deberíamos abandonar nuestros modos habituales de hablar o reinterpretarlos en términos de poco intuitivas paráfrasis? Finalmente, en muchos casos no está claro si hay una estrategia general para parafrasear o es preciso recurrir a paráfrasis distintas según el caso[7]. En síntesis, considero que el antirrealista está comprometido con un problemático revisionismo semántico generalizado, en la medida en que involucra la discutible adopción de una estrategia descriptivista y/o el igualmente discutible recurso a la paráfrasis para todos los usos de enunciados que contienen nombres de ficción.

En cuanto a la segunda razón para optar por la opción realista, considero que el abstractismo no es una posición especialmente controvertida, dado que los filósofos se han comprometido con la existencia de entidades abstractas desde tiempos inmemoriales. Está claro, sin embargo, que la inclusión de los personajes literarios en el reino de las entidades abstractas requiere una re-definición del concepto. Los personajes literarios, a diferencia de entidades abstractas paradigmáticas, como, por ejemplo, los números y los conjuntos, dependen para existir de las obras literarias que les dieron origen y, por tanto, puede pensarse que poseen una existencia *temporal* que depende a su vez de la de aquéllas. En este ensayo no voy a involucrarme en profundidad con la cuestión ontológica, más allá de adherir a la teoría abstractista en las versiones de Kripke (2011, 2013 basado en 1973) y Thomasson (1999), quienes conciben a tales entidades como un tipo de artefacto, es decir, algo que no es natural sino un producto

---

[7] Véase, paradigmáticamente, el texto de Sainsbury (2005). Para una crítica de su propuesta, véase mi 2008. Una propuesta antirrealista distinta, que evitaría esta objeción, puede encontrarse en Caso (2016).

de la actividad creativa de los seres humanos[8,9].

Solo destaco entonces que la introducción de este nuevo tipo de entidades abstractas involucra un cambio en la concepción tradicional, dado que a diferencia de las tradicionales, como los llamados 'universales' (un modo muy tradicional de concebir a las propiedades), los números y los conjuntos, los personajes de ficción tienen una ubicación *temporal* –si bien se trata claramente de una ubicación no estándar: existen siempre y cuando existan las correspondientes obras literarias que les dieron origen. Por otro lado, son *particulares*, no universales, como las Ideas platónicas y las esencias aristotélicas. Más aun, los personajes literarios han sido concebidos como los objetos teóricos de la teoría literaria y los estudios culturales, así como la Revolución Francesa es considerado un objeto teórico de la historia[10]. En este sentido, constituyen los referentes de nombres característicos de esas disciplinas, del mismo modo en que las especies constituyen los referentes de nombres característicos de la biología –en relación con este punto, es interesante tomar en cuenta que para algunos filósofos de la biología (Ghiselin 1974, Hull 1976, 1978) las especies, si bien son entidades abstractas, también deben ser concebidas como individuos *particulares*, con una peculiar localización temporal.

En síntesis, sobre la base de los argumentos de Kripke (1972) según los cuales ningún nombre propio común y corriente (y así suelen ser los que aparecen en narraciones de ficción, como es el caso de 'Ulises') es semánticamente equivalente a una descripción así como de su propuesta original en favor del carácter abstracto de los personajes de ficción (1973, 2011, 2013), daré por sentada la tesis de que la naturaleza semántica de los nombres de ficción no puede ser descriptiva: se trata de designadores rígidos que designan entidades abstractas en todos los mundos posibles en los que existen. Ahora bien, suele pensarse que la opción realista, especialmente en su versión abstractista, podría ser plausible en relación con los usos metafictivos pero que sin duda no lo es en relación con los fictivos (y parafictivos). Una de las razones es el hecho de que resulta difícil explicar cómo un objeto abstracto (o inexistente o meramente posible, si se prefiere alguna de las otras versiones) puede poseer el tipo de propiedades que se le adscriben en las narraciones de ficción, tales como dormir en

---

[8] Véase Thomasson (1999) para una concepción de los personajes de ficción como entidades abstractas que dependen para existir de obras literarias abstractas. En contraste, hay autores que objetan la posibilidad de concebir a las obras literarias (y musicales) en términos de entidades abstractas. García Ramírez y Mayerhofer (2015), por ejemplo, consideran que ninguna entidad abstracta puede ser creada, dado que las entidades abstractas existen fuera del tiempo, lo cual equivale a rechazar la posibilidad de que existan entidades abstractas de un tipo *sui generis*. Si bien se trata de una cuestión controvertida, sospecho que algunos tipos pueden ser creados: creamos líneas de colectivo, palabras y, en general, lenguajes.

[9] Es oportuno indicar que una excepción a esta línea originada en el manuscrito de Kripke de 1973 la constituye la teoría abstractista de Zalta (1983), según la cual los personajes de ficción son semejantes a entidades platónicas, esto es, entidades que no han sido creadas por los seres humanos (posición que me resulta sumamente anti-intuitiva).

[10] Van Inwagen 1977 es el *locus classicus* de esta tesis, que es presentada como el argumento principal para ser realista respecto de los usos metafictivos.

una playa o ser un hombre y no un dios, en el caso de Ulises (mientras que no resulta tan complejo explicar cómo es posible predicar de ellos que son personajes literarios, por ejemplo). *Prima facie el abstractista parece tener un problema para explicar la verdad aparente o ficticia de los usos ficticios: en la medida en que 'Ulises' refiere a una entidad abstracta, no hay ninguna lectura posible de (1) que lo haga verdadero, dado que ninguna entidad abstracta puede dormir en una playa*[11]. (Más aun, parece ser no solo falsa sino categorialmente falsa, esto es, involucrar un error categorial.)[12]

En lo que sigue, me ocuparé de este problema: más específicamente, me propongo explicar en el marco del realismo abstractista cuál es el contenido singular de los usos ficticios de enunciados como (1) y en qué sentido se puede decir que esos usos son verdaderos. Mi objetivo es mostrar que el marco del realismo abstractista hace posible compatibilizar el carácter singular de la proposición involucrada en cada caso con el carácter ficticio de las propiedades adscriptas. En términos del ejemplo anterior, *explicaré cómo es posible que Ulises, un personaje literario entendido como una entidad abstracta, esté durmiendo en una playa*.

2. Los nombres de ficción como designadores rígidos

2.1. Crítica de algunas posiciones estándar

La respuesta abstractista estándar al problema señalado en el apartado anterior apela a la idea de que, a pesar de las apariencias, las oraciones como (1), (2) y (4) son ambiguas, dado que hay dos sentidos en los cuales es posible adscribir una propiedad a un personaje de ficción: este puede o bien *instanciar* o bien *codificar* una determinada propiedad (Zalta 1983)[13]. De este modo, aun cuando sea falso que Ulises instancia la propiedad de estar dormido, es verdad que la codifica, dado que se trata de una propiedad que posee en la historia narrada por la *Odisea*. Ahora bien, Crane señala que las propiedades presuntamente codificadas por los personajes deben entenderse como propiedades *dependientes de la representación*: "La mejor manera de pensar en los objetos no existentes como codificando propiedades es pensar en ellos como siendo

---

[11] Véase, por ejemplo, la siguiente cita de Sainsbury: "Ningún objeto abstracto es un detective, toca el violín o vuela, de modo que las oraciones comunes de la ficción (tales como 'Holmes es un detective', 'Holmes toca el violín', 'Pegaso vuela') recibirán el mismo valor de verdad, la falsedad, tanto en la explicación de RWR como en aquella en la cual esos usos reciben personajes ficticios robustos como referentes." (2005: 211)

[12] En mi opinión, este es uno de los dos problemas principales que enfrenta el abstractista. El otro es el de explicar cómo pueden ser verdaderos algunos enunciados existenciales negativos que intuitivamente parecen serlo, tales como "Ulises no existe". Por razones de espacio, no me ocuparé de este problema en este ensayo. Puede verse mi 2014a para un intento de solución, así como Predelli (2002).

[13] Para una distinción semejante, véase van Inwagen (2003). Los neo-meinongianos (Parsons 1980, Priest 2005), por su parte, proponen una distinción paralela entre tipos de propiedades: de acuerdo con ella, Ulises tiene, por un lado, la propiedad extra-nuclear de ser un personaje literario y, por otro, la propiedad nuclear de estar durmiendo en la playa de Ítaca.

2.1 Crítica de algunas posiciones estándar

representados con esas propiedades" (Crane 2013: 70). Pero esto no ofrece una respuesta a nuestro problema inicial: el hecho de que el fenómeno de la representación de un objeto concreto por uno abstracto permita afirmar que este último tiene las propiedades del primero es el punto de partida y precisamente lo que hay que explicar. De este modo, la distinción entre *tener propiedades de suyo* y *tener propiedades por ser una representación de otro objeto* (atribuida a algo cuya naturaleza es ser básicamente una representación de otro objeto) resulta tan poco iluminadora como la distinción entre instanciar y codificar una propiedad: ambas parecen ser maneras alternativas de *enunciar* la posibilidad de las dos lecturas básicas de las que son pasibles los enunciados como (1), a saber, la fáctica y la fictica.

Otro acercamiento tradicional considera que (1) es, bajo su interpretación ficticia, equivalente al siguiente enunciado, que contiene un *operador de ficción*, interpretado por lo general como un operador intensional:

(7) De acuerdo con la *Odisea*, Ulises duerme en la playa de Ítaca,

que resulta literal y estrictamente verdadero (Lewis 1978). 'De acuerdo con la *Odisea*', como todo operador intensional, cambia la circunstancia de evaluación de la oración incrustada del mundo real a un mundo alternativo, el mundo tal como es descripto en la *Odisea*[14]. Sin embargo, como se explicará más abajo, el abstractista no puede considerar que el mundo en cuestión (un mundo en el que Ulises es una persona real que navega de vuelta hacia Ítaca después de haber combatido en la guerra de Troya) es un mundo posible: en la medida en que 'Ulises' es considerado un designador rígido, se supone que designa la misma entidad abstracta en todos los mundos posibles en los que esta existe (y nada en los que no existe), por tanto, en ningún caso a una persona capaz de navegar, combatir en una guerra o quedarse dormida en la playa. Hay dos consecuencias principales que quisiera destacar: (i) por un lado, dado que en todos los mundos posibles Ulises es una entidad abstracta, no hay mundo posible en el cual sea un héroe de guerra, y en general, no hay mundo posible en el que sea una persona real; (ii) por otro lado, los mundos en los que muere en la guerra de Troya o ahogado en el mar tampoco son posibilidades reales. En lo que sigue trataré de explicar con mayor detalle ambas afirmaciones.

En primer lugar, si 'Ulises' designa al personaje literario Ulises, una entidad abstracta, en todos los mundos posibles en los que existe, no hay mundo posible en el que Ulises sea una persona real: lo que pudo haber existido es un mundo en el cual existe una persona real con todas las propiedades que le son atribuidas a Ulises en la *Odisea* pero esa persona no sería idéntica a Ulises, dado que no sería un personaje literario creado en la antigua Grecia sino que tendría un origen distinto[15]. Una posibilidad real

---

[14] Por una cuestión de simplicidad, daré por sentado que las circunstancias de evaluación están constituidas solo por mundos posibles, si bien, como es sabido, usualmente se considera que incluyen asimismo otros parámetros, como, por ejemplo, el tiempo.

[15] ¿Qué ocurre con personajes tales como el Napoleón de *La guerra y la paz*, los cuales, a diferencia de

es, por ejemplo, el mundo en el cual Ulises no es un personaje famoso porque en él no hay rastros de la civilización griega. En general, las posibilidades *reales* son los mundos en los Ulises es un personaje literario al que le faltan algunas de las propiedades que tiene en el mundo real (como la de ser famoso) o tiene propiedades que son distintas de las que de hecho tiene (tal como la de ser ignorado por la mayoría de los críticos literarios).

La segunda afirmación establece que los mundos en los que Ulises muere en la guerra de Troya o ahogado en el mar tampoco son posibilidades reales. También se sigue de la tesis según la cual 'Ulises' designa rígidamente el personaje literario de Ulises, esto es, una entidad abstracta: si Ulises es una entidad abstracta en todos los mundos en los que existe, no hay mundo posible en el cual tenga propiedades tales como morir en la guerra de Troya o ahogarse en el mar, esto es, propiedades que solo pueden ser poseídas por entidades concretas (y, más específicamente, seres vivos). Es importante tomar en cuenta que esto no implica que todas las propiedades que se adscriben a un personaje en una narración de ficción son constitutivas de la identidad del personaje: este podría haber tenido en la historia de ficción propiedades diferentes de las que de hecho tiene. En términos del ejemplo anterior, la *Odisea* podría haber sido tal que Ulises resultara muerto durante la guerra de Troya, es decir, Ulises podría haber muerto en Troya en la historia de ficción. Pero, en la medida en que esas son posibilidades *dentro de la historia de ficción o internas a(l marco de) la narración ficticia, las considero puramente ficticias y no reales*. De acuerdo con esto, una narración de ficción sirve para introducir un contexto ficticio, con su correspondiente conjunto de mundos alternativos (o posibilidades ficticias), las cuales son concebibles o imaginables pero no son metafísicamente posibles (dado que lo metafísicamente posible está, como es sabido, determinado por el mundo real: no hay mundo metafísicamente posible en el que Ulises no sea lo que de hecho es, esto es, un personaje literario, aunque, por supuesto, hay mundos metafísicamente posibles en los que tiene propiedades que de hecho no tiene, tales como ser ignorado por los críticos literarios, etc.) Mientras que los mundos metafísicamente posibles hacen verdaderos a enunciados contrafácticos tales como

(8) Si no hubieran quedado rastros de la civilización griega antigua, Ulises habría sido ignorado por los críticos literarios,

los *mundos ficticiamente posibles* o *mundos posibles meramente ficticios* hacen verdaderos a contrafácticos como

---

Ulises, pueden ser concebidos como entidades reales? No quiero tomar partido en esta cuestión controvertida. Si es verdad que tales personajes son importados del mundo real (y, por tanto, son idénticos a entidades reales), mi afirmación deberá restringirse a los que entonces podrían considerarse personajes de ficción en sentido estricto; de lo contrario, podrá considerarse que comprende a todos los personajes pertenecientes a una narración de ficción. Para la cuestión de la importación de personajes, véase, por ejemplo, Lewis (1978).

(9) Si Ulises hubiera sido seducido por el canto de las sirenas, nunca habría vuelto a Itaca.

Por tanto, para un abstractista, hay cierta tensión entre, por un lado, considerar que los nombres de ficción son designadores rígidos y, por otro, comprometerse con la existencia de mundos posibles en donde nuestro nombre 'Ulises' designe algo diferente de lo que designa en el mundo real, tal como una persona que puede, entre otras cosas, quedarse dormida en la playa[16] [17].

2.2. La idea del 'cambio de contexto' y los mundos posibles ficticios

Mi propósito es entonces adherir a un enfoque alternativo, el Enfoque del Cambio de Contexto (*Context-Shift View*), propuesto por Predelli (1997, 2005 y 2008) y luego Recanati (2010), al cual quisiera agregar algunos elementos. Según este enfoque, los usos ficticios sugieren, sin la presencia de operador intensional alguno, un cambio de contexto. La propuesta es considerar que, bajo la interpretación ficticia, las circunstancias con respecto a las cuales un enunciado que contiene un nombre de ficción debe evaluarse como verdadero o falso no son, como es usual, las del contexto de emisión; más claramente, el contexto apropiado para evaluar el enunciado no incluye como valor al mundo real sino al mundo del relato de ficción. Bajo la interpretación fáctica, en cambio, el contexto de evaluación apropiado incluye al mundo real[18]. De este modo, según mi lectura de esta propuesta, dar cuenta de la intuición según la cual (1) es verdadero de acuerdo con el relato de la *Odisea* involucra postular *un cambio de contexto*, lo cual significa que el enunciado debe ser evaluado en relación no con el mundo real en el cual la emisión tiene lugar, ni tampoco con un mundo alternativo metafísicamente posible, sino en relación con *el mundo en el que transcurre la historia de ficción, un mundo puramente ficticio* –en el sentido antes explicado según el cual algunas posibilidades son puramente ficticias. Ahora bien, si bien acuerdo fuertemen-

---

[16]Nótese que este problema no surge si los nombres de ficción no son considerados designadores rígidos, como es el caso de Lewis (1978), quien piensa que designan a aquellos individuos, no importa cuán diferentes sean entre sí, que satisfacen las descripciones ficticias correspondientes en cada mundo; desde el punto de vista ontológico, se los considera, en cada caso, contrapartes del mismo individuo.

[17]Para otras críticas a este enfoque, véase, por ejemplo, Predelli (2005: 69).

[18]Véanse, por ejemplo, las siguientes citas de Predelli:
"Supóngase que estamos discutiendo la película *Amadeus*, de Milos Forman, y que digo (6) Salieri encomendó el *Requiem*. Mi emisión es aparentemente verdadera: en la película, el compositor Antonio Salieri es la misteriosa figura que encomienda anónimamente la Misa de los Muertos [...]" (2005: 66)
"En este caso, el parámetro que diferencia al índice apropiado del ingenuo es la coordenada de mundo posible: el índice con respecto al cual mi emisión es evaluada no contiene al mundo en el que tuvo lugar, esto es, el mundo real, sino al 'mundo' de la historia." (2005: 70)
Vale la pena aclarar que 'índice' es la palabra usada por Predelli para lo que usualmente se denomina 'contexto'.

te con el núcleo de esta propuesta, quisiera destacar algunos puntos fundamentales en los que me distancio de ella.

Predelli piensa que los indéxicos que aparecen en un cierto enunciado de ficción toman sus referentes en relación con (no el contexto real de emisión sino) un contexto que podría llamarse 'ficticio'. La siguiente cita, aunque un poco larga, sirve para aclarar el punto:

> Supongamos que, al hablar de la película [*Amadeus*], digo:
>
> (8) A pesar de que Mozart pensó que la misteriosa figura era el fantasma de su padre, quien de hecho le encomendó el *Requiem* fue Salieri.
>
> Dado cómo las cosas son descriptas en la película *Amadeus*, esta emisión es verdadera. Pero, para obtener el comportamiento semántico correcto de la expresión 'quien de hecho le encomendó el *Requiem*', es preciso evaluar (8) *con respecto a un contexto que, a diferencia del contexto de emisión, contenga al mundo ficticio de Amadeus como su parámetro de mundo posible*. Con respecto a tal contexto, 'quien de hecho le encomendó el *Requiem*' denota a Salieri, y la oración "Quien de hecho le encomendó el *Requiem* fue Salieri" resulta verdadera; con respecto al contexto de emisión, por otro lado, 'quien de hecho le encomendó el *Requiem*' denota al Conde Walsegg, y "Quien de hecho le encomendó el *Requiem* fue Salieri" es evaluada como falsa (Predelli 1997: 74; la traducción y el énfasis son míos).

Del mismo modo, parece pensar que las descripciones y los nombres deben ser interpretados en relación con el contexto ficticio, esto es, el único contexto relevante para la interpretación de los usos ficticios.

En la línea de Walton (1990), pienso que los usos ficticios involucran fundamentalmente una indicación o prescripción acerca de lo que debemos imaginar que *existe aun cuando sabemos que no es así*[19]. Realizarlos y comprenderlos implica involucrarse en un proceso interpretativo que a su vez comprende un uso de la imaginación característicamente desconectado de nuestra percepción del mundo externo y nuestra acción en él. Ahora bien, considero que ese proceso interpretativo tiene como punto de partida nuestra interacción con la obra literaria: más específicamente, en un primer estadio, interpretamos la narración literaria como tal (a diferencia de una crónica periodística, por ejemplo), lo cual significa que consideramos que los nombres de ficción que aparecen en ella refieren a personajes literarios (y no a gente real). Esta es la base en la cual anclan nuestros poderes imaginativos: entonces, en un segundo estadio, imaginamos que esos personajes literarios son personas reales que viven sus propias

---

[19] La parte de la oración que ha sido destacada tiene importancia, dado que señala la diferencia entre el uso de la imaginación en relación con las narraciones de ficción y su uso en relación con las narraciones fácticas. Véase Matravers (2014) para la tesis según la cual la imaginación no es una facultad ni exclusiva ni característica de nuestra interacción con la ficción.

2.2 La idea del 'cambio de contexto' y los mundos posibles ficticios 293

vidas. (Esto parece reproducir, en la esfera de la competencia semántica, la concepción metafísica de Walton de los juegos de *make-believe* en términos de *props* que funcionan como base del proceso imaginativo.)

Por lo tanto, a diferencia de Predelli, considero que los indéxicos, las descripciones definidas y los nombres que aparecen en una narración de ficción toman sus referentes del contexto real de emisión, el cual es luego transformado por la imaginación en algo completamente distinto de lo que de hecho es. Es precisamente el trabajo de nuestra imaginación sobre los objetos provistos por el contexto real de emisión lo que determina el cambio de contexto. En otras palabras, cambiamos de un contexto a otro porque somos capaces de transformar, imaginativamente, lo que nos es dado originalmente, en virtud de nuestra interacción interpretativa con la narración literaria. Esta última juega el papel de un manual de instrucciones: si no fuera por las descripciones de los personajes, lugares y tiempos que contiene, no seríamos capaces de imaginar absolutamente nada al enfrentarnos con el uso ficticio de nombres e indéxicos. En términos de nuestro ejemplo anterior, la interpretación de 'Ulises' involucra la referencia a un personaje literario que imaginamos ser una persona real, a partir del perfil provisto por la *Odisea*. Asimismo, la interpretación de 'de hecho' involucra la referencia al mundo real que imaginamos ser un mundo de ficción, diferente de cómo el mundo de hecho es; e igualmente con el resto de los indéxicos[20].

Ahora bien, una vez que el contexto ha cambiado en virtud de nuestra imaginación, la emisión correspondiente será evaluada como verdadera o falsa con respecto al mundo del nuevo contexto (esto es, la nueva circunstancia de evaluación del contexto). En términos del ejemplo anterior, una vez que construimos imaginativamente una persona a partir de nuestra comprensión de los rasgos de personalidad y las acciones de Ulises, todo uso ficticio de (1) resultará verdadero con respecto al mundo ficticio de la *Odisea*. Del mismo modo, una vez que construimos imaginativamente una persona a partir de nuestra comprensión de los rasgos personales y las acciones de Salieri, todo uso ficticio de (8) en la cita anterior resultará verdadero con respecto al mundo ficticio

---

[20]En este punto la propuesta presenta cierta semejanza con la de Recanati: "Los rasgos objetivos del contexto de emisión son realmente 'dados' y, por lo tanto, no pueden cambiarse. Pero lo que el hablante puede hacer es simular o fingir (*pretend*) que el contexto es diferente del que en verdad es. Si la simulación es mutuamente manifiesta, será parte de lo que el hablante significa el que la oración es emitida en un contexto diferente del contexto real c. En tal situación sí tiene lugar un cambio de contexto: hay dos contextos, el contexto real c en el que la emisión es producida, y el contexto simulado c' en el que la emisión se presenta a sí misma como siendo producida."(2010: 193) Sin embargo, hay ciertos aspectos en los que también me distancio de esta posición. En primer lugar, creo que los rasgos del contexto de emisión por lo general (no sólo en el caso de los usos ficticios) no están 'dados' sino que involucran un proceso de interpretación por parte del oyente, en el que pone en juego este conocimiento pragmático de la situación de uso -el tipo de conocimiento que le permite, por ejemplo, interpretar un determinado uso de 'acá' en una oración como "Acá hace frío" como haciendo referencia a una ciudad entera y no a la habitación de la casa en la que se encuentra. En segundo lugar, no creo que sea necesario imaginar que la oración es emitida en un contexto diferente: como destaco en el texto principal, la mayor parte de las veces, es suficiente con imaginar que el personaje es una persona real, el mundo real, uno de ficción, etc.

de *Amadeus*[21].

En resumen,

(i) los indéxicos, las descripciones definidas y los nombres son interpretados en el contexto real de emisión;

(ii) la acción de nuestra imaginación determina lo que puede ser caracterizado (tal vez metafóricamente) como 'un cambio de contexto';

(iii) las emisiones e inscripciones son evaluadas como verdaderas o falsas con respecto al mundo del nuevo contexto.

Hay ciertos aspectos de esta propuesta que quisiera enfatizar.

Ante todo, el mundo ficticio de la *Odisea*, si bien es relevante para *evaluar* los usos ficticios de (1), no juega ningún papel en fijar el referente de 'Ulises', constituido por el personaje literario del *mundo real* en el cual fue originalmente fundado o anclado por el autor de la *Odisea*. Al igual que cualquier otro nombre, sea o no ficticio, 'Ulises' es un designador rígido: designa al mismo personaje literario en todos los mundos en los que existe. Lo que determina la identidad de un personaje a través de los mundos posibles no es tan diferente de lo que determina la identidad de una persona real: en ambos casos parece tratarse de un rasgo histórico, pero mientras que en el caso de una persona es su código genético, la identidad de un personaje remite al proceso de creación de la correspondiente narración literaria, de cuya existencia depende[22].

En segundo lugar, como vimos, la imaginación determina un cambio de contexto, según el cual el mundo relevante para la evaluación de un uso ficticio resulta ser

---

[21] La presente propuesta resulta en este aspecto semejante a la teoría abstractista de Salmon (1998). Mi contribución se centra específicamente en el ofrecimiento de una explicación específica acerca de cómo los usos ficticios pueden ser considerados verdaderos en la historia de ficción o relativamente al marco provisto por una narración de ficción, desde una perspectiva abstractista semejante a la de Salmon.

[22] Podría pensarse, en la línea de Currie (1990: 12-18), que los usos fictivos no involucran aserciones, ni preguntas, ni ningún otro tipo de acto de habla usual (lo cual no sería el caso ni de los paraficticios ni de los metaficticios). De acuerdo con esta perspectiva, la actitud de fingimiento o simulación que acompaña a los usos ficticios daría lugar a un tipo de fuerza ilocucionaria distinta, que podría llamarse 'simulativa', sin afectar entonces al contenido proposicional. Simular afirmar que Ulises duerme en la playa de Ítaca sería realizar un acto de habla simulativo con un genuino contenido proposcional, <Ulises duerme en la playa de Ítaca>. En otras palabras, simular afirmar que Ulises duerme en la playa de Ítaca sería equivalente a sostener (1) como verdadera con respecto al mundo ficticio de la *Odisea*. En mi opinión, la virtud que tiene este enfoque es también la de compatibilizar el componente de simulación con la función referencial de los nombres: la simulación afecta a la dimensión ilocucionaria de un acto de habla, su fuerza, mientras que la referencia concierne a la dimensión locucionaria, su contenido proposicional. Mi impresión es que Walton parece pensar que el reconocimiento del componente de simulación conduce a parafrasear los enunciados que contienen nombres de ficción en posiciones referenciales en términos de otros en donde eso no ocurre -como si la presencia de un componente de simulación fuera lo que determinara que los nombres de ficción no pueden tener una función referencial (véase también Evans 1982 para una posición similar). Sin embargo, la propuesta defendida en este trabajo es independiente de la idea de que los usos ficticios tienen una fuerza ilocucionaria especial, no asertórica.

un mundo puramente ficticio. Tal mundo, como se mencionó anteriormente, no es un mundo metafísicamente posible: es el mundo tal como es descripto en una narración literaria, el producto de la imaginación de cierto escritor. Es un mundo concebible, epistémicamente posible, un modo en el cual podríamos haber descubierto que es el mundo si este hubiera sido tal como lo imaginó un cierto escritor. Las posibilidades epistémicas incluyen mundos que son compatibles con lo que podemos saber *a priori*, esto es, antes de llevar a cabo una investigación empírica –por lo que sabemos *a priori*, el agua podría ser XYZ y Héspero podría ser diferente de Fósforo. Del mismo modo, por lo que sabemos *a priori*, Ulises podría ser una persona real en lugar de un personaje literario[23].

Por último, quisiera mencionar el siguiente problema: ¿no es acaso implausible e incluso absurdo pensar que los lectores de obras literarias de ficción son llamados a imaginar que entidades abstractas son cosas concretas o seres humanos que habitan mundos ficticios?[24] Creo que esta objeción confunde a los *personajes* con sus *descripciones metafísicas*: al interactuar con las narraciones literarias y, en particular, al comprender los nombres de ficción, los lectores captan personajes literarios, los cuales, desde el punto de vista abstractista, son interpretados en términos de entidades abstractas. De la misma manera, al comprender un texto matemático, uno capta números, los cuales pueden ser metafísicamente interpretados de diferentes maneras (en términos de capacidades psicológicas, como entidades abstractas, etc.) En tanto lectores, no somos conscientes del estatuto metafísico de los referentes de los términos que utilizamos. Por lo que sé, algo semejante podría decirse acerca de nuestra interacción con la gente: por lo general, ignoramos por completo qué tipo de entidades se considera que somos desde un punto de vista metafísico[25].

3. Conclusión

En este trabajo me he ocupado del análisis semántico de los enunciados que contienen nombres de ficción. En la primera parte, después de contrastar los diferentes usos, fictivos, parafictivos y metafictivos, que puede hacerse de ellos, he destacado que una teoría semántica adecuada debe dar cuenta tanto de su contenido proposicional como de la intuición de verdad/falsedad asociada con ellos. A continuación, he caracterizado los dos tipos principales de propuestas ofrecidas: las realistas, según las

---

[23] Dejo de lado el hecho, señalado por Lewis (1978), de que algunas narraciones de ficción parecen involucrar mundos lógicamente imposibles, es decir, mundos que no son ni metafísica ni epistémicamente posibles. De todos modos, el punto básico según el cual el mundo relevante para la evaluación no es uno metafísicamente posible queda en pie.

[24] Esta objeción ha sido presentada por Sainsbury (2005) a la versión del abstractismo propuesta por Salmon (1998).

[25] Desde un punto de vista abstractista, puede considerarse que los autores de las narraciones de ficción también interactúan con personajes, cuyo estatuto metafísico de entidades abstractas suele ser completamente ignorado por aquéllos (y está a menudo muy lejos de sus intereses).

cuales los nombres de ficción son, como el resto de los nombres, expresiones genuinamente referenciales que designan rígidamente objetos ontológicamente *sui generis*, los objetos ficticios; y las antirrealistas, según las cuales los enunciados que contienen nombres de ficción deben parafrasearse y reemplazarse por otros que no implican compromiso ontológico alguno con objetos ficticios. La opción realista involucra preservar la simplicidad de la semántica a riesgo de incrementar de manera inconveniente el compromiso ontológico, mientras que, a la inversa, la opción antirrealista propone preservar una ontología austera a riesgo de complicar excesivamente la semántica. Una vez presentada esta disyuntiva, he dado algunas razones para preferir la opción realista, en su versión abstractista. Por un lado, la revisión semántica involucrada por las posiciones antirrealistas comprende la adopción de una estrategia descriptivista para explicar el significado de los nombres de ficción que los distinguiría tajantemente de los restantes nombres (a los cuales, como es sabido, no es plausible pensar que pueda aplicarse). Por otro lado, la inclusión de objetos ficticios entre las entidades abstractas no parece constituir un incremento temerario del compromiso ontológico. La posición resultante enfrenta, sin embargo, un problema fundamental que queda sintetizado en el siguiente interrogante: ¿cómo puede una entidad abstracta poseer las propiedades que habitualmente se adscriben a los personajes literarios en las narraciones de ficción? En términos de nuestro ejemplo, si Ulises es una entidad abstracta, ¿cómo puede decirse con verdad que se quedó dormido en la playa de Ítaca? En la segunda parte del trabajo desarrollo una respuesta a este problema que apela a una elaboración del Enfoque del Cambio de Contexto: desde mi perspectiva, los usos ficticios deben evaluarse como verdaderos o falsos con respecto al mundo posible de la historia de ficción, esto es, un mundo posible en la ficción o relativamente al marco introducido por la narración de ficción, una posibilidad ficticia. Tomar en cuenta las posibilidades ficticias, producto de nuestra transformación imaginativa del mundo real, es lo que nos permite decir que es ficticiamente verdadero o verdadero en la ficción que Ulises se ha quedado dormido.

Agradecimientos

Agradezco mucho a Max Freund y a Marco Ruffino por editar e invitarme a colaborar en este volumen, y a Eduardo García Ramírez, Alfonso Losada, Andrés Saab y Laura Skerk, por sus agudos comentarios a una primera versión de este trabajo. Este artículo fue escrito en virtud del subsidio PIP N° 633 otorgado por el CONICET y el subsidio PICT N° 02457 otorgado por la ANPCyT.

Referencias

[Bonomi(2008)] A. Bonomi. Fictional Contexts. *P. Bouquet, L. Serafini y R. Thomason (eds.), Perspectives on Context, (Stanford: CSLI Publications)*, pages 213–248, 2008.

[Braun(2005)] David Braun. Empty Names, Fictional Names, Mythical Names. *Noûs*, 39:596–631, 2005.

[Caso(2016)] Ramiro Caso. "Un antirrealismo sin paráfrasis para las entidades de ficción". A medio siglo de Formas lógicas, realidad y significado. Homenaje a Thomas Moro Simpson, etc. Publicado en 2016 eds. A. Moretti, E. Orlando & N. Stigol (Buenos Aires: Eudeba), 2016.

[Crane(2013)] Tim Crane. *The Objects of Thought*. Oxford: Oxford University Press, 2013.

[Currie(1990)] Gregory Currie. *The Nature of Fiction*. Cambridge: Cambridge University Press, 1990.

[Evans(1982)] Gareth Evans. Existential Statements. *The Varieties of Reference*, ed. J. McDowell, (Oxford: Oxford University Press), pages 343–372, 1982.

[Frege(1892)] Gottlob Frege. On Sinn and Bedeutung. *The Frege Reader*, ed. M. Beaney, (Oxford: Blackwell), pages 151–171, 1892.

[García Ramírez and Mayerhofer(2015)] E. García Ramírez and I. Mayerhofer. A Plea for Concrete Universals. *Crítica, Revista Hispanoamericana de Filosofía*, 47 (139):3–46, 2015.

[Ghiselin(1974)] Michael T. Ghiselin. A Radical Solution to the Species Problem. *Systematic Zoology*, 23:536–544, 1974.

[Hull(1978)] D. Hull. A Matter of Individuality. *Philosophy of Science*, 45:335–360, 1978.

[Hull(1976)] David Hull. Are Species Really Individuals? *Systematic Zoology*, 25:174–191, 1976.

[Kaplan(1977)] David Kaplan. Demonstratives. *Themes from Kaplan, eds. J. Almog, J. Perry & H. Wettstein, (New York and Oxford: Oxford University Press)*, pages 481–563, 1977.

[Kripke(2011)] S. Kripke. Vacuous Names and Fictional Entities. *Philosophical Troubles. Collected Papers, (Oxford: Oxford University Press)*, 1:52–74, 2011.

[Kripke(2013)] S. Kripke. *Reference and Existence*. The John Locke Lectures. (Oxford: Oxford University Press). (Manuscrito de 1973), 2013.

[Kripke(1972)] Saul Kripke. *Naming and Necessity*. Cambridge: Harvard University Press, 1972.

[Lewis(1978)] David Lewis. Truth in Fiction. *American Philosophical Quarterly*, 15: 37–46, 1978.

[Lo Guercio(2016)] Nicolás Lo Guercio. *Archivos mentales: una solución al problema de las emisiones metafictivas*. A medio siglo de Formas lógicas, realidad y significado. Homenaje a Thomas Moro Simpson, eds. A. Moretti, E. Orlando & N. Stigol, (Buenos Aires: Eudeba), 2016.

[Matravers(2014)] D. Matravers. *Fiction and Narrative*. Oxford: Oxford University Press, 2014.

[Orlando(2008)] E. Orlando. Fictional Terms without Fictional Entities. *Crítica. Revista Hispanoamericana de Filosofía*, 40(120):111–127, 2008.

[Orlando(2014a.)] E. Orlando. Ficción y compromiso ontológico. *Quaderns de Filosofia*, 1(1):39–54, 2014a.

[Orlando(2014b)] E. Orlando. Ficción y referencia: notas en contra de la estrategia descriptivista en su versión fregeana clásica. *Cuadernos de Filosofía*, 61:37–48, 2014b.

[Parsons(1980)] Terence Parsons. *Nonexistent Objects*. New Haven: Yale University Press, 1980.

[Parsons(1982)] Terence Parsons. Are There Nonexistent Objects? *American Philosophical Quarterly*, 19:365–371, 1982.

[Predelli(2017)] S. Predelli. *Proper Names*. A Millian Account. (Oxford: Oxford University Press), 2017.

[Predelli(1997)] Stefano Predelli. Talk about Fiction. *Erkenntnis*, 46:60–77, 1997.

[Predelli(2002)] Stefano Predelli. 'Holmes' and Holmes: a Millian Analysis of Names from Fiction. *Dialectica*, 56:261–279, 2002.

[Predelli(2005)] Stefano Predelli. *Contexts. Meaning, Truth and the Use of Language*. Oxford: Oxford University Press, 2005.

[Predelli(2008)] Stefano Predelli. Modal Monsters and Talk about Fiction. *Journal of Philosophical Logic*, 37:277–297, 2008.

[Priest(2005)] Graham. Priest. *Towards Non-Being. The Logic and Metaphysics of Intentionality*. Oxford: Clarendon Press, 2005.

[Recanati(2010)] Francois Recanati. *Truth-Conditional Pragmatics*. Oxford: Clarendon Press, 2010.

[Russell(1905)] Bertrand Russell. On Denoting. *Mind*, 14:479–493, 1905.

[Sainsbury(2010a)] M. Sainsbury. Fiction and acceptance-relative truth, belief and assertion. *Truth in Fiction, ed. F. Lihoreau, (Frankfurt: Ontos Verlag)*, pages 137–152, 2010a.

[Sainsbury(2010b)] M. Sainsbury. *Fiction and Fictionalism*. (Londres: Routledge), 2010b.

[Sainsbury(2005)] Mark Sainsbury. *Reference without Referents*. Oxford: Clarendon Press, 2005.

[Salmon(1998)] Nathan Salmon. Nonexistence. *Nôus*, 32:277–319, 1998.

[Salmon(2002)] Nathan Salmon. Mythical Objects. *Meaning and Truth. Investigations in Philosophical Semantics, eds. J. Campbell, M. O'Rourke & D. Shier, (New York: Seven Bridges Press)*, pages 105–123, 2002.

[Simpson(1964)] Thomas M. Simpson. *Formas lógicas, realidad y significado*. Buenos Aires: Eudeba, 1964.

[Taylor(2000)] Kenneth Taylor. Emptiness without Compromise. *Empty Names, Fiction and the Puzzles of Non-Existence, eds. A. Everett & T. Hofweber, (Stanford: CSLI Publications)*, pages 17–36, 2000.

[Thomasson(1999)] Amie Thomasson. *Fiction and Metaphysics*. Cambridge: Cambridge University Press, 1999.

[van Inwagen(1977)] Peter van Inwagen. Creatures of Fiction. *American Philosophical Quarterly*, 14:299–308, 1977.

[van Inwagen(2003)] Peter van Inwagen. Existence, ontological commitment and fictional entities. *The Oxford Handbook to Metaphysics*, eds. M. Loux & D. Zimmerman, (Oxford: Oxford University Press), pages 131–160, 2003.

[von Meinong(1904)] Alexis von Meinong. Teoría del objeto. *traducción castellana de E. García Máynez, en Cuadernos de Crítica* (México: Instituto de Investigaciones Filosóficas, UNAM, 1981), 13:5–57, 1904.

[Walton(1990)] Kendall Walton. *Mimesis as Make-Believe. Foundations of the Representational Arts*. Cambridge: Harvard University Press, 1990.

[Walton(2000)] Kendall Walton. Existence as Metaphor. *Empty Names, Fiction and the Puzzles of Non-Existence*, eds. A. Everett & T. Hofweber, (Stanford: CSLI Publications), pages 69–94, 2000.

[Zalta(1983)] Edward Zalta. *Abstract Objects: An Introduction to Axiomatic Metaphysics*. Dordrecht: D. Reidel, 1983.

www.ingramcontent.com/pod-product-compliance
Lightning Source LLC
Chambersburg PA
CBHW070721160426
43192CB00009B/1273